# SOMATOSTATIN

ENDOCRINE UPDATES
*Shlomo Melmed, M.D., Series Editor*

---

J.A. Fagin (ed.): Thyroid Cancer. 1998.ISBN: 0-7923-8326-5
J.S. Adams and B.P. Lukert (eds.): Osteoporosis: Genetics,
Prevention and Treatment. 1998. ISBN: 0-7923-8366-4.
B.-Å. Bengtsson (ed.): Growth Hormone. 1999. ISBN: 0-7923-8478-4
C. Wang (ed.): Male Reproductive Function. 1999. ISBN: 0-7923-8520-9
B. Rapoport and S.M. McLachlan (eds.): Graves' Disease:
Pathogenesis and Treatment. 2000. ISBN: 0-7923-7790-7.
W. W. de Herder (ed.): Functional and Morphological Imaging
of the Endocrine System. 2000. ISBN 0-7923-7923-9
H.G. Burger (ed.): Sex Hormone Replacement Therapy. 2001.
ISBN 0-7923-7965-9
A. Giustina (ed.): Growth Hormone and the Heart. 2001.
ISBN 0-7923-7212-3
W.L. Lowe, Jr. (ed.): Genetics of Diabetes Mellitus. 2001.
ISBN 0-7923-7252-2
J.F. Habener and M.A. Hussain (eds.): Molecular Basis of Pancreas
Development and Function. 2001. ISBN 0-7923-7271-9
N. Horseman (ed.): Prolactin. 2001. ISBN 0-7923-7290-5
M. Castro (ed.): Transgenic Models in Endocrinology. 2001
ISBN 0-7923-7344-8
R. Bahn (ed.): Thyroid Eye Disease. 2001. ISBN 0-7923-7380-4
M.D. Bronstein (ed.): Pituitary Tumors in Pregnancy
ISBN 0-7923-7442-8
K. Sandberg and S.E. Mulroney (eds.): RNA Binding Proteins:
New Concepts in Gene Regulation. 2001. ISBN 0-7923-7612-9
V. Goffin and P. A. Kelly (eds.): Hormone Signaling. 2002
ISBN 0-7923-7660-9
M. C. Sheppard and P. M. Stewart (eds.): Pituitary Disease. 2002
ISBN 1-4020-7122-1
N. Chattopadhyay and E.M. Brown (eds.): Calcium-Sensing Receptor.
2002. ISBN 1-4020-7314-3
H. Vaudry and A. Arimura (eds.): Pituitary Adenylate Cyclase-
Activating Polypeptide. 2002. ISBN: 1-4020-7306-2
Gaillard, Rolf C. (ed.): The ACTH AXIS: Pathogenesis, Diagnosis
and Treatment. 2003. ISBN 1-4020-7563-4
Beck-Peccoz, Paolo (ed.): Syndromes of Hormone Resistance on the
Hypothalamic-Pituitary-Thyroid Axis. 2004. ISBN 1-4020-7807-2
Ghigo, Ezio (ed.): Ghrelin. 2004. ISBN 1-4020-7770-X
Srikant, Coimbatore B. (ed.): Somatostatin. 2004.
ISBN 1-4020-7799-8

# SOMATOSTATIN

*edited by*

## Coimbatore B. Srikant
*Department of Medicine*
*McGill University Heath Center*
*Montréal, Québec, Canada*

**KLUWER ACADEMIC PUBLISHERS**
**Boston / New York / Dordrecht / London**

**Distributors for North, Central and South America:**
Kluwer Academic Publishers
101 Philip Drive
Assinippi Park
Norwell, Massachusetts 02061 USA
Telephone (781) 871-6600
Fax (781) 681-9045
E-Mail: kluwer@wkap.com

**Distributors for all other countries:**
Kluwer Academic Publishers Group
Post Office Box 322
3300 AH Dordrecht, THE NETHERLANDS
Telephone 31 786 576 000
Fax 31 786 576 254
E-Mail: services@wkap.nl

 Electronic Services <http://www.wkap.nl>

**Library of Congress Cataloging-in-Publication Data**

A C.I.P. Catalogue record for this book is available
from the Library of Congress.

Somatostatin
Coimbatore B. Srikant
ISBN 1-4020-7799-8
e-Book ISBN 1-4020-8033-6

*Printed on acid-free paper.*
Printed in the United States of America.

*The Publisher offers discounts on this book for course use and bulk purchases.*
*For further information, send email to <Laura.Walsh@wkap.com>.*

**Dr. Yogesh C. Patel (1942-2003)**

Dr. Yogesh C. Patel made seminal contributions to every aspect of somatostatin research including its gene expression, biosynthesis, post-translational processing, secretion, metabolism and signaling. He was the first to develop a sensitive radioimmunoassay to quantitate somatostatin in tissue extracts and plasma, document its molecular heterogeneity and map its tissue distribution. He went on to delineate the biosynthesis and processing of prosomatostatin and the pharmacology of its receptor subtypes and demonstrated the existence of a novel dimension of signaling diversity due to molecular oligomierization of these subtypes amongst themselves and with other GPCRs. He also provided strong leadership in the field through his numerous scholarly reviews and by organizing several international symposia. Dr. Patel was a colleague, collaborator or mentor and a dear friend to many of us. Dedication of this monograph on somatostatin to his memory is hence a fitting tribute to his outstanding achievements and contributions.

C.B. Srikant

# CONTENTS

## EXPRESSION AND FUNCTION OF SOMATOSTATIN RECEPTORS

**APPLICATIONS IN DISEASE**

**THE FUTURE**

# PREFACE

Thirty years have passed since somatostatin was discovered and its hormonal function defined [1,2]. The wide range of anatomical distribution and actions of somatostatin and its receptors stimulated intense scientific and clinical interest, and fostered extensive research into all aspects of its biology including regulation of its gene expression, processing of its precursor, biological and cellular actions relating to regulation cell secretion, motility and proliferation. The development of metabolically stable peptide analogs helped define its usefulness in the treatment of endocrine diseases and cancer.

The heterogeneity of somatostatin receptors recognized in 1981 [3] was later shown to result from the existence and variable expression of five distinct subtypes that constitutes two distinct subfamilies with distinguishing pharmacological characteristics [4,5]. The molecular cloning of these receptor subtypes reignited the interest in this field leading to a major increase in our insight into the biology of somatostatin and its receptor subtypes, and has led to the design and development of subtype-selective peptide and non-peptide agonists and antagonists. Significant progress has been made in both basic and clinical research on somatostatin. Noteworthy developments include the role of somatostatin in neuronal patterning, delineation of subtype-selective cytostatic / cytotoxic antiproliferative actions, manipulation of cellular signaling by oligomeric association of somatostatin receptor subtypes amongst themselves or with other G protein-coupled receptors, application of radionuclide-tagged derivatives to localize somatostatin receptor positive tumors and their treatment through targeted delivery of radionuclide or chemotoxin tagged agonists through receptor-mediated endocytosis. These advances have spurred ongoing basic and clinical research geared towards the development of rationalized somatostatin therapy in endocrine disorders, cancer, gastrointestinal bleeding and immune disorders. This volume provides an update of specific advances written by leading researchers in the field and will serve as the single source of ready reference to the diverse aspects of the research on the biology of somatostatin and its receptors.

# REFERENCES

1.      Krulich, L, Dhariwal, AP, and McCann, SM, *Stimulatory and inhibitory effects of purified hypothalamic extracts on growth hormone release from rat pituitary in vitro.* Endocrinology, 1968. 83: 783-790.
2.      Brazeau, P, Vale, W, Burgus, R, Ling, N, Butcher, M, Rivier, J, and Guillemin, R, *Hypothalamic polypeptide that inhibits the secretion of immunoreactive pituitary growth hormone.* Science, 1973. 179: 77-79.
3.      Srikant, CB and Patel, YC, *Receptor binding of somatostatin-28 is tissue specific.* Nature, 1981. 294: 259-260.
4.      Patel, YC and Srikant, CB, *Somatostatin and its receptors.* Adv Mol Cell Endocrinol, 1999. 3: 43-73.
5.      Moller, LN, Stidsen, CE, Hartmann, B, and Holst, JJ, *Somatostatin receptors.* Biochim Biophys Acta, 2003. 1616: 1-84.

# NOTE ON NOMENCLATURE

Several abbreviations have been used for somatostatin in the past. These include SRIF (somatotrophin release inhibiting factor) and somatostatin (S, SS or SST). Consequently, the two biologically active forms of somatostatin have been referred to as SRIF-14 / S-14 / SS-14 / SST-14 and SRIF-28 / S-28 / SS-28 / SST-28 respectively. In this book, the most widely accepted nomenclature of SST-14 and SST-28 are used to describe the two naturally occurring forms of somatostatin. The IUPAC had recommended the abbreviation sst for the somatostatin receptor [1]. Given that SST is the preferred abbreviation for somatostatin, SSTR is a more logical abbreviation for its receptors and continues to be widely used in current literature. In this book both abbreviations have been employed to denote somatostatin receptors.

## REFERENCES

1.  Hoyer, D, Bell, GI, Berelowitz, M, Epelbaum, J, Feniuk, W, Humphrey, PP, O'Carroll, AM, Patel, YC, Schonbrunn, A, Taylor, JE, and et al., *Classification and nomenclature of somatostatin receptors.* Trends Pharmacol Sci, 1995. 16: 86-88.

# CONTRIBUTORS

**Christian Bruns**
Head of Unit 1
Transplantation
S-386-646
Novartis Pharma AG
CH-4002 Basel, Switzerland
E-mail:

**Annamaria Colao**
Dept. of Molecular & Clinical
Endocrinology & Oncology
Federico II University of Naples
Via S Pansini 5
80131 Naples
Italy
E-mail: rpivone@tin.it

**Luis de Lecea**
Depts. of Molecular Biology and
Neuropharmacology MB-10
The Scripps Research Institute
10550 N. Torrey Pines Rd.
La Jolla, CA 92037
E-mail: llecea@scripps.edu

**David Elliott**
4611 JCP
University of Iowa
College of Medicine
200 Hawkins Drive
Iowa City, Iowa 52242, USA
E-mail:

**Jacques Epelbaum**
Directeur de Recherches INSERM
U. 159 INSERM
2ter rue d=Alesia
75014 Paris, France
E-mail: epelbaum@broca.inserm.fr

**Pekka Hayry**
Transplantation Laboratory
Haartman Institute

P.O. Box 21 (Haartmaninkatu 3)
FIN 00014 Univ. of Helsinki
Finland
E-mail: pekka.hairy@helsinki.fi

**Ujendra Kumar**
Department of Medicine
Fraser Laboratories for Diabetes
Research
Royal Victoria Hospital
687 Pine Avenue West
Montreal, Quebec H3A 1A1
E-mail:
ujendra.kumar@muhc.mcgill.ca

**Eric P. Krenning**
Department Nucleaire
Geneeskunde
Dijkzigt Hospital
P.O. Box 2040
3000 CA Rotterdam
The Netherlands
E-mail:krenning@nuge.azr.nl

**Dik Kwekkeboom**
University Hospital Rotterdam
Department of Nuclear Medicine
40 Dr Molewaterplein
3015 GD Rotterdam
Holland
E-mail:
djkwekkeboom@hotmail.com

**Malcolm Low**
Vollum Institute, L-474
Oregon Health Sciences Univ.
3181 S.W. Sam Jackson Park Rd
Portland, OR 97201-3098
E-mail:low@ohsu.edu

**Kjell Oberg**
Chairman
Dept. of Medical Sciences

Uppsala University
Dept. Of Internal Medicine
University Hospital
SE751 85
Uppsala, Sweden
E-mail: kjell.oberg@medicin.uu.se

**Jean-Claude Reubi**
Division of Cell Biology and
Experimental Cancer Research
Institute of Pathology
University of Berne
Murtenstrasse 31, P.O. Box 62
CH-3010, Berne, Switzerland
E-mail: reubi@pathology.unibe.ch

**Dietmar Richter**
Institut fur Zellbiochemie und
Klinische Neurobiologie,
Universitatsklinikum Hamburg-
Eppendorf (UKE),
Universitat Hamburg,
Suderfeldstrasse 24
22529 Hamburg
Germany
E-mail: richter@uke.uni-
hamburg.de

**Lois E.H. Smith**
Department of Ophthalmology
Children=s Hospital
300 Longwood Avenue
Boston, MA 02114
U.S.A.
E-mail:
lois.smith@childrens.Harvard.edu

**Coimbatore B. Srikant**
Fraser Laboratories for Diabetes
Research
Department of Medicine
McGill University Health Science
Center - RVH
687 Pine Avenue West
Montreal, Quebec H3A 1A1
E-mail:

coimbatore.srikant@mcgill.ca

**Christiane Susini**
Directeur de Recherche,
Unite de Recherche U531,
Biologie et pathologie Digestive
IFR 31
Chu Rangueil, Bt L3,
1 avenue Jean Poulhes
31403 Toulouse Cedex 4, France
E-mail:
susinich@toulouse.inserm.fr

**Mario Vallejo**
Instituto de Investigaciones
Biomedicas (CSIC)
Calle Arturo Duperier, 4
28029 Madrid
Spain
E-mail: mvallejo@iib.uam.es

**Hubert Vaudry**
European Institute for Peptide
Research
Laboratory of Cellular &
Molecular Neuroendocrinology
INSERM U413, UA CNRS
University of Rouen
76821 Mont-Saint-Aignan Cedex
France
E-mail: Hubert.vaudry@univ-
rouen.fr

# 1

# SOMATOSTATIN GENE STRUCTURE AND REGULATION

Mario Vallejo

*Instituto de Investigaciones Biomédicas Alberto Sols, Consejo Superior de Investigaciones Científicas y Universidad Autónoma de Madrid, 28029 Madrid, Spain.*

## 1. INTRODUCTION

Work carried out in the late sixties and early seventies led to the discovery of somatostatin (SST) as a small hypothalamic polypeptide acting on the pituitary to inhibit growth hormone secretion (Brazeau et al., 1973; Krulich et al., 1968). Subsequent work established that SST is present in other extrahypothalamic regions of the central nervous system, in cells of the peripheral nervous system, and in non-neural tissues including thyroid, gut and endocrine pancreas (Arimura et al., 1975; Hökfelt et al., 1975; Nordeen et al., 1977; Patel and Reichlin, 1978; Peinado et al., 1989; Reichlin, 1983; Rorstad et al., 1979). In each of these locations SST acts as a neurotransmitter or hormone with different functions (Reichlin, 1983).

Studies to elucidate the biosynthesis of the two forms of SST, namely somatostatin-14 (SST-14) and somatostatin-28 (SST-28), were carried out by analyzing specific products detected after radioactive labeling of hypothalamic or pancreatic cells, or synthesized *in vitro* from mRNA obtained from tissues enriched in SST (Goodman et al., 1980 and 1981; Joseph-Bravo et al., 1980; Oyama et al., 1980; Patzelt et al., 1980; Shields, 1980; Zingg and Patel, 1982). Those studies revealed that both forms of SST derive from a larger inactive precursor, preproSST, which is processed by post-translational enzymatic cleavage to yield the active polypeptide forms.

The development of recombinant DNA technology allowed the isolation and cloning of rat and human cDNAs encoding preproSST (Funckes et al., 1983; Goodman et al., 1982; Shen et al., 1982). This work established that preproSST is a polypeptide of 116 amino acids, and opened the road for the elucidation of the sequence and structure of the SST gene. The chromosomal location of the human gene was mapped to 3q28 (Zabel et al., 1983).

# 2.   STRUCTURE OF THE SOMATOSTATIN GENE

## 2.1   Coding Region

The elucidation of the nucleotide sequence of the human (Shen and Rutter, 1984), rat (Montminy et al., 1984; Tavianini et al., 1984), and mouse (Fuhrmann et al., 1990) genes encoding preproSST revealed a similar structure.

The polypeptidic SST precursor is encoded by two different exons separated by a relatively small intron (Figure 1). The first exon contains a 5' untranslated region of approximately 100 nucleotides in length. This is followed by a stretch of nucleotides encoding approximately the amino-terminal half of the preprohormone, which includes the signal peptide. The carboxy-terminal half of the polypeptide is encoded by the second exon. Within the region encoded by this exon there is a segment that corresponds to amino acids 63 to 77 of the preprohormone, thus close to the amino terminal end of SST-28. In the rat, specific immunoreactivity for this fragment was found in different areas of the central nervous system, but it has not been demonstrated that it corresponds to any biologically active hormone or neuromodulator (Lechan et al., 1983). The second exon also includes a 3' untranslated region with a typical AATAAA polyadenylation site.

The intron separates codon CAG, encoding glutamine at position 46 of the preprohormone from codon GAA, encoding glutamic acid at position 47. These two amino acids belong to the region that separates the signal peptide (amino acids 1 to 24) from the mature SST polypeptides, located in the carboxy-terminal end of the precursor. This region is not processed to generate other biologically active peptides. Therefore, the intron does not really separate regions encoding functional domains corresponding to peptides with different biological activities.

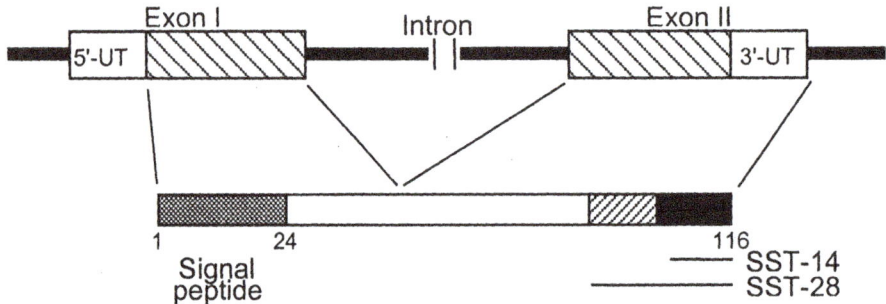

*Figure 1. Schematic depiction of the coding region of the somatostatin gene.*
*A single intron interrupts the region encoding preproSST. The two forms of somatostatin (SST-14 and SST-28) generated by post-translational enzymatic processing are located at the carboxy-terminus of the polypeptide. UT, untranslated regions of the primary RNA transcript.*

The intron between the two coding exons is longer in humans (877 base-pairs) than in rats and mice (630 and 665 base-pairs, respectively), due in part to the existence of short homopolymeric sequences and a segment containing a dinucleotide TG repeat (Shen and Rutter, 1984). This dinucleotide repeat has been found to be polymorphic, (Weber and May, 1989), generating six different alleles with a calculated heterogeneity index of 0.48.

## 2.2    Regulatory Region

The existence of DNA *cis*-regulatory elements controlling the transcription rates of the SST gene was first investigated by determining the activity of chloramphenicol acetyltransferase (CAT) generated in cells that had been transfected with SST-CAT reporter plasmids. These plasmids were constructed by sequential 5'-end deletions of a region of the promoter of the rat SST gene spanning 900 nucleotides. Transfections were carried out in different cell lines including SST-expressing thyroid and pancreatic cells (Andrisani et al., 1987; Powers et al., 1989). Those studies indicated that cell-specific expression of the SST gene requires the integrity of a proximal region of the promoter located next to the TATA box, that was also identified as a cAMP-response element (CRE) responsible for cAMP-dependent stimulation of SST gene transcription (Montminy et al., 1986a).

The octameric palindrome of the CRE (nucleotides –48 to –41 relative to the transcriptional initiation site) is separated from the TATA box, which has the variant sequence TTTAAAA, by a stretch of GA repeats (nucleotides –40 to –30). This GA-rich cassette was found to be essential for cell-specific expression in cell lines derived from pancreatic islets (Powers et al., 1989). In the light of our recent knowledge about transcriptional coactivators and TATA box binding factors that couple transcription factors to RNA polymerase II, it is possible to envisage the existence of a multiprotein complex that assembles on a region of the SST gene spanning the CRE, the GA box, and the TATA box. It is likely that this complex includes cell-specific components required for the correct spatial coupling of RNA polymerase II-activating proteins.

Upstream from the CRE, and in close proximity, there is a regulatory element that was found to be necessary for pancreatic islet D-cell-specific expression of the SST gene. This element, spanning nucleotides –114 to –78, and named SST upstream enhancer (SST-UE), acts in synergy with the adjacent CRE (Vallejo et al., 1992a) (Figure 2). The SST-UE is a tripartite element consisting of functionally interdependent domains known as domains A, B, and C (Vallejo et al., 1992a; Vallejo et al., 1992b).

The domain B of the SST-UE includes in its sequence a TAAT motif typical of a consensus binding site for homeodomain transcription factors. This sequence, also referred to as tissue-specific enhancer I (TSE-I), is known to be important

4

for the regulation of SST gene expression by homeodomain proteins (Leonard et al., 1992 and 1993; Schwartz and Vallejo, 1998; Vallejo et al., 1992a and 1992b).

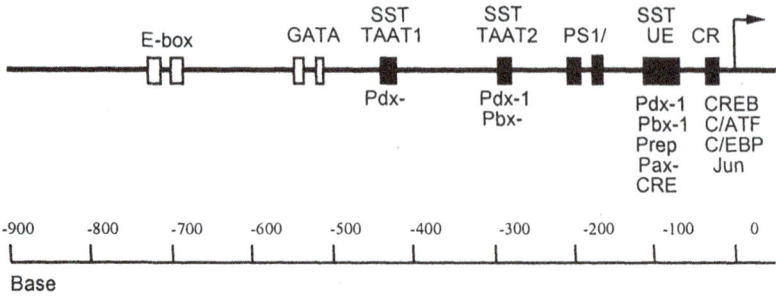

*Figure 2*. Diagram showing the 5'-flanking region of the rat somatostatin gene. Black boxes represent DNA cis-regulatory elements characterized by transfection, mutation and DNA-protein binding studies. White boxes represent sequences that contain consensus E-box or GATA motifs, but have not been demonstrated to be functionally active. Identified transcription factors known to recognize some of the regulatory elements are indicated. The scale represents the number of base-pairs relative to the transcriptional initiation site, indicated by an arrow.

There are other DNA *cis*-regulatory elements in the SST promoter that have a consensus TAAT motif and act as targets for regulation by homeodomain proteins. These are SST-TAAT1 (nucleotides –449 to –445), (Miller et al., 1994), SST-TAAT2 (nucleotides –303 to –280) also known as TSE-II (Leonard et al., 1993; Miller et al., 1994), and SST-TAAT3 (nucleotides –368 to –365) (Schwartz and Vallejo, 1998).

Transient transfection studies in pancreatic and neural cells have yielded evidence for the existence of negative acting regulatory elements distributed throughout the promoter region that repress SST gene transcription (Powers et al., 1989; Schwartz and Vallejo, 1998; Vallejo et al., 1992a). In cells derived from pancreatic islets two of these promoter silencer elements, known as proximal silencers 1 and 2 (PS1 and PS2; nucleotides –237 to –220, and –208 to –189, respectively), have been characterized (Vallejo et al., 1995a). In contrast, in thyroid cells, transcriptional regulation of the SST gene does not seem to involve active silencer elements (Andrisani et al., 1987; Medina et al., 1999). Taken together, these data indicate that cell-specific expression of the SST gene in pancreatic and neural cells is regulated by both positive- and negative-acting elements that function coordinately to modulate appropriate expression levels. Studies carried out in cells derived from embryonic rat cortex and hippocampus revealed that expression of the SST gene in neural cells is regulated by the strong positive activity of the CRE acting as an enhancer, even in the absence of stimulation by cAMP (Schwartz and Vallejo, 1998). This strong activity appears to be counterbalanced by the influence of negative-acting regulators located upstream. At least three of these elements correspond to pancreatic enhancers (see below). Thus, it appears that expression of the SST gene occurs via transcriptional mechanisms that differ depending on the specific cell type.

## 3. REGULATION BY TRANSCRIPTION FACTORS ACTING ON DNA *CIS*-REGULATORY ELEMENTS

### 3.1 bZip Proteins and Cyclic AMP Response Element (CRE)

Once the SST gene was cloned and the sequence of the 5'-flanking region determined, a search for a DNA *cis*-regulatory element required for cAMP-dependent regulation at the transcriptional level was triggered by two sets of observations. First, SST release is increased by agents that stimulate adenylylcyclase activity, suggesting that SST biosynthesis could be coupled to its release. Second, cAMP is able to stimulate SST biosynthesis by inducing the accumulation of SST mRNA levels (Montminy et al., 1986a). That work led to the characterization of the CRE as an octameric palindrome whose characteristic sequence, TGACGTCA, is conserved in many other genes that respond to cAMP stimulation with increases in their transcription rates (Montminy et al., 1986b). Thus, the CRE acts both as a cell-specific enhancer in basal conditions (see above), and as an inducible cAMP-dependent regulatory element to mediate increases in gene transcription in response to the appropriate extracellular signals.

DNA binding assays using nuclear extracts from different types of cells indicate that the CRE is able to recruit a number of protein complexes to the SST promoter (Andrisani et al., 1988; Montminy and Bilezikjian, 1987; Vallejo et al., 1995b). All proteins identified so far as SST-CRE-binding transcription factors belong to the bZip class of transcription factors (Vallejo, 1994). The bZip is a conserved domain present in all of these proteins, which contains two adjacent regions that subserve different functions. A basic region that makes direct contact with the DNA, and a leucine zipper involved in protein dimerization, a requisite for binding to DNA for this type of proteins.

The first protein to be identified as a target for the SST-CRE was the transcription factor CREB (Gonzalez et al., 1989; Hoeffler et al., 1988). CREB belongs to the bZip family of transcription factors characterized by the existence of a basic DNA-binding domain located adjacent to a so called leucine zipper domain, which is critical for dimer formation. The elucidation of the crystal structure of the CREB bZip dimer bound to the SST-CRE revealed the existence of hydrogen bonds in the basic region and leucine zipper domain between residues that are conserved in other members of the family (CREM and ATF-1) and that are critical for dimerization (Schumacher et al., 2000). In addition, it was discovered that there is a cavity between the dimeric basic region and the SST-CRE occupied by a $Mg^{2+}$ ion that enhances the binding to DNA significantly.

CREB is a substrate for phosphorylation by cAMP-dependent protein kinase A (PKA) at a specific serine residue present in the so called kinase inducible domain (KID) (Gonzalez and Montminy, 1989). In basal conditions, unphosphorylated CREB is bound to the CRE but does not activate transcription. Activation by phosphorylation only occurs after nuclear translocation of the catalytic subunit of PKA triggered by cAMP (Hagiwara et al., 1993). Once phosphorylated, the KID of CREB bound to the CRE interacts with the coactivator proteins CBP and p300, which in turn activate RNA polymerase II by direct contact with proteins of the transcriptional activation complex assembled on the TATA box (Kwok et al., 1994). In addition, CBP and p300 are histone acetyltransferases that induce histone acetylation, a critical step for CREB-induced transcription (Korzus et al., 1998).

The CBP/p300 domain that interacts with CREB is known as KIX. The interaction between KID and KIX can be inhibited by methylation of KIX induced by coactivator-associated arginine methyltransferase 1 (CARM1), thus inhibiting cAMP/CREB-dependent activation of the SST gene (Xu et al., 2001).

Recruitment of CBP by phosphorylated CREB is essential but not sufficient for transcriptional activation, because a constitutive CREB activation domain rich in glutamine residues termed Q2 is also necessary for activity. The Q2 domain stimulates transcription by interacting with specific components of TFIID on the basal transcriptional machinery assembled on the TATA box. Thus, CREB phosphorylation triggers the establishment of cooperativity between the KID and the Q2 domains in a chromatin-dependent manner, leading to activation of SST gene expression (Asahara et al., 2001).

The SST-CRE is also recognized by other bZip transcription factors in pancreatic islet-derived cells. These factors include JunD, C/ATF and C/EBPβ (Ubeda et al., 1999; Vallejo et al., 1995b). JunD is an AP-1-like protein of the Jun/Fos family of transcription factors, and can activate the SST promoter in response to cAMP stimulation (Ubeda et al., 1999). On the other hand C/EBPβ is a member of the CAAT-box/enhancer binding proteins (Williams et al., 1991), and acts as a transcriptional transactivator protein in basal conditions (Vallejo et al., 1995b). The other protein, C/ATF, is a C/EBP-related activating transcription factor that can bind the CRE as a homodimer, but is also able to form heterodimers with C/EBP proteins to recruit them to the CRE (Vallejo et al., 1993).

In SST-producing RIN-1027-B2 cells, derived from a rat insulinoma, C/ATF and C/EBPβ act as potent transcriptional transactivators of the SST gene in a cAMP-independent manner (Vallejo et al., 1995b). In these cells the SST gene maintains readily detectable basal levels of expression, but it is not responsive to cAMP stimulation (Patel et al., 1991; Powers et al., 1989; Vallejo et al., 1992a and 1995b). This circumstance appears to be due to the expression of a heat-stable inhibitor of PKA that prevents phosphorylation of CREB (Vallejo et al., 1995b). As a consequence, dephosphorylated CREB represses transcription of the SST gene by competing with C/ATF and C/EBPβ for binding to the CRE. Thus, in this case, CREB acts as a repressor rather than an activator of SST gene transcription. Interestingly, a different SST producing cell line derived from pancreatic islets, Tu6 cells, also exhibit a defect that prevents phosphorylation of CREB. In these cells, however, CREB is able to transactivate the SST gene by interacting with the homeodomain protein Isl-1 bound in its proximity (Leonard et al., 1992).

## 3.2 Homeodomain Proteins and TAAT Elements

Most DNA cis-regulatory elements that recognize homeodomain transcription factors are characterized by the presence of a canonical TAAT motif. This TAAT motif is present in several elements located in the SST gene 5'-flanking region. These include SST-UE, SST-TAAT1, SST-TAAT2 and SST-TAAT3 (Leonard et al., 1992 and 1993; Miller et al., 1994; Schwartz and Vallejo, 1998; Vallejo et al., 1992a and 1992b).

The first homeodomain transcription factor to be identified as a transactivator of the SST gene was Pdx-1 (also known as STF-1, IDX-1, and IPF-1) (Leonard et al., 1993; Miller et al., 1994; Ohlsson et al., 1993). Pdx-1 regulates the expression of several pancreatic genes besides SST (Carty et al., 1997; Ohlsson et al., 1993; Waeber et al., 1996). In addition, it is required for the formation of the pancreas during development and for the correct function of pancreatic islets in the adult (Ahlgren et al., 1998; Jonsson et al., 1994; Offield et al., 1996; Stoffers et al., 1997a and 1997b). Pdx-1 regulates SST gene transcription in pancreatic cells by binding to SST-TAAT1, SST-TAAT2 and SST-UE-B (Andersen et al., 1999; Goudet et al., 1999; Leonard et al., 1993; Lu et al., 1996; Miller et al., 1994; Peers et al., 1995).

Recently, Pdx-1 was found to be expressed in the central nervous system during embryonic development (Pérez-Villamil et al., 1999), and to regulate SST gene transcription in neural cells (Schwartz et al., 2000). Interestingly, Pdx-1 does not recognize SST-TAAT1 or SST-TAAT2 in neural cells. In these cells, SST gene regulation by Pdx-1 is accomplished by binding to SST-TAAT3 and SST-UE-B (Schwartz et al., 2000). Thus, the observation that Pdx-1 binds to SST-TAAT1 and SST-TAAT2 only in pancreatic cells suggests that differences in the protein environment within the nucleus of neural or pancreatic cells may promote cell-specific protein-protein interactions generating heterodimeric complexes with different DNA binding specificity.

Not much information is available on the role of specific homeodomain proteins as direct regulators of SST gene expression in the central nervous system. Most of the available evidence has been provided by work carried out using genetically engineered mice in which specific regulatory genes have been deleted. In those animals, different populations of cells that normally express SST are not present, but it is not possible to establish whether lack of SST gene expression is selective and due to the absence of a specific homeodomain transcription factor, or whether that transcription factor is required for the generation of cells that secondarily would express the SST gene. Thus, Marín et al. (2000) observed that striatal interneurons expressing SST are absent in mice lacking the homeodomain gene *Nkx2.1*. In a different study, it was observed that deletion of *Orthopedia*, another homeodomain-encoding gene, results in absence of SST-expressing neurons in the hypothalamic periventricular and arcuate nuclei, although other types of hypothalamic cells that express different neuropeptides were also affected by this mutation.

In pancreatic cells, Pdx-1 is known to interact with other homeodomain proteins including the atypical homeodomain Pbx-1 and Prep1 transcription factors (Peers et al., 1995, Goudet et al., 1999) and the Pair-domain homeoprotein Pax-6 (Andersen et al., 1999). The interaction between Pdx-1 and Pbx-1 is cooperative, and the resulting heterodimer recognizes the SST-TAAT2 element. This interaction requires the integrity of the Pdx-1 homeodomain and of a pentapeptide motif (Phe-Pro-Trp-Met-Lys) located in the amino-terminal region of the protein, separated from the homeodomain by 22 residues (Peers et al., 1995). This motif is present in a significant number of homeodomain proteins that are known to interact with Pbx-like proteins, all of which belong to the three amino acid loop extension-class of atypical homeodomain proteins whose prototype is the *Drosophila* Extradenticle (Mann and Chan, 1996).

As mentioned earlier, Pdx-1 also binds to the SST-UE-B. The adjacent SST-UE-A contains the sequence TGATTGA, which corresponds to a perfect consensus binding site for Pbx proteins. In this case, however, Pbx-1 interacts directly with another atypical homeoprotein, Prep1 (Berthelsen et al., 1998). Thus, the heterodimeric Pbx-1/Prep1 complex binds to the SST-UE-A and forms a ternary complex with Pdx-1 bound on the adjacent SST-UE-B (Goudet et al., 1999).

Finally, the homeodomain protein Pax-6, which is one of the Pair-class transcription factors (Mansouri et al., 1996), binds to the SST-UE-C and appears to form a complex with Pdx-1 (Andersen et al., 1999). In addition, the SST-UE-C was found to be recognized by the transcription factor CREB despite the lack of sequence similarity with the consensus CRE (Vallejo et al., 1992b).

Taken together, these data indicate that the SST-UE is an important regulatory element for the assembly of a complex containing different types of interacting homeodomain proteins (Figure 3). At least in pancreatic cells, the SST-UE acts in synergy with the almost adjacent CRE. Therefore, both elements appear to act as components of an important enhancer unit that is critical for cell-specific expression of the SST gene.

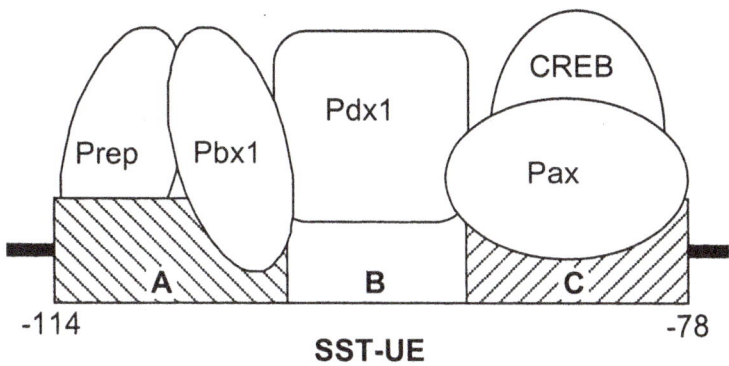

*Figure 3.* Schematic representation of identified transcription factors binding to the SST-UE. The three different domains of SST-UE, A, B and C, are shown. Evidence for interactions between Prep1 and Pbx-1, Pbx-1 and Pdx-1, and Pdx-1 and Pax-6 have been reported. Interactions between CREB and any of these homeodomain transcription factors have not been studied. Numbers indicate the boundaries of the SST-UE in terms of nucleotide position relative to the transcriptional initiation site.

## 3.3    Regulation by Promoter Silencer Elements

The SST gene 5'-flanking region contains DNA *cis*-regulatory elements that repress gene transcription in pancreatic and neural cells. However, the specific functions of these elements and their distribution differ between both types of cells (Powers et al., 1989; Schwartz and Vallejo, 1998; Vallejo et al., 1992a). For example, transient transfection experiments indicate the existence of repressor elements located in the region spanning nucleotides –750 to – 550, but these appear to be functional only in neural cells.

In pancreatic cells, active negative control elements appear to be located downstream from nucleotide –425. Thus, distal repressors acting in neural cells may correspond to regulatory elements important for the modulation of adequate levels of expression of the SST gene specifically in the nervous system (Schwartz and Vallejo, 1998).

Differences in the function of regulatory elements are not only evident between pancreatic and neural cells, but also exist between neural cells derived from different regions of the brain. Thus, the pancreatic proximal silencers PS1 and

PS2 (Vallejo et al., 1995a) are active in cerebrocortical cells but not in hippocampal cells (Schwartz and Vallejo, 1998).

Transcription factors that recognize PS1 and PS2 in the pancreas have not been identified. PS2 has an E-box (CATGTG) typical of binding sites for transcription factors of the basic-helix-loop-helix (bHLH) family, known to regulate transcription of pancreatic genes (Dumonteil et al., 1998). On the other hand, PS1 contains a TAAT motif typical of homeodomain-binding sites.

Notably, the pancreatic enhancers SST-TAAT1, SST-TAAT2 and SST-UE-B act as silencer elements in neural cells, since mutation of each one of these elements results in increased levels of SST promoter activity (Schwartz and Vallejo, 1998). It remains to be determined whether homeodomain proteins that recognize these elements in neural cells act as transcriptional repressors, or whether they act as transactivators that compete for binding with brain-specific repressor proteins to the same DNA sites.

## 3.4    Regulation by bHLH Proteins

In addition to the E-box present within the PS2 element mentioned in the previous section, the SST promoter contains at least two other E-box elements located approximately 700 base-pairs upstream from the transcription start site (see Figure 2). To date, no bHLH transcription factors that bind these sites or regulate SST gene expression directly in endocrine or neural cells have been described. However, genetic evidence indicates that expression of certain bHLH-encoding genes during development is important for the differentiation of different endocrine cells that express SST.

Genetic deletion of the bHLH *neurogenin 3* (*ngn3*) gene, which is transiently expressed in endocrine progenitor cells of the developing pancreas, results in failure to generate all four islet cell types (Gradwohl et al., 2000). In addition, SST-expressing cells of the stomach and intestine require expression of *ngn3* for their differentiation (Lee et al., 2002; Jenny et al., 2002). In the pancreas, expression of the bHLH transcription factor NeuroD/BETA2 is also required for the proper development of the endocrine cells of islets, but deletion of the *NeuroD/BETA2* gene does not alter SST gene expression in intestinal cells (Naya et al, 1997).

In the central nervous system, absence of hypothalamic SST-producing cells has been documented in mice lacking the bHLH-encoding genes *Sim1* (Michaud et al., 1998) or *Arnt2* (Hosoya et al., 2001).

# 4. REGULATION OF SST EXPRESSION BY EXTRACELLULAR SIGNALS

The effects of a number of extracellular-acting agents on SST gene expression have been investigated for a number of years. This work has been carried out generally using neural cells, although a report presenting evidence that SST gene transcription is regulated by TSH in thyroid cells has been published recently (Medina et al., 1999). In general, a correlation between the effects induced by extracellular agents and the DNA *cis*-regulatory elements that could act as targets for the intracellular signaling pathways activated by those agents have not been documented.

In cells from the central nervous system, TGFβ and leptin were shown to decrease SST mRNA levels (Quintela et al., 1997a and 1997b). Although a direct mechanism repressing gene transcription cannot be ruled out, it appears that TGFβ acts primarily by inducing the degradation of SST mRNA (Quintela et al., 1997b). Thyroid and steroid hormones have also been reported to induce changes in SST mRNA levels in hypothalamic and cerebrocortical cells (De los Frailes et al., 1988; Lorenzo et al., 1995; Werner et al., 1988). However, direct regulation of SST gene transcription by nuclear receptors has not been reported.

Finally, depolarization of fetal neurons induced by potassium results in an increase in the levels of SST mRNA (Tolón et al., 1994). This is a calcium-dependent effect that appears to be related with a stimulation of adenylate cyclase activity (Tolón et al., 2000), and therefore it is possibly mediated by the CRE.

# 5. CONCLUDING REMARKS

Since the elucidation of the sequence of the SST gene in the mid-eighties, a great deal of progress has been accomplished in our understanding of the molecular mechanisms that regulate the expression of the SST gene at the transcriptional level. However, important questions remain to be solved. Positive- and negative-acting DNA *cis*-regulatory elements responsible for the cell-specific expression of the gene in neural and other tissues need to be identified and fully characterized. Additional transcription factors that specifically recognize SST DNA *cis*-regulatory elements have to be isolated and their mechanism of action on the basal transcriptional machinery elucidated. Recently, the notion that chromatin structure and regulation are clearly part of the molecular mechanisms that regulate gene transcription has emerged as an important component of control. This aspect needs to be explored in detail in neural, pancreatic, thyroid, and other cells to understand how the SST gene operates in different cell types. Clearly, a substantial amount of exciting work lies ahead of us.

# REFERENCES

1. Acampora, D., Postiglione, M.P., Avantaggiato, V., Di Bonito, M., Vaccarino, F.M., Michaud, J. and Simeone, A. (1999) Progressive impairment of developing neuroendocrine cell lineages in the hypothalamus of mice lacking the *Orthopedia* gene. Genes Dev. 13: 2787-2800.
2. Ahlgren, U., Jonsson, J., Jonsson, L., Simu, K. and Edlund, H. (1998) β-Cell-specific inactivation of the mouse Ipf1/Pdx1 gene results in loss of the β-cell phenotype and maturity onset diabetes. Genes Dev. 12: 1763-1768.
3. Andersen, F.G., Jensen, J., Heller, R.S., Petersen, H.V., Larsson, L.I., Madsen, O.D. and Serup, P. (1999) Pax6 and Pdx1 form a functional complex on the rat somatostatin gene upstream enhancer. FEBS lett. 445: 315-320.
4. Andrisani, O.U., Hayes, T.E., Roos, B. and Dixon, J.E. (1987) Identification of the promoter sequences involved in the cell-specific expression of the rat somatostatin gene. Nucleic Acids Res. 15: 5715-5728.
5. Andrisani, O.M., Pot, D.A., Zhu, Z. and Dixon, J.E. (1988) Three sequence-specific DNA-protein complexes are formed with the same promoter element essential for expression of the rat somatostatin gene. Mol. Cell. Biol. 8: 1947-1956.
6. Arimura, A., Sato, H., Dupont, A., Nishi, N. and Schally, A. (1975) Somatostatin: abundance of immunoreactive hormone in rat stomach and pancreas. Science 189: 1007-1009.
7. Asahara, H., Santoso, B., Guzman, E., Du, K., Cole, P.A., Davidson, I. and Montminy, M. (2001) Chromatin-dependent cooperativity between constitutive and inducible activation domains in CREB. Mol. Cell. Biol. 21: 7892-7900.
8. Berthelsen, J., Zappavigna, V., Ferretti, E., Mavilio, F. and Blasi, F. (1998) The novel homeoprotein Prep1 modulates Pbx-Hox protein cooperativity. EMBO J. 17: 1434-1445.
9. Brazeau, P., Vale, W., Burgus, R., Ling, N., Butcher, M., Rivier, J. and Guillemin, R. (1973) Hypothalamic polypeptide that inhibits secretion of immunoreactive pituitary growth hormone. Science 179: 77-79.
10. Carty, M.D., Lillquist, J.S., Peshavaria, M., Stein, R. and Soeller, W.C. (1997) Identification of cis- and trans-active factors regulating human islet amyloid polypeptide gene expression in pancreatic β-cells. J. Biol. Chem. 272: 11986-11993.
11. De los Frailes, M.T., Cacicedo, L., Lorenzo, M.J., Fernández, G. and Sánchez-Franco, F. (1988) Thyroid hormone action on biosynthesis of somatostatin by fetal rat brain cells in culture. Endocrinology 123: 898-904.
12. Dumonteil, E., Laser, B., Constant, I. and Phillipe, J. (1998) Differential regulation of the glucagon and insulin I gene promoters by the basic helix-loop-helix transcription factors E47 and BETA2. J. Biol. Chem. 273: 19945-19954.
13. Fuhrmann, G., Heiling, R., Kempf, J. and Ebel, A. (1990) Nucleotide sequence of the mouse preprosomatostatin gene. Nucleic Acids Res. 18: 1287.
14. Funckes, C.L., Minth, C.D., Deschenes, R., Magazin, M., Tavianini, M.A., Sheets, M., Collier, K., Weith, H.L., Aron, D.C., Ross, B.A. and Dixon, J.E. (1983) Cloning and characterization of a mRNA-encoding rat preprosomatostatin. J. Biol. Chem. 258: 8781-8787.
15. Gonzalez, G.A. and Montminy, M.R. (1989) Cyclic AMP stimulates somatostatin gene transcription by phosphorylation of CREB at serine 133. Cell 59: 675-680.
16. Gonzalez, G.A., Yamamoto, K.K., Fischer, W.H., Karr, D., Menzel, P., Biggs, W., Vale, W.W. and Montminy, M.R. (1989) A cluster of phosphorylation sites on the cyclic AMP-regulated nuclear factor CREB predicted by its sequence. Nature 337: 747-752.

17. Goodman, R.H., Lund, P.K., Jacobs, J.W. and Habener, J.F. (1980) Pre-prosomatostatins. Products of cell-free translation of messenger RNA from anglerfish islets. J. Biol. Chem. 255

18. Goodman, R.H., Lund, P.K., Barnett, F.H. and Habener, J.F. (1981) Intestinal pre-prosomatostatin. Identification of mRNA coding for a precursor by cell-free translation and hybridization with a cloned islet cDNA. J. Biol. Chem. 256: 1499-1501.

19. Goodman, R.H., Jacobs, J.W., Dee, P.C. and Habener, J.F. (1982) Somatostatin-28 encoded in a cloned cDNA obtained from a rat medullary thyroid carcinoma. J. Biol. Chem. 257: 1156-1159.

20. Goudet, G., Delhalle, S., Biemar, F., Martial, J.A. and Peers, B. (1999) Functional and cooperative interactions between the homeodomain PDX1, Pbx, and Prep1 factors on the somatostatin promoter. J. Biol. Chem. 274: 4067-4073.

21. Gradwohl, G., Dierich, A., LeMeur,, M. and Guillemot, F. (2000) neurogenin3 is required for the development of the four endocrine cell lineages of the pancreas. Proc. Natl. Acad. Sci. USA 97: 1607-1611.

22. Hagiwara, M., Brindle, P., Harootunian, A., Armstrong, R., Rivier, J., Vale, W.W., Tsien, R. and Montminy, M.R. (1993) Coupling of hormonal stimulation and transcription via the cyclic AMP-responsive factor CREB is rate limited by nuclear entry of protein kinase A. Mol. Cell. Biol. 13: 4852-4859.

23. Hoeffler, J.P., Meyer, T.E., Yun, Y., Jameson, J.L. and Habener, J.F. (1988) Cyclic AMP responsive DNA-binding protein: structure based on a cloned placental cDNA. Science 242: 1430-1433.

24. Hökfelt, T., Efendic, S., Hellerström, C., Johansson, O., Luft, R. and Arimura, A. (1975) Cellular localization of somatostatin in endocrine-like cells and neurons of the rat with special reference to de $A_1$-cells of the pancreatic islets and to the hypothalamus. Acta Endocrinol. 80 (suppl. 200): 1-41.

25. Hosoya, T., Yoshihito, O., Takahashi, S., Morita, M., Kawauchi, S., Ema, M., Yamamoto, M. and Fujii-Kuriyama, Y. (2001) Defective development of secretory neurones in the hypothalamus of Arnt2-knockout mice. Genes Cells 6: 361-374.

26. Jonsson, J., Carlsson, L., Edlund, T. and Edlund, H. (1994) Insulin-promoter-factor 1 is required for pancreas development in mice. Nature 371: 606-609.

27. Joseph-Bravo, P., Charli, J.L., Sherman, T., Boyer, H., Bolivar, F. and McKelvy, J.F. (1980) Identification of a putative hypothalamic mRNA coding for somatostatin and of its product in cell-free translation. Biochem. Biophys. Res. Commun. 94: 1004-1012.

28. Korzus, E., Torchia, J., Rose, D.W., Xu, L., Kurokawa, R., McInerney, E.M., Mullen, T.M., Glass, C.K. and Rosenfeld, M.G. (1998) Transcription factor–specific requirements for coactivators and their acetyltransferase functions. Science 279: 703-707.

29. Krulich, L., Dhariwal, A.P.S. and McCann, S.M. (1968) Stimulatory and inhibitory effects of purified hypothalamic extracts on growth hormone release from rat pituitary in vitro. Endocrinology 83: 783-790.

30. Kwok, R.P.S., Lundblad, J.P., Chrivia, J.C., Richards, J.P., Bachinger, P., Brennan, R.G., Roberts, S.E.G., Green, M.R. and Goodman, R.H. (1994) Nuclear protein CBP is a coactivator for the transcription factor CREB. Nature 370: 223-226.

31. Lechan, R.M., Goodman, R.H., Rosenblat, M., Reichlin, S. and Habener, J.F. (1983) Prosomatostatin-specific antigen in rat brain: Localization by immunocytochemical staining with an antiserum to a synthetic sequence of preprosomatostatin. Proc. Natl. Acad. Sci. USA 80: 2780-2784.

32. Leonard, J., Serup, P., Gonzalez, G., Edlund, T. and Montminy, M. (1992) The LIM family transcription factor Isl-1 requires cAMP response element binding protein to promote somatostatin expression in pancreatic islet cells. Proc. Natl. Acad. Sci. USA 89: 6247-6251.

33. Leonard, J., Peers, B., Johnson, T., Ferreri, K., Lee, S. and Montminy, M. (1993) Characterization of somatostatin transactivating factor-1, a novel homeobox factor that stimulates somatostatin expression in pancreatic islet cells. Mol. Endocrinol. 7: 1275-1283.

34. Lorenzo, M.J., Cacicedo, L., Tolón, R.M., Balsa, J.A. and Sánchez-Franco, F. (1995) Triiodothyronine regulates somatostatin gene expression in cultured fetal rat cerebrocortical cells. Peptides 16: 249-253.
35. Lu, M., Miller, C.P. and Habener, J.P. (1996) Functional regions of the homeodomain protein IDX-1 required for transactivation of the rat somatostatin gene. Endocrinology 137: 2959-2967.
36. Mann, R.S. and Chan, S.K. (1996) Extra specificity from extradenticle: The partnership between Hox and Exd/Pbx proteins. Trends Genet. 12: 258-262.
37. Mansouri, A., Hallonet, M. and Gruss, P. (1996) Pax genes and their roles in cell differentiation and development. Curr. Opin. Cell Biol. 8: 851-857.
38. Marín, O., Anderson, S.A. and Rubenstein, J.L.R. (2000) Origin and molecular specification of striatal interneurons. J. Neurosci. 20: 6063-6076.
39. Medina, D.L., Velasco, J.A. and Santisteban, P. (1999) Somatostatin is expressed in FRTL-5 thyroid cells and prevents thyrotropin-mediated down-regulation of the cyclin-dependent kinase inhibitor p27$^{kip1}$. Endocrinology 140: 87-95.
40. Michaud, J.L., Rosenquist, T., May, N.R. and Fan, C.M. (1998) Development of neuroendocrine lineages requires the bHLH-PAS transcription factor SIM1. Genes Dev. 12: 3264-3275.
41. Miller, C.P., McGehee, R.E. and Habener, J.F. (1994) IDX-1: a new homeodomain transcription factor expressed in rat pancreatic islets and duodenum that transactivates the somatostatin gene. EMBO J. 13: 1145-1156.
42. Montminy, M.R. and Bilezikjian, L.M. (1987) Binding of a nuclear protein to the cyclic-AMP response element of the somatostatin gene. Nature 328: 175-178.
43. Montminy, M.R., Goodman, R.H., Horovitch, S.J. and Habener, J.F. (1984) Primary structure of the gene encoding rat preprosomatostatin. Proc. Natl. Acad. Sci. USA 81: 3337-3340.
44. Montminy, M.R., Low, M.J., Tapia-Arancibia, L., Reichlin, S., Mandel, G. and Goodman, R.H. (1986a) Cyclic AMP regulates somatostatin mRNA accumulation in primary diencephalic cultures and in transfected fibroblast cells. J. Neurosci. 6: 1171-1176.
45. Montminy, M.R., Sevarino, K., Wagner, J.A., Mandel, G. and Goodman, R.H. (1986b) Identification of a cyclic-AMP-responsive element within the rat somatostatin gene. Proc. Natl. Acad. Sci. USA 83: 6682-6686.
46. Naya, F.J., Huang, H.P., Qiu, Y., DeMayo, F.J., Leiter, A.B. and Tsai, M.J. (1997) Diabetes, defective pancreatic morphogenesis, and abnormal enteroendocrine differentiation in BETA2/NeuroD-deficient mice. Genes Dev. 11: 2323-2334.
47. Nordeen, S.V., Polak, J.M. and Pearse, A.G.E. (1977) Single cellular origin of somatostatin and calcitonin in the rat thyroid gland. Histochemistry 53: 243-247.
48. Offield, M.F., Jetton, T.L., Labosky, P.A., Ray, M., Stein, R.W., Magnuson, M.A., Hogan, B.L. and Wright, C.V. (1996) PDX-1 is required for pancreatic outgrowth and differentiation of the rostral duodenum. Development 122: 983-995.
49. Ohlsson, H., Karlsson, K. and Edlund, T. (1993) IPF-1, a homeodomain-containing transactivator of the insulin gene. EMBO J. 12: 4251-4259.
50. Oyama, H., O'Connell, K. and Permutt, A. (1980) Cell-free synthesis of somatostatin. Endocrinology 107: 845-847.
51. Patel, Y.C. and Reichlin, S. (1978) Somatostatin in hypothalamus, extrahypothalamic brain and peripheral tissues of the rat. Endocrinology 102: 523-530.
52. Patel, C.Y., Papachristou, D.N., Zingg, H.H. and Farkas, E.M. (1991) Regulation of islet somatostatin secretion and gene expression: selective effects of adenosine 3',5'-monophosphate and phorbol esters in normal islets of Langerhans and in a somatostatin-producing rat islet clonal cell line 1027-B2. Endocrinology 128: 1754-1762.
53. Patzelt, C., Tager, H.S., Carroll, R.J. and Steiner, D.F. (1980) Identification of prosomatostatin in pancreatic islets. Proc. Natl. Acad. Sci. USA 77: 2410-2414.
54. Peers, B., Sharma, S., Johnson, T., Kamps, M. and Montminy, M. (1995) The pancreatic islet factor STF-1 binds cooperatively with Pbx to a regulatory element in the somatostatin

promoter: Importance of the FPWMK motif and the homeodomain. Mol. Cell. Biol. 15: 7091-7097.

55. Peinado, M.A., Viader, M., Puig-Domingo, M., Hernández, G., Reiter, R.J. and Webb, S. (1989) Regional distribution of immunoreactive somatostatin in the bovine pineal gland. Neuroendocrinology 50: 550-554.

56. Pérez-Villamil, B., Schwartz, P. and Vallejo, M. (1999) The pancreatic homeodomain transcription factor IDX1/IPF1 is expressed in neural cells during brain development. Endocrinology 140: 3857-3860.

57. Powers, A.C., Tedeschi, F., Wright, K.E., Chan, J.S. and Habener, J.F. (1989) Somatostatin gene expression in pancreatic islet cells is directed by cell-specific DNA control elements and DNA-binding proteins. J. Biol. Chem. 264: 10048-10056.

58. Quintela, M., Señaris, R.M., Heiman, M.L., Casanueva, F.F. and Dieguez, C. (1997a) Leptin inhibits in vitro hypothalamic somatostatin secretion and somatostatin mRNA levels. Endocrinology 138: 5641-5644.

59. Quintela, M., Señaris, R.M. and Dieguez, C. (1997b) Transforming growth factor-βs inhibit somatostatin messenger ribonucleic acid levels and somatostatin secretion in hypothalamic cells in culture. Endocrinology 138: 4401-4409.

60. Reichlin, S. (1983) Somatostatin. N. Engl. J. Med. 309: 1495-1501.

61. Rorstad, O.P., Epelbaum, J., Brazeau, P. and Martin, J.B. (1979) Chromatographic and biological properties of immunoreactive somatostatin in hypothalamic and extrahypothalamic brain regions of the rat. Endocrinology 105: 1083-1090.

62. Schumacher, M.A., Goodman, R.H. and Brennan, R.G. (2000) The structure of a CREB bZIP-somatostatin CRE complex reveals the basis for selective dimerization and divalent cation-enhanced DNA binding. J. Biol. Chem. 275: 35242-35247.

63. Schwartz, P.T. and Vallejo, M. (1998) Differential regulation of basal and cyclic adenosine 3',5'-monophosphate-induced somatostatin gene transcription in neural cells by DNA control elements that bind homeodomain proteins. Mol. Endocrinol. 12: 1280-1293.

64. Schwartz, P.T., Pérez-Villamil, B., Rivera, A., Moratalla, R. and Vallejo, M. (2000) Pancreatic homeodomain transcription factor IDX1/IPF1 expressed in developing brain regulates somatostatin gene transcription in embryonic neural cells. J. Biol. Chem. 275: 19106-19114.

65. Shen, L.P., Pictet, R.L. and Rutter, W.J. (1982) Human somatostatin I: Sequence of the cDNA. Proc. Natl. Acad. Sci. USA 79: 4575-4579.

66. Shen, L.P. and Rutter, W.J. (1984) Sequence of the human somatostatin I gene. Science 224: 168-171.

67. Shields, D. (1980) In vitro biosynthesis of somatostatin. J. Biol. Chem. 255: 11625-11628.

68. Stoffers, D.A., Ferrer, J., Clarke, W.L. and Habener, J.F. (1997a) Early-onset type-II diabetes mellitus (MODY4) linked to IPF-1. Nat. Genet. 17: 138-139.

69. Stoffers, D.A., Zinkin, N.T., Stanojevic, V., Clarke, W.L. and Habener, J.F. (1997b) Pancreatic agenesis attributable to a single nucleotide deletion in the human IPF1 coding region. Nat. Genet. 15: 106-110.

70. Tavianini, M.A., Hayes, T.E., Magazin, M.D., Minth, C.D. and Dixon, J.E. (1984) Isolation, characterization, and DNA sequence of the rat somatostatin gene. J. Biol. Chem. 259: 11798-11803.

71. Tolón, R.M., Sánchez-Franco, F., De los Frailes, M.T., Lorenzo, M.J. and Cacicedo, L. (1994) Effect of potassium-induced depolarization on somatostatin gene expression in cultured fetal rat cerebrocortical cells. J. Neurosci. 14: 1053-1059.

72. Tolón, R.M., Sánchez-Franco, F., Villuendas, G., Vicente, A.B., Palacios, N. and Cacicedo, L. (2000) Potassium depolarization-induced cAMP stimulates somatostatin mRNA levels in cultured diencephalic neurons. Brain Res. 868: 338-346.

73. Ubeda, M., Vallejo, M. and Habener, J.F. (1999) CHOP enhancement of gene transcription by interactions with Jun/Fos AP-1 complex proteins. Mol. Cell. Biol. 19: 7589-7599.

74. Vallejo, M. (1994) Transcriptional control of gene expression by cAMP-response element binding proteins. J. Neuroendocrinol. 6: 587-596.

16

75. Vallejo, M., Miller, C.P. and Habener, J.F. (1992a) Somatostatin gene transcription regulated by a bipartite pancreatic islet D-cell-specific enhancer coupled synergetically to a cAMP response element. J. Biol. Chem. 267: 12868-12875.

76. Vallejo, M., Penchuk, L. and Habener, J.F. (1992b) Somatostatin gene upstream enhancer element activated by a protein complex consisting of CREB, Isl-1-like, and α-CBF-like transcription factors. J. Biol. Chem. 267: 12876-12884.

77. Vallejo, M., Ron, D., Miller, C.P. and Habener, J.F. (1993) C/ATF, a member of the activating transcription factor family of DNA-binding proteins, dimerizes with CAAT/enhancer-binding proteins and directs their binding to cAMP response elements. Proc. Natl. Acad. Sci. USA 90: 4679-4683.

78. Vallejo, M., Miller, C.P., Beckman, W. and Habener, J.F. (1995a) Repression of somatostatin gene transcription mediated by two promoter silencer elements. Mol. Cell. Endocrinol. 113: 61-72.

79. Vallejo, M., Gosse, M., Beckman, W. and Habener, J.F. (1995b) Impaired cyclic AMP-dependent phosphorylation renders CREB a repressor of C/EBP-induced transcription of the somatostatin gene in an insulinoma cell line. Mol. Cell. Biol. 15: 415-424.

80. Weber, J.L. and May, P.E. (1989) Abundant class of human DNA polymorphiSST which can be typed using polymerase chain reaction. Am. J. Hum. Genet. 44: 388-396.

81. Waeber, G., Thompson, N., Nicod, P. and Bonny, C. (1996) Transcriptional activation of the GLUT2 gene by the IPF-1/STF-1/IDX-1 homeobox factor. Mol. Endocrinol. 10:1327-1334.

82. Werner, H., Koch, Y., Baldino, F. and Gozes, I. (1988) Steroid regulation of somatostatin mRNA in the rat hypothalamus. J. Biol. Chem. 263: 7666-7671.

83. Williams, S.C., Cantwell, C.A. and Johnson, P.F. (1991) A family of C/EBP-related proteins capable of forming covalently linked leucine zipper dimers in vitro. Genes Dev. 5: 1553-1567.

84. Xu, W., Chen, H., Du, K., Asahara, H., Tini, M., Emerson, B.M., Montmini, M. and Evans, R.M. (2001) A transcriptional switch mediated by cofactor methylation. Science 294: 2507-2511.

85. Zabel, B.U., Naylor, S.L., Sakaguchi, A.Y., Bell, G.I. and Shows, T.B. (1983) High resolution chromosomal location of human genes for amylase, proopiomelanocortin, somatostatin, and a DNA fragment (D3S1) by in situ hybridization. Proc. Natl. Acad. Sci. USA 80: 6932-6936.

86. Zingg, H.H. and Patel, Y.C. (1982) Biosynthesis of immunoreactive somatostatin by hypothalamic neurons in culture. J. Clin. Invest. 70:1101-1109.

# 2
# PROCESSING AND INTRACELLULAR TARGETING OF SOMATOSTATIN

Rania Mouchantaf, Yogesh C. Patel and Ujendra Kumar
*Fraser Laboratories, Departments of Medicine, Pharmacology and Therapeutics, Neurology and Neurosurgery, McGill University, Royal Victoria Hospital and Montreal Neurological Institute, Montreal, Quebec, Canada H3A 1A1.*

## INTRODUCTION

Since the discovery in 1973 of the tetradecapeptide somatostatin 14 (SST-14) (1), which is the growth hormone inhibitory factor derived from the mammalian hypothalamus, considerable progress has been made in understanding the enzymes and secretory pathways implicated in somatostatin (SST) production. Mammalian preprosomatostatin (PPSST) is synthesized as an inactive precursor consisting of 119 amino acids which is cleaved post-translationally to yield several mature products (1, 2). Processing principally occurs at the C-terminal segment generating the biologically active forms of the precursor. In addition, a second cleavage takes place at the amino terminus of the molecule resulting in a product of unknown function. However high degree of sequence homology throughout vertebrate evolution within that region argues that it may harbor information for properly directing PSST within the secretory pathway (3).

## 1.    Prosomatostatin Processing

By means of genetic recombinant techniques the entire sequence of PPSST has been elucidated for several tissues (4, 5). In common with other peptide-synthesizing systems entering the secretory pathway, SST is initially synthesized as part of an inactive precursor molecule, PPSST, on ribosomes in the rough endoplasmic reticulum (RER). Once translated, rapid cleavage via a signal peptidase produces the prohormone form, prosomatostatin (PSST), that is later processed enzymatically to yield several mature products (6). Processing of PSST principally occurs at the C-terminal end generating the two bioactive forms SST-14 and SST-28 (Figure 1). SST-14 is produced through dibasic cleavage at an Arg-Lys residue, whereas endoproteolysis at a monobasic Arg site generates SST-28. Mammalian tissues contain variable mixtures of SST-14 and SST-28 due to differential processing of PSST (7). In the pancreatic islets, stomach, and neural tissues SST-14 predominates; it is

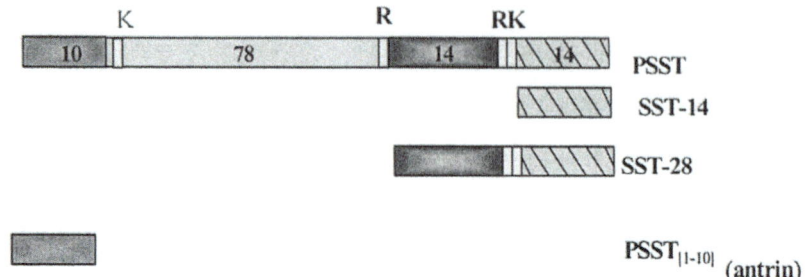

*Figure 1. Diagram depicting mammalian PSST and its cleavage products.*

virtually the only form in retina, peripheral nerves and enteric neurons. The SST-28 form accounts for 20-30% of total SST-like immunoreactivity in the brain. In the periphery, SST28 is synthesized predominantly in the intestinal mucosa as a terminal processed product of PSST (7, 8).

Enzymes implicated in PSST processing belong to a family of mammalian subtilisin/kexin-related, $Ca^{2+}$-dependent serine proteinases (collectively termed precursor convertases or PCs) (9, 10). Currently there are seven known mammalian PCs that have been identified comprising PC1 (PC3), PC2, furin (PACE), PC4, PC5 (PC6), PACE4, and PC7. The tissue distribution and subcellular localization of the PCs is diverse and, can be subdivided into four categories. Furin, PC5B, and PC7 are found in varying amounts in a wide variety of tissues, generally retained in the *trans*-Golgi (TGN) by virtue of their transmembrane domain providing them with a strategic location to cleave substrates entering the constitutive secretory pathway (CSP) *en route* to the cell surface. PC1 and PC2 are the major forms expressed only in the neuroendocrine system and brain acting on precursors entering the regulated secretory pathway (RSP). The short soluble isoform of PC5 (PC5A) also cleaves precursors entering the RSP but contrary to PC1 and PC2 it has a broader tissue distribution in both endocrine and non-endocrine cells. PACE4 is expressed in both endocrine and non-endocrine cells conceivably processing precursors in both the CSP and RSP. Finally, PC4 mRNA expression is exclusive to the testis. These enzymes cleave a variety of precursors at a general consensus $(\mathbf{R/K})$-$(Xaa)_n$-$(\mathbf{K/R})\downarrow$ sequence, where Xaa is any amino acid except Cys and n = 0,2,4 or 6.

Initial work toward the elucidation of specific proteases particularly involved in PSST processing began in 1984 by incubating synthetic peptides containing the necessary predicted cleavage sites with extracts of rat hypothalamic tissue (11). The enzyme isolated was capable of processing the synthetic peptide and SST-28 to SST-14 *in vitro*. Later, in 1987, direct biochemical evidence demonstrated the presence of two different types of endopeptidases implicated in cleaving PSST to either SST-14 or SST-28 (12). In 1991, following the identification of multiple mammalian cDNAs coding for protein PCs (9, 10), opportunities were created allowing the isolation of specific enzymes capable of cleaving PSST at basic sites. Through Edman degradation, the partial sequence of an enzyme purified from anglerfish (AF) islets with capability in cleaving PSST to SST-14 demonstrated homology to both human and mouse PC2 of more than 64% and 57% respectively (13). Such was the first reported finding associating PSST processing with the subtilisin/kexin family. Direct roles for PC1 and PC2 in SST-14 production were characterized by studying heterologues processing of rat PPSST in both endocrine (AtT-20, GH3) and, constitutive (COS-7) cells using cDNA transfection experiments (14, 15). Co-transfection experiments of PPSST with PC1 or PC2 cDNAs demonstrated that both are candidate proteases for dibasic cleavage required for SST-14 formation. In both constitutive as well as regulated secretory cells, PC1 is active, whereas PC2 is only active in neuroendocrine cells. These differences can be explained by the poor ability of constitutively secreting cells to process proPC2 to the active PC2 form compared to regulated secretory cells. The pattern of expression of PC1 and PC2 in normal tissues known to process PSST efficiently to SST-14 also implicates both enzymes in this conversion. For example in islet $\delta$ cells, PC2 is expressed and is likely the candidate islet SST-14 covertase (16). Similarly, in experiments designed to study the effects of PC2 gene disruption on prohormone processing in the islets of Langerhans, complete absence of SST-14 production was observed in extracts derived from PC2-/- mice (17).

As for SST-28 production, furin was found to be the most likely candidate. Such an observation stemmed from structure function studies revealing the substrate specificity of furin to be composed of the following motif: R-X-R/K-R↓ (10, 18). Such a specificity-profile implied considerable overlap with that for processing at the monobasic Arg cleavage site predicted for SST-28 containing the **R**-L-E-L-Q-**R** motif. Experimental evidence implicating furin in SST-28 production was demonstrated through heterologues processing of rat PPSST in COS-7, PC12, AtT-20, and GH3 cells; all of which express the enzyme (9, 10). Furthermore, direct evidence for the role of furin was provided by overexpressing the enzyme in COS-7 cells in a dose dependent manner using vaccinia virus infection experiments (19). An increase in SST-28 production correlated with the levels of furin expression. Additionally, PACE4 processed PSST to SST-28, authenticating this protease as a second SST-28 convertase (20).

The presence of basic amino acids although necessary, might not be sufficient for recognition by the PC endoproteases. Secondary structure motifs surrounding the cleavage site such as β-turns or α-helical structures have recently been shown to play an important role by discriminating between functional dibasic residues and, those that are not cleaved (21, 22). In the case of PPSST the region occupying SST-28 (1-12), the $NH_2$-terminal sequence of SST-28, separating the two C-terminal basic cleavage sites has been shown to play a key role in prohormone processing as predicted through secondary structure predictions (23). Therefore, the global conformation along with the basic residues are both important for PPSST recognition by PCs.

## 2.    Amino Terminal Processing of PSST

In addition to C-terminal processing of the SST precursor, another product originally isolated from the endocrine portion of the stomach corresponding to the first 10 amino acids of PSST was isolated from rat in 1987 (24). The highest concentration of the decapeptide at the time was found in the gastric antrum and hence named antrin. Subsequently, all SST producing tissues have been shown to be rich with it (25, 26). Scanning the $NH_2$-terminus for basic residues, $PSST_{[1-10]}$ would be produced by initial cleavage of PSST at the - $Phe^{10}$-$Leu^{11}$-$Gln^{12}$-**$Lys^{13}$** ↓ **$Ser^{14}$** peptide bond to generate $PSST_{[1-13]}$. Following the action of a PC endopeptidase, a carboxypeptidase would then be required to trim out the C-terminal residues from the processing intermediate resulting in $PSST_{[1-10]}$. Initial attempts to characterize the enzyme responsible for antrin production began by studying the ability of furin to carry on such a role due to its preferential cleavage of substrates C-terminal to monobasic residues (19). However, overexpression of furin in COS-7 cells did not produce any changes in the pattern of PSST processing to antrin. Similarly infecting LoVo cells, a cell line expressing an inactive form of furin, with PPSST did not hinder antrin production (19). Hence, a more stringent approach to test the requirement for the monobasic $Lys^{13}$ cleavage site was carried out by mutating it to an Ala residue (27); a residue not accepted by PCs (9, 10). Surprisingly $PSST_{[1-10]}$ cleavage was not blocked, producing no significant changes in antrin processing. Therefore other enzymes having non-basic cleavage specificity, most likely mediate amino terminal processing of PPSST.

Efforts to identify such proteinases led to the recent cloning of a new type-I membrane bound subtilase called subtilisin-kexin-isozyme 1 (SKI-1) or site-1 protease (S1P), whose amino acid sequence is highly conserved among rodent and human species (9, 10). It is the first known mammalian subtilisin/kexin-like enzyme capable of cleaving proproteins at nonbasic residues. SKI-1 is a $Ca^{2+}$–dependent subtilase that is widely expressed with possible substrate cleavage specificity C-terminal to Thr, Leu, Phe and Lys residues. In addition to autocatalytic processing, four substrates have been identified for SKI-1 and include: serum responsive element binding protein (SREBP) which regulates cholesterol metabolism (28), human pro-brain-derived neurotrophic factor (hpro-BDNF) (29), ATF6 transcription factor (30), and the surface fusion glycoprotein (GP) of the Lassa virus (31) Table 1. Since the consensus sequence at the NH$_2$-terminus of PSST qualifies as a SKI-1 substrate the enzyme became an attractive candidate for producing antrin. Overexpression experiments of SKI-1 in HEK-293 and COS-cells significantly increased the amount of antrin detected in the medium, making it a likely candidate enzyme (27). Additionally, detailed mutagenesis experiments demonstrated that efficient processing of PSST requires the -$R^6$-$L^7$-$R^8$-$Q^9$-$F^{10}$-$L^{11}$↓ recognition motif. Interestingly in mutant CHO cells, lacking SKI-1, reduced but detectable levels of antrin were generated with reversal of the defect observed upon reintroduction of the enzyme (27). Hence for certain substrates the requirement for SKI-1 is not absolute, with the possibility of other enzymes cleaving at hydrophobic residues highly likely. The need for the identification of more proteinases responsible for such processing is important, especially, since they could play major roles in regulating neurodegenerative disorders such as Alzheimer's disease.

*Table 1* Substrates cleaved by SKI-1

| Precursor protein | Cleavage site sequence P8-P7-P6-P5-P4-P3-P2-P1↓ |
|---|---|
| hProBDNF | K -A- G- S- R- G- L- T↓ |
| hSREBP-2 | S- G- S- G- R- S- V- L↓ |
| LassaV-gp | I- Y- I- S- R- R- L- L↓ |
| hATF6 | A- N- Q- R- R- H- L- L↓ |
| rPSST | D- P- R-L- R- Q- F- L↓ |

## 3.  Role Of Compartmentalization In Prosomatostatin Cleavage

During intracellular transport from the ER to vesicles prohormones need to be processed to their mature forms. Originally it was proposed that these proteolytic events can only occur once the precursor is packaged into granules. Using immunoelectron microscopy techniques it was demonstrated that proinsulin (a precursor hormone with a similar targeting route as SST) cleavage is initiated in acidic clathrin-coated vesicles that bud from the TGN (32). As for the compartment in which PSST is processed, the weight of evidence indicates that significant processing can occur in a compartment other than secretory granules. Based on the enzymes acting on PSST one would expect that SST-28 production takes place in the Golgi, whereas the organelle in which the cleavage of the precursor to SST-14 remains to be precisely identified. The first attempt to elucidate the subcellular localization involved in generating SST-28 was studied using AF PPSST II (33). By using a combination of immunofluorescence and immunogold staining, SST-28 was clearly observed in the TGN network of the cells. Obviously, high concentration of the immunolabelling was also demonstrated in dense core granules emerging from the Golgi. Later, studies aimed at identifying the subcellular compartments in which cleavage at both the monobasic and dibasic cleavage sites were carried out through intracellular fractionation of rat cortical brain (34). Three main secretory compartments were isolated which corresponded to the ER, Golgi and granules. By using a specific antibody that distinguishes SST-14 from SST-28, immunocytochemsitry, electron microscopy and HPLC results all demonstrated that both forms were enriched in the Golgi. Additionally, the utilization of temperature block and monensin treatment, known to inhibit forward transport from the Golgi, did not inhibit either cleavage (35, 36). Furthermore, the kinetic relationship between the biosynthesis of PPSST and its cleavage products was studied using rat hypothalamic and cortical neurons (37). Pulse-chase studies demonstrated that PPSST processing to SST-14 and SST-28 can start as early as 5 minutes after pulse labeling.

The best *in vivo* example demonstrating the ability of PPSST to be cleaved in the absence of granule requirement stems from its expression in immune cells, well known not to contain a RSP or even specialized secretory granules. The SST content in lymphoid organs (spleen and thymus) of male rats was studied by using an anti-SST antibody recognizing both peptide forms SST-14 and SST-28 (38). mRNA and sequential immunostaining of rat spleen using SST antibody and subsequently a monoclonal antibody against a rat B-cell surface antigen revealed the presence of SST immunoreactivity in some, but not all, B cells. Sequential immunostaining of rat thymus revealed the presence of SST immunoreactivity in a small population of T lymphocytes in the medulla. Additionally activated macrophages isolated from granulomas of *Schistosoma mansoni*-infected mice produced SST-14 (39).

## 4. Sorting and Intracellular Targeting

Generally neuroendocrine cells have two secretory pathways the RSP and the CSP Figure 2 (40, 41). The CSP is used for membrane renewal and passive secretion of proteins that are not responsive to secretagogues. The RSP is used to store PSST and other proteins in secretory granules to be subsequently released in response to appropriate stimuli. Both pathways initially meet in the RER where proteins are synthesized and eventually transported to the TGN compartment. Here proteins segregate to different pools of vesicles by as yet a poorly defined mechanism. One theory postulates the possible existence of a sorting receptor for the RSP that interacts with regulated secretory proteins. The membrane bound form of carboxypeptidase E (CPE) has been has been shown to exhibit characteristic features of a sorting receptor and is implicated in sorting of a number of neuroendocrine hormones (42, 43). A second theory proposes that regulated secretory proteins possess an intrinsic ability to form aggregates leading to packaging of condensed products into secretory granules thereby sorting them away from soluble proteins that are carried off by bulk flow in small vesicles (44, 45). Additionally, it is believed that one of the requirements for a soluble protein such as SST to enter the RSP is some sort of a signal most likely present within its amino acid sequence. The need to postulate the presence of a sorting signal to carry proteins into RSP came from investigations demonstrating that not all proteins are detected in secretory granules (44, 45). Later, amino acid analysis of 15 prohormones that have been shown to be correctly sorted into the RSP in AtT-20 cells, lead to the discovery of a motif that is shared by all of them (46).

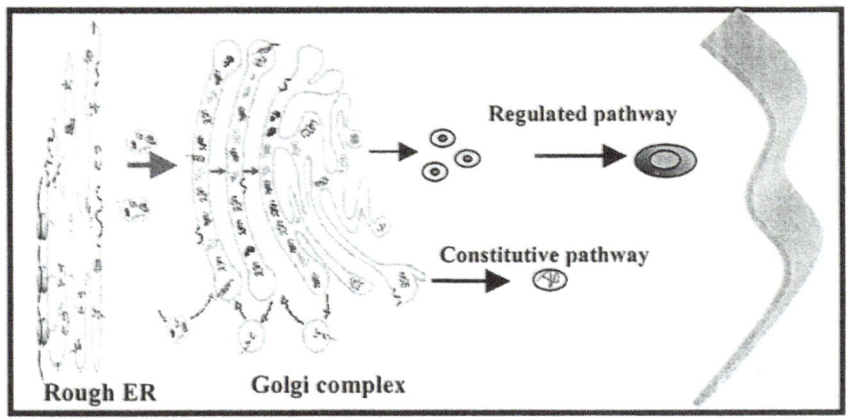

*Figure 2* Compartmental organization of the secretory pathway.

In the case of PSST it is well established that it undergoes regulated release. Among the intracellular mediators known to modulate SST secretion include ions, cAMP, and activators of protein kinase-C (47, 48). Since the amino terminus of PSST has been highly conserved throughout vertebrate evolution its role in targeting the precursor to the RSP was tested in a number of studies in order to identify the exact sorting motif. The importance of this region was initially addressed by studying differences in targeting capabilities of AF PPSST I and AF PPSST II (49). Heterologues expression of both precursors independently in AtT-20 cells and RIN 5F cells resulted in regulated release of the PPSST I precursor product and constitutive release of PPSST II product. The major differences between the two AF PPSST precursors lies in their pro-region. Therefore, the pro-region was tested for its ability for targeting to the RSP by creating a fusion gene containing the leader sequence and, the $NH_2$-terminal 54 a.a of rPPSST linked to the COOH-terminal 48 a.a of AF PPSST II. Transfected RIN 5F cells secreted the processed product upon stimulation with cAMP demonstrating that the N-terminal sequences of PPSST rerouted the hybrid protein towards the RSP. Later experiments in which the first 82 a.a. containing the pro-region of PPSST were fused to $\alpha$ globin protein, a protein that does not undergo regulated release, resulted in its successful targeting to the RSP (50). Additionally, construction of mutant precursors of rPPSST that either lack, or have replaced portions of the pro-region revealed that the $NH_2$-terminus not only contains signals for regulated secretion but that there could be multiple signals each of which could independently cause sorting to two different pools of RSVs (51). Such a conclusion was due to differences in cellular stimulation produced by treating the cells with two different secretagogues. A direct role for the $NH_2$-terminus in PSST targeting was carried out through extensive alanine scanning and deletional mutagenesis of the PSST[3-15] region (52). Mutants created were stably expressed in AtT-20 cells. Regulated secretion was studied by analyzing basal and stimulated release of SST-14 LI, and by immunocytochemistry for staining of SST-14 LI in punctate granules. The region $Pro^5$ to $Gln^{12}$ was identified as being important in precursor targeting with $Leu^7$ and $Leu^{11}$ being critical. Molecular modelling demonstrated that these two residues are located in close proximity on a hydrophobic surface of an amphipathic α-helical structure.

## SUMMARY

The intense interest in SST as a model to study neuropeptide synthesis has provided a framework for understanding the biosynthetic pathways, the putative role of PCs, and the secretory pathways required for PSST maturation to its active products. Kinetic studies have demonstrated that PSST is rapidly and independently processed to SST-14, SST-28, and PSST[1-10]. Both PC1 and PC2 are capable of processing PSST to SST-14. Furin and PACE 4 effect monobasic cleavage and are candidate SST-28 convertases. Contrary to previous belief, cleavage at the NH2-terminus has been shown not require the monobasic $Lys^{13}$ residue with SKI-1 acting as the most likely convertase. Secretory granules are not a requirement for PSST maturation. Additionally, the conserved amino terminal segment of PSST serves as a sorting for the regulated secretory pathway. Beyond the pure biochemical and basic research interest in defining the specific intracellular events implicated in processing and sorting of PSST, combining the knowledge gained from the processing and targeting aspects to treatments of various diseases would be the ultimate satisfaction of any scientist. Such an approach was elegantly attempted by Rivera and colleagues who used the ER as the storage compartment for genetically engineered secretory proteins such as insulin and growth hormone (53).

## REFERENCES

1.   Brazeau P, Vale W, Burgus R, Ling N, Butcher M, Rivier J, Guillemin R. Hypothalamic polypeptide that inhibits secretion of immunoreactive pituitary growth hormone. *Science* 1973; 179: 77-79.

2.   Goodman RH, Aron DC, Roos BA. Rat preprosomatostatin: structure and processing by microsomal membranes. *J Biol Chem* 1982; 258: 5570-5573.

3.      Goodman RH, Jacobs JW, Chin W, Lund PK, Dee PC, Habener JF. Nucleotide sequence of a cloned structural gene coding for a precursor somatostatin. *Proc Natl Acad Sci USA* 1980; 77: 5869-5873.

4.      Hobart P, Crawford R, Shen L, Pictet R, Rutter WJ. Cloning and sequence analysis of cDNAs coding for two distinct somatostatin precursors found in the endocrine portion of anglerfish. *Nature* 1980; 288: 137-141.

5.      Warren TG, Shields D. Cell-free biosynthesis of somatostatin precursors: evidence for multiple forms of preprosomatostatin. *Proc Natl Acad Sci* USA 1982; 79: 3729-3733.

6.      Shields D, Warren TG, Green RF, Roth SE, Brenner MJ. The primary events in the biosynthesis and post-translational processing of different precursors to somatostatin. In: Rich DH, Gross E, eds. Peptides: synthesis-structure-function: proceedings of the seventh American Peptide symposium. Rockford, III. Pirce Chemical Company, 1981: 471-479.

7.      Patel YC, Wheatley T, Ning C. Multiple forms of immunoreactive somatostatin: comparison of distribution in neural and nonneural tissues and portal plasma of the rat. *Endocrinology* 1981; 109: 1943-1949.

8.      Baskin DG, Ensinck JW. Somatostatin in epithelial cells of intestinal mucosa present is present primarily as somatostatin-28. *Peptides* 1984; 5: 615-621.

9.      Zhou, A., Webb, G., Zhu, X., and Steiner, D.F. Proteolytic processing in the secretory pathway. *J Biol Chem* 1999; 274: 20745-20748.

10.     Seidah NG, Chretein M. Proprotein and prohormone convertases: a family of subtilases generating diverse bioactive polypeptides *Brain Res* 1999; 848: 45-62.

11.     Gluschankof P, Morel A, Gomez S, Nicolas P, Fahy C, Christine F, Cohen P. Enzymes processing somatostatin precursors: an Arg-Lys esteropeptidase from the rat brain cortex converting somatosatin-28 into somatostatin-14. *Proc Natl Sci USA 1984*; 81: 6662-6666.

12.     Mackin RB, Noe BD. Direct evidence for two distinct prosomatostatin converting enzymes. *J Biol Chem* 1987; 262: 6453-6456.

13.     Mackin RB, Noe BD, Spiess J. Identification of a somatostatin-14 generating propeptide cnverting enzyme as a member of the kes2/furin/PC family. *Endocrinology* 1991; 129: 2263-2265.

14.     Galanopoulou AS, Kent G, Rabanni SN, Seidah NG, Patel YC. Heterologous processing of prosomatostatin in constitutive and regulated secretory pathways: Putative role of the endoproteases furin, PC1, and PC2. *J Biol Chem* 1993; 268:6041-9.

15.     Galanopoulou AS, Seidah NG, Patel YC. Heterologous processing of rat prosomatostatin to somatostatin-14 by PC2: requirement for secretory cell but not the secretion granule. *Biochem J* 1995; 311: 111-118.

16.     Marcinkeiwicz M, Ramala D, Seidah NG, Chretiem M. Developmemtal expression of the prohormoe convertases PC1 and PC2 in mouse pancreatic islets. *Endocrinology* 1994; 135: 1651-1660.

17.     Furuta M, Yano H, Zhou A, Roulle Y, Holst JJ, Carrol R, Ravazzola M, Orci L, Furuta H, Steiner DF. Defective prohormone processing and altered pancreatic islet morphology in mice lacking active SPC2. *Proc Natl Acad Sci USA* 1997; 94: 6646-6651.

18.     Watanabe T, Nakagawa T, Lkemizu J, Nagahama M, Murakami K, Nakayama K. Sequence requirements for precursor cleavage within the constitutive secretory pathway. *J Biol Chem* 1992; 267: 8270-8274.

19.     Galanopoulou AS, Seidah NG, Patel YC. Direct role of furin in mammalian prosomatostatin processing. *Biochem J* 1995; 309: 33-40.

20.     Brakch N, Galanopoulou AS, Patel YC, Boileau G, Seidah NG. Comparative processing of rat prosomatostatin by the convertases PC1, PC2, furin, PACE4 and PC5 in constitutive and regulated secretory pathways. *FEBS Lett* 1995; 362: 143-146.

21.     Brakch N, Rholam M, Boussetta A, Cohrn P. Role of beta-turn in proteolytic processing of peptide hormone precursors at dibasic sites *Biochemistry* 1993; 432: 4925-4930.

26

22. Rholam M, Nicholas P, Cohen P. Precursors for peptide hormones share common secondary structures forming features at the proteolytic processing sites. *FEBS Lett* 1986; 207: 1-6.

23. Brakch N, Lazar N, Panchal M, Allemandou F, Boileau G, Cohen P, Rholam M The Somatostatin-28(1-12)-NPAMAP sequence: an essential helical-promoting motif governing prosomatostatin processing at mono- and dibasic sites. *Biochemistry* 2002; 41: 1630-1639.

24. Benoit R, Ling N, Esch F. A new prosomatostatin-derived peptide reveals a pattern for prohormone cleavage at monobasic sites. *Science* 1987; 238: 1126-1129.

25. Rabbani SN, Patel YC. Peptides derived by processing of rat prosomatostatin near the amino-terminus: characterization, tissue distribution and release. *Endocrinology* 1990; 126: 2054-2061.

26. Ravazzola M, Benoit R, Orci L. Prosomatostatin-derived antrin is present in gastric D cells and in portal circulation. *J Clin Invest* 1989; 83: 362-366.

27. Mouchantaf R, Sulea T, Seidah NG, and Kumar U. Prosomatostatin is proteolytically processed at the amino terminal segment by the subtilase SKI-1. *Regulatory Peptides* 2004 (submitted).

28. Sakai J, Rawson RB, Espenshade PJ, Cheng D, Seegmiller AC, Goldstein JL, Brown MS. Molecular identification of the sterol-regulated luminal protease that cleaves SREBPs and controls lipid metabolism. *Mol Cell* 1998; 2:505-514.

29. Seidah NG, Mowla SJ, Hamelin J, Mamarbachi AM, Benjannet S, Toure BB, Basak A, Munzer JS, Marcinkiewicz J, Zhong M, Barale JC, Lazure C, Murphy RA, Chretien M, Marcinkiewicz M. Mammalian subtilisin/kexin isozyme SKI-1: a widely expressed proprotein convertase with a unique cleavage specificity and cellular localization. *Proc Natl Acad Sci USA* 1999; 96: 1321-1326.

30. Ye J, Rawson RB, Komuro R, Chen X, Dave UP, Prywes R, Brown MS, Goldstein JL. ER stress induces cleavage of membrane-bound ATF6 by the same proteases that process SREBPs. *Mol Cell* 2000; 6: 1355-1364.

31. Lenz O, Meulen JT, Klenk HD, Seidah NG, garten W. The Lassa virus glycoprotein precursor GP-C is proteolytically processed by subtilase SKI-1/S1P. *Proc Natl Acad Sci USA* 2001; 98: 12710-12705.

32. Orci L, Ravazzola M, Storch MJ, Anderson RG, Vassalli JD, Perrelet A. Proteolytic maturation of insulin is a post-Golgi event which occurs in acidifying clathrin-coated secretory vesicles. *Cell* 1987; 49: 865-868.

33. Bourdais, J, Devillers G, Girard R, Morel A, Benedett L, Cohen P. Porsomatostatin II processing in the trans-Golgi network of anglerfish pancreatic cells. *Biochem Biophys Res Commun* 1990; 170: 1263-1271.

34. Lepage-Lezin A, Joseph-Bravo P, Devilliers G, Benedetti L, Launay J, Gomez S, Cohen P. Prosomatostatin is processed in eth Golgi apparatus of rat neural cells. *J Biol Chem* 1991; 266: 1679-1688.

35. Xu H and Shields D. Prohormone processing in the trans-Golgi network: endoproteolytic cleavage of prosomatostatin and formation of nascent secretory vesicles in permeabilized cells. *J Cell Biol* 1993; 122: 1169-1184.

36. Patel YC, Galanopoulou AS, Rabanni SN, Liu JL, Ravazzola M, Amherdt M. Somatostatin-14, somatostatin-28, and prosomatostatin[1-10] are independently and efficiently processed from prosomatostatin in the constitutive secretory pathway in islet somatostatin tumour cells (1027B2). *Mol Cell Endocrinol* 1997; 131: 183-194.

37. Zingg HH, Patel YC. Biosynthesis of immunoreactive somatostatin by hypothalamic neurons in culture. *J Clin Invest* 1982; 70: 1101-1109.

38. Aguila MC, Dees WL, Haensly WE, McCann SM. Evidence that somatostatin is localized and synthesized in lymphoid organs. *Proc Natl Acad Sci USA* 1991; 88: 11485-11489.

39. Weinstock JV, Blum A, Malloy T. Macrophages within the granulomas of murine schistosomiasis mansoni are a source of a somatostatin 1-14 like molecule. *Cell Immunol* 1990; 131: 381-388.

40. Kelly RB. Pathways of protein secretion in eukaryotes. *Science* 1985; 230:25-32.

41.   Rothman JE, Orci L. Molecular dissection of the secretory pathway. *Nature* 1992; 355: 409-415.

42.   Cool DR, Normant E, Shen FS, Chen HC, Pannell L, Zhang Y, Loh YP. Carboxypeptidase E is a regulated secretory pathway sorting receptor: genetic obliteration leads to endocrine disorders in Cpe$^{fat}$ Mice. *Cell* 1997; 88: 73-83.

43.   Normant E, Loh YP. Depletion of carboxypeptidase E, a regulated secretory pathway sorting receptor, causes misrouting and constitutive secretion of proinsulin and proenkephalin, but not chromogranin A. *Endocrinology* 1998; 139: 2137-2145.

44.   Tooze SA. Biogenesis of secretory granules in the trans-Golgi network of neuroendocrine and endocrine cells. *Biochem Biophys Acta* 1998; 1404: 231-244.

45.   Chung KN, Walter P, Aponte GW, Moore HP. Molecular sorting in the secretory pathway. *Science* 1988; 243: 192-197.

46.   Kizer JS, Tropsha A. A motif found in propeptides that may target them to secretory vesicles. *Biochem Biophys Res Commun* 1991; 174: 586-592.

47.   Patel YC. General aspects of the biology and function of somatostatin. In: Basic and Clinical Aspects of Neuroscience. Edited by C., Weil EE Muller, and MO Thorner. Berlin: Springer-Verlag 1992; 4:1-16.

48.   Reichlin S, Saperstein R, Jackson IMD, Boyd AE, Patel YC. Hypothalamic hormone. *Annu Rev Physiol* 1976; 38: 389-424.

49.   Sevarino KA, Stork P, Ventimiglia R, Mandel G, Goodman RH. Amino-terminal sequences of prosomatostatin direct intracellular targeting but not processing specificity. *Cell* 1989; 57: 11-19.

50.   Shields D, Stoller T. The propeptide of preprosomatostatin mediates intracellular transport and secretion of α-globin from mammalian cells. *J Cell Biol* 1989; 108: 1647-1655.

51.   Sevarino KA, Stork P. Multiple preprosomatostatin sorting signals mediate secretion via discrete cAMP- and tetradecoylphorbolacetate-responsive pathways. *J Biol Chem* 1991; 266: 18507-18513.

52.   Mouchantaf R, Kumar U, Sulea T, and Patel YC. A conserved α helix at the amino terminal of prosomatostatin serves as a sorting signal for the regulated secretory pathway. *J Biol Chem* 2001; 276: 23308-23316.

53.   Rivera VM, Wang X, Wardwell S, Courage NL, Volchuk A, Keenan T, Holt DA, Gilman M, Orci L, Cerasoli F, Rothman JE, Clackson T. Regulation of protein secretion through controlled aggregation in the endoplasmic reticulum. *Science* 2000; 287: 826-830.

# 3
# CORTISTATIN- A NOVEL MEMBER OF THE SOMATOSTATIN GENE FAMILY

Véronique Fabre, Avron D. Spier, Raphaëlle Winsky-Sommerer, José R. Criado and Luis de Lecea
*Department of Molecular Biology. The Scripps Research Institute, La Jolla, CA.*

## INTRODUCTION

Cortistatin (CST) is a recently discovered neuropeptide from the somatostatin (SST) gene family named after its predominantly cortical expression and ability to depress cortical activity [1]. CST shows many remarkable structural and functional similarities to its related neuropeptide SST. However, the many physiological differences between CST and SST are just as remarkable as the similarities. CST-14 shares 11 of its 14 amino acids with SST-14, including the FWKT tetramer thought to be responsible for SST's receptor interactions and the pair of cysteine residues that likely render the peptides cyclic [2]. Yet the nucleotide sequences and chromosomal localizations of these genes clearly indicate they are products of separate genes and CST's activity in the brain is widely distinct from that of SST [3].

Now cloned from human, mouse and rat sources, *in vitro* assays show CST is able to bind all five cloned SST receptors (SSTR) and shares many pharmacological and functional properties with SST, including the depression of neuronal activity via activation of the M-current. However, distinct from SST, CST has been shown to induce slow-wave sleep, reduce locomotor activity, and activate cation selective currents not responsive to, or antagonized by, SST [4]. The effects of CST-14 on EEG activity appear to be mediated by antagonism of the excitatory effects of acetylcholine in cortical activity. The expression of mRNA encoding CST follows a circadian rhythm and is upregulated on deprivation of sleep, suggesting cortistatin is a sleep modulatory factor. Many lines of evidence also indicate that CST, like SST, is involved in the processes of learning and memory. Recent studies have shown that CST overexpressing mice have impaired long-term potentiation (LTP) and injection of CST-14 into the rat brain causes performance impairment in learning paradigms.

In this review we address the discovery and characterization of this novel SST-like neuropeptide, including its cloning, proteolytic processing, expression in the brain, electrophysiology and pharmacology. We will discuss the role of

CST in hippocampal and cortex associated functions including regulation of sleep and cognition. We also examine the evidence pointing towards a specific receptor for this novel neuropeptide member of the SST gene family.

## CLONING OF PREPROCORTISTATIN

A partial cDNA clone of preprocortistatin was first isolated from a hippocampal subtracted brain cDNA library [1]. The nucleotide sequence of the full-length cDNA clone suggested that it encoded a novel putative 112 amino acid protein, whose C-terminal revealed a strong similarity with preproSST. The protein encoded by the cDNA was named preproCST, in recognition of its strong similarity with preproSST and its predominantly cortical expression. The name cortistatin also reflects its inhibition of cortical activity (see below).

Preprocortistatin begins with a 27-residue apparent secretion signal sequence. Cleavage of the preprospecies to procortistatin would produce a protein that could be processed at either of two tandem basic amino acid pairs to produce CST-29 and CST-14, analogous to the cleavage of preproSST at residues 28 and 14 [5], or at both basic pairs to additionally produce cortistatin-13 (CST-13). Whereas CST-13 is unrelated to known species, CST-14 shares 11 of 14 residues with SST-14, including two cysteine residues that are likely to render the peptide cyclic and the FWKT motif that is critical for SST-14 binding to its receptors (Figure 1) [2].

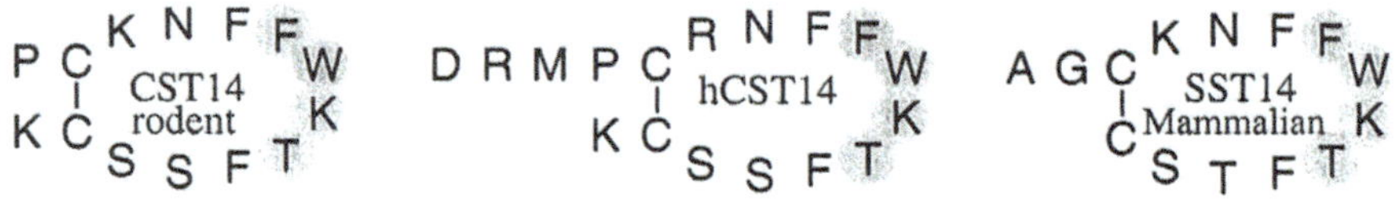

*Figure 1. Predicted secondary structures of rodent cortistatin-14, human cortistatin-17 and mammalian SST-14.*

CST-14 and SST-14 are permuted by one amino acid; the alignment of CST-14 begins at residue 2 of SST-14, and CST-14 terminates with a lysine residue beyond the C-terminal cysteine of SST-14. Although the C-terminal lysine would be susceptible to cleavage by carboxypeptidases, release experiments have demonstrated that this residue is present in the endogenous peptide [6].

A full-length cDNA encoding preprocortistatin has also been obtained from mouse and human [3, 7]. Excluding the introduction of two gaps, mouse and rat nucleotide sequences are 86% identical. Comparison of the two sequences indicates that the mouse preprocortistatin mRNA contains 108 amino acids instead of 112 for the rat. The presence of only two cleavage sites in the mouse deduced amino acid sequence shows that processes of the prepropeptide would give rise to two peptides, mCST-44 and mCST-14, analogous to rCST-14. The human nucleotide sequence shows a much lower degree of identity to the rat sequence (71%). The human preprocortistatin deduced amino acid sequence has 114 residues and begins with a 29- amino acid hydrophobic putative secretory signal sequence. Analysis of the putative processing sites in human preprocortistatin revealed that it may be cleaved at two RR sites, giving rise to hCST-29 and a C-terminal seventeen residue peptide (hCST-17) that shares 13 of the last 14 residues with rat and mouse CST-14. The Lys-Lys pair that lies just N-terminal to CST-14 in rat and mouse is not conserved in the human sequence. Other possible products that follow the signal sequence (hCST-21 and hCST-31) are not conserved across species, although rCST-31 and hCST-31 share 13 residues clustered in their N-terminal regions that are conserved among the rat, mouse and human prohormone sequences.

Chromosomal mapping of the mouse preprocortistatin gene on distal chromosome 4 confirmed that CST-14 and SST-14 are the products of different genes [3], as the gene encoding preproSST is located on mouse chromosome 16 [8]. The human CST gene has recently been mapped to 1p36.3-1p36.2 in close proximity to a region associated with neuroblastoma [9]. However, the identification of cort locus as a tumor suppressor gene is unlikely, as no mutations have been detected in the preprocortistatin human gene in neuroblastoma cells [9].

## PREPROCORTISTATIN PROCESSING AND RELEASE

Analysis of the deduced preprocortistatin protein indicates the presence of several dibasic residues that may be substrates for prohormone convertases [10]. The enzymes that process preprocortistatin are still unknown, however the cell line used to study preprocortistatin cleavage (AtT20) contains the convertases PC1 and PC2 [11]. Interestingly these enzymes are both highly expressed in the cerebral cortex [12], a brain region presenting the highest level of cortistatin expression (see below). Moreover, these enzymes have also been implicated in preprosomatostatin processing [11, 13] and hence are putative candidates to process preprocortistatin *in vivo*. Co-localization studies of preprocortistatin mRNA with several processing enzymes visualized by immunohistochemistry (PC1, PC2 and 7B2) on brain sections revealed that preprocortistatin mRNA co-localized in every instance with PC1 immunoreactivity, in contrast to PC2 which only overlapped partially. This study suggests that PC1 prohormone convertase is responsible for procortistatin processing *in vivo* (N. Seidah and L. de Lecea, unpublished results).

Recently, Puebla *et al* [6] have shown that rat preprocortistatin is cleaved at the two C-terminal dibasic cleavage sites, KK and KR, to produce rCST-14 and rCST-29 analogously to SST-14 and SST-28. The lack of evidence for a cleavage at both C-terminal dibasic sites to produce rCST-14 and rCST-13 indicates that rCST-14 and rCST-29 are the major products of the preprocortistatin processing. The ratio of rCST-14:rCST-29:preprocortistatin in tissue culture cells was found to be 41:55:4.5 and the ratio of secreted peptides rCST-14:rCST-29 was shown to be approximately 2:1. Thus, although the two peptides are produced approximately equally, the rCST-14 form is preferentially released compared to rCST-29. More recently, the presence of

CST-14 has been demonstrated *in vivo* in preproSST knock-out mice. HPLC purification of peptide extracts from SST knock-out mice revealed a peak of SST-like immunoreactivity, which coincided with synthetic CST-14 [14].

## EXPRESSION OF PREPROCORTISTATIN IN THE BRAIN

Preprocortistatin appears to be brain-specific, as judged by Northern blot analysis. A single band of approximately 600 nucleotides was detected in samples prepared from brain but not adrenal gland, liver, spleen, thymus, ovary, testes or anterior pituitary [1, 3]. Cortistatin cDNA sequences have been cloned from human peripheral tissues including fetal heart, fetal lung, prostate, colon, kidney and many tumors, as revealed by their presence in the expressed sequence tag database (Genbank), although the functional relevance of this expression remains to be determined.

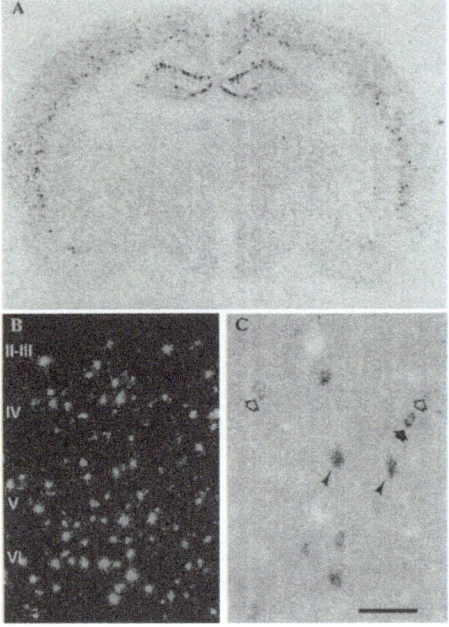

*Figure 2. Distribution of preprocortistatin mRNA-positive cells. A. Coronal section of rat brain hybridized in situ with a preprocortistatin probe. Note that the signal is restricted to scattered cells in the neocortex and hippocampus. B. Dark field image showing preprocortistatin-expressing cells in all cortical layers. C. Preprocortistatin positive cells (empty arrows) partly colocalize (filled arrows) with SST-positive neurons (arrowheads).*

The distribution pattern of preprocortistatin mRNA, determined by *in situ* hybridization on rat brain sections, indicates that preprocortistatin mRNA is expressed in scattered neurons throughout the cerebral cortex and hippocampus (Figure 2). However, other areas of the brain also show preprocortistatin mRNA expression, although at much lower levels. For example, in the olfactory bulb, granule GABAergic neurons are positive for preproCST mRNA. In the striatum, a small number of positive cells can be detected, resembling cholinergic or GABAergic interneurons. A few cells in the hypothalamus, corresponding to the periventricular nucleus, are positive for preprocortistatin mRNA. No signals could be detected in the thalamus, mesencephalon, brainstem, cerebellum or spinal cord [3, 7, 15].

In the cortex, CST-positive cells are especially abundant in layers II-III and VI (Figure 2B). Interestingly, the distribution of CST positive cells is not uniform in all cortical areas, the visual cortex displays about twice as many preprocortistatin mRNA-containing neurons as the somatosensory cortex. In the hippocampal formation, cortistatin mRNA expression is found in a small subset of non-pyramidal cells in the subiculum and in the stratum oriens of the CA1 and CA3 fields. No preprocortistatin mRNA was detected in the dentate gyrus. Double *in situ* hybridization experiments have shown that preprocortistatin mRNA co-localized in every instance with either GAD65 or GAD67, demonstrating the GABAergic nature of cortistatin-expressing cells [15].

Since SST is also expressed in GABAergic interneurons, the question arises whether CST-14 and SST are expressed in the same cells. Using combined immunocytochemistry to SST-14 and *in situ* hybridization to preprocortistatin

*Table I.* Distribution of cortistatin-expressing cells relative to their somatostatin content (% labeled cells).

|  | % SST | % CST | % SST/CST | Total |
|---|---|---|---|---|
| **Neocortex** (parietal area 1) |  |  |  |  |
| Layer II-III | 40 | 60 | 0 | 100 |
| Layer VI | 20 | 35 | 45 | 100 |
| **Hippocampal formation** |  |  |  |  |
| Subiculum | 40 | 30 | 30 | 100 |
| CA1 | 60 | <5 | 37 | 100 |
| CA3 | 80 | 0 | 20 | 100 |
| Hilus | 100 | 0 | 0 | 100 |

mRNA de Lecea *et al* [15] showed that the two peptides are expressed in distinct, though partially overlapping, sets of neurons. PreproCST mRNA and SST immunoreactivity do not co-localize in layer II-III of the cerebral cortex (Figure 2C). No CST-positive neurons are present in the hilar region of hippocampus, where SST-14 is expressed (see Table I).

The developmental expression of preproCST mRNA has been characterized by northern blot and by *in situ* hybridization on brain sections at different developmental stages [15]. PreproCST mRNA appears at day post-natal 15 and reaches a maximum level during the second post-natal week (between P15 and P20). This pattern of expression correlates with the maturation of cortical interneurons.

The regulation of preprocortistatin mRNA accumulation has been investigated in various paradigms underlying again the differences between the two peptides. PreproSST mRNA increases its steady-state concentration four fold upon kainate injection in the dentate gyrus and CA1, whereas preprocortistatin mRNA does not respond to kainate, indicating that CST and SST respond to different signals [16]. Also, preprocortistatin mRNA is regulated by the light/dark cycle, with maximum levels before sleep onset [17].

To determine whether cortistatin products could be detected in the brain, we designed MAP (multiple antigen peptides) to contain the sequence of the cortistatin precursor immediately N-terminal to mature CST-14 (rat CST1-13: $NH_2$- QERPPLQQPPHRD –COOH) which is divergent between rodents and humans and is unrelated to any neuropeptide species. Anti-MAP sera was processed for immunocytochemical staining.

The most prominent CST immunoreactivity was found in the upper layers (II-III) of the neocortex, including motor, somatosensory and visual cortices (Fig 3A). Cortistatin-positive cells in the cortex were non-pyramidal and resembled local circuit GABA ergic interneurons (Fig 3 A,D,E). In the upper layers most cst positive neurons were fusiform and bipolar (Fig 3D). In the deeper cortical layers, positive neurons displayed multiple non-pyramidal morphologies (Fig 3E). The localization of cst-positive neurons was consistent with that found by in situ hybridization [15], although fewer cells could be detected with the antibody, probably those which showed higher concentration of preproCST mRNA. Staining in the hippocampal formation revealed a population of interneurons that are reminiscent of SST-immunopositive GABAergic interneurons in the CA1 region. No staining was detected in the dentate gyrus or hiluar region. It is noteworthy that cortistatin-immunoreactive neurons and fibers were detected in the periventricular hypothalamus, a population of cells reminiscent of SST-immunopositive neurons. This finding may be important to understand the hypothalamic actions of CST and SST.

## ELECTROPHYSIOLOGICAL PROPERTIES OF CST-14

Considering the high level of CST expression in the hippocampus and the hyperpolarizing effect of SST on hippocampal neurons, the possible involvement of CST in the physiology of the hippocampus was first investigated by means of current- and voltage-clamp recordings in hippocampal slice preparations [1]. Superfusion of CST-14 hyperpolarizes pyramidal neurons. Unlike SST-14, the CST-14 effect develops slowly, reaching a maximum steady effect six to eight minutes after the onset of the response. This effect contrasts with a much shorter (2-3 min) time-to-peak of

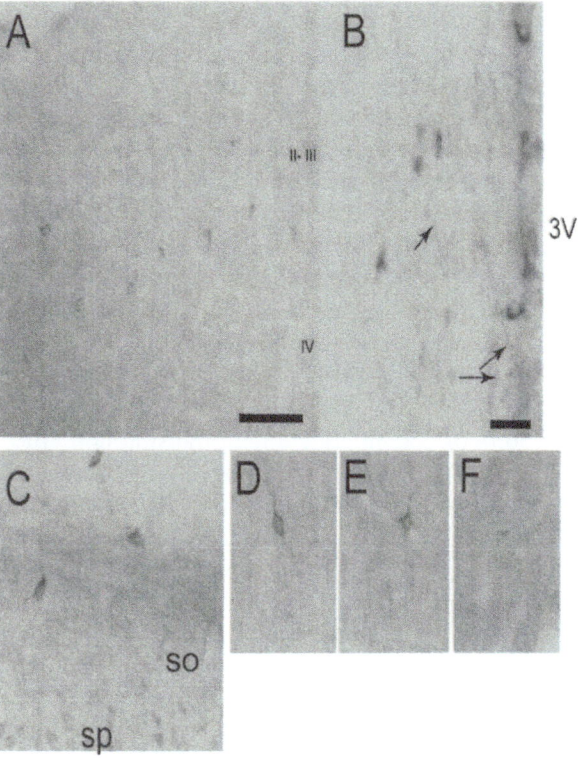

*Figure 3. Cortistatin-like immunoreactivity in the rat brain. Antisera to the rat cortistatin precursor was incubated with rat brain sections. The distribution of CST-immunopositive neurons is fully consistent with the in situ hybridization results in the neocortex (A), namely,scattered, nonpyramidal cells in the layers II-III and VI. B. A few CST neurons were also detected in the periventricular hypothalamus. C In the hippocampal formation, CST-immunoreactive neurons were non-pyramidal, and concentrated in the stratum oriens of the CA1. D-F. Heterogeneous morphologies of CST-immunopositive neurons in the cortex (D), amygdala (E) and striatum (F).*

the effect of SST-14 on hippocampal neurons under the same experimental conditions.

To determine the mechanism of the CST-induced inhibition, the effect of CST-14 on the M current ($I_M$), a non-inactivating voltage-dependent potassium current seen in hippocampal neurons, has been studied [18]. As previously described for SST-14 [18, 19], CST-14 superfusion increases the amplitude of the $I_M$ relaxation concomitantly with an outward steady-state current, with recovery to control levels upon washout. The same effects were observed following treatment with CST-29.

In the hippocampus, the nature of cells sensitive to cortistatin has been identified *in vitro* and *in vivo* as glutamatergic neurons. Thus, application of CST-14, as SST-14, reduces evoked excitatory postsynaptic currents (EPSPs) mediated by NMDA and non-NMDA glutamate receptors in CA1 neurons *in vitro*, whereas neither peptide alters inhibitory postsynaptic currents (IPSPs) [20, 21]. *In vivo*, iontophoretic application of CST-14 in anesthetized rats decreases the firing rate of CA1 pyramidal cells induced by NMDA pulses. The modulation of glutamatergic neurons by CST-14 has been observed in extracortical areas. Application of CST-14 in hypothalamic neurons inhibits glutamate-induced responses, probably through the activation of type 2 SST receptors [22]. Together these results suggest that CST generates physiological responses by reducing the activity of excitatory neurons.

The modulation of cholinergic neurotransmission by CST, and more specifically its ability to antagonize the effects of acetylcholine (ACh) on cortical excitability, is an important feature to dissect the mechanism by which CST regulates cortical activity and sleep/wakefulness rhythms (see below). Iontophoretic application of ACh significantly reduces paired-pulse (PP) inhibition in CA1 neurons *in vivo*, an effect that is mediated in part by hippocampal interneurons. However, while CST-14 itself produces no effect on PP responses, it completely antagonizes the effects of ACh. In contrast, local application of SST-14 on CA1 neurons enhances ACh release and potentiates ACh responses. Further experiments using iontophoretic administration of CST-14 and ACh indicated that the interaction between the two transmitters is not limited to an inhibitory role of CST on cholinergic transmission. Single unit recordings in the hippocampus revealed that iontophoretic administration of CST-14 enhances the ACh-induced increase in firing rate (Figure 4). Considering the high concentration of ACh used in this preparation, the modulation of cholinergic activity by CST-14 has been investigated in a more physiological preparation. The ability of the peptide to modulate ACh effects has been assessed in the hippocampus by stimulating cholinergic afferents from the medial septum with a train of stimuli known to facilitate the population spike amplitudes in the CA1. Addition of CST-14 dramatically reduced the facilitation induced by the septal stimulation (Figure 4), which is consistent with inhibitory properties of CST on ACh responses observed in the PP inhibition paradigm. The bimodal action of CST-14 on cholinergic neurotransmission may have important implications in the sleep-inducing properties of CST-14 (see below).

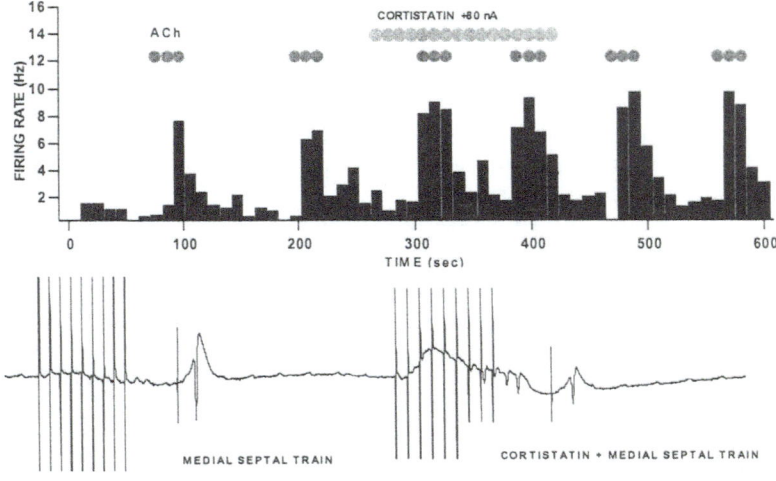

**Figure 4**. *Interaction between CST-14 and ACh. CST-14 enhances the firing rate of pyramidal neurons that have been administered with high concentrations of Ach. (top). In contrast, a mild stimulation of cholinergic afferents is antagonized by CST-14 (bottom).*

## PHARMACOLOGICAL PROFILE OF CST-14

The strong structural similarity between CST-14 and SST suggested that CST-14 might bind to somatostatin receptors. SST binds to five different known receptors, named SSTR1-5, which are members of the 7-transmembrane G-protein coupled receptor superfamily and display distinct, but overlapping expression patterns [23]. rCST-14 and hCST-17 have been shown to displace $^{125}$I-Tyr-SST-14 binding to each of the 5 cloned SSTRs expressed in transfected cell lines, with affinities in the low nanomolar range similar to those of SST-14 [7, 24, 25] (Table II). CST-14 has also been shown to be an effective agonist to SSTR(s) expressed by GH4 cells, as measured by inhibition of vasoactive intestinal peptide (VIP) or thyroid-releasing hormone (TRH) induced cAMP accumulation, with indistinguishable efficacy from SST-14 [1]. GH4 cells are thought to primarily express SSTR1, and cortistatin has further been shown to agonize SSTRs 2-5 in transfected CHO cells [7].

The anatomical distribution of CST-14 binding sites has been analyzed using autoradiography with $^{125}$I Tyr$^{10}$CST on mouse brain sections. Similarly to SST-14 [26], CST-14 shows binding sites throughout the cerebral cortex, especially in the deep layers, hippocampal formation and medial habenula and basolateral amygdala [27]. Most of the SST-14 binding sites in the brain are competed with unlabeled SST-14 (100 nM). But, interestingly, 100 nM CST-14 does not compete $^{125}$I-Tyr-SST-14 labeling in the cortex or amygdala, whereas 100nM

SST-14 fully displaces the signal, thus indicating the presence of receptors with different affinities for SST-14 and CST-14.

Several studies have examined the peptide structures necessary for cortistatin binding to SSTRs. As with SST-14, CST-14 does not show any preferred conformation in solution, as determined by circular dichroism and nuclear magnetic resonance [24], suggesting that the conformation of the peptides is not an important factor for binding to SSTRs. Fukusumi et al [7] showed that hCST-17, hCST-15 and hCST-13 all displaced $^{125}$I-Tyr-SST-14 binding to SSTR1-5 with similar efficacy. Surprisingly this indicates that the four N-terminal extracyclic residues of hCST-17 are not important for mediating a detectable pharmacological difference between hCST-17 and SST-14. Using the rationale that the residues contained within the Cys-Cys loop of cortistatin and somatostatin only differ by one amino acid whereas the extracyclic residues are distinct, Criado et al [24] examined the role of the extracyclic residues in cortistatin pharmacology. With the cyclic moiety of the peptide utilizing a sequence obtained from octreotide, a potent SST analogue, the extracyclic residues were systematically modified. It was shown that both the N-terminal Pro and C-terminal Lys are necessary to elicit cortistatin's unique physiological effects on sleep and locomotor activity. With only the N-terminal Pro present a cortistatin-like effect on locomotor activity but not on sleep was observed and if only the C-terminal Lys is present the compound behaves like somatostatin. Analog compounds with the N-terminal proline bound with nanomolar affinities to SSTR3 and 5, but not to SSTRs 1, 2 and 4 (Table II).

*Table II.* Affinities of SST-14, CST-14 and analogues to the five human SSTRs. Values are Kd [nM] (adapted from Criado et al [24]).

| Peptide | hSSTR1 | hSSTR2 | hSSTR3 | hSSTR4 | hSSTR5 | Sequence |
|---|---|---|---|---|---|---|
| SST-14 | 2.3±0.47 | 0.23±0.04 | 1.17±0.23 | 1.7±0.3 | 1.4±0.3 | Ala-Gly-c[Cys-Lys-Asn-Phe-Phe-Trp-Lys-Thr-Phe-Thr-Ser-Cys]-OH |
| Octreotide | 875±180 | 0.57±0.08 | 26.8±7.7 | >1000 | 6.8±1.0 | D-Phe-c[Cys-Phe-D-Trp-Lys-Thr-Cys]-Thr-ol |
| 96145 | >1000 | 115±18 | 934 | >1000 | 26.6±8.8 | D-Phe-c[Cys-Tyr-D-Trp-Lys-Val-Cys]-Lys-NH$_2$ |
| 96166 | >1000 | 930±69 | 27±1.0 | >1000 | 93±39 | Pro-c[Cys-Tyr-D-Trp-Lys-Val-Cys]-NH$_2$ |
| 96165 | >1000 | >1000 | 105±13 | >1000 | 60.4±19 | Pro-c[Cys-Tyr-D-Trp-Lys-Val-Cys]-Lys-NH$_2$ |
| 96149 | >1000 | >1000 | >1000 | >1000 | >1000 | Pro-c[Cys-Tyr-D-Trp-Lys-Cys]-Lys-NH$_2$ |
| CST-14 | 2.1±0.8 | 0.5±0.1 | 3.8±0.9 | 18.2±2.5 | 0.9±0.2 | Pro- c[Cys-Lys-Asn-Phe-Phe-Trp-Lys-Thr-Phe-Ser-Ser-Cys]-Lys |

Interestingly, SSTR5 mRNA is expressed only at low levels in the brain, while SSTR3 mRNA is expressed abundantly in the cortex and hippocampus. This leads to the possibility that cortistatin-like effects *in vivo* may be mediated through SSTR3.

## CST-14 MODULATES SLEEP/WAKEFULNESS RHYTHMS

To investigate the physiological function of CST, a first set of experiments investigated the ability of CST to affect the general behavior of rat. For this purpose, the effects of CST-14 on locomotor activity were examined in cannulated, freely moving rats. Intracerebroventricular injection of different amounts (0.1, 0.5 and 1 μg) of synthetic cyclic CST-14 markedly decreased locomotor activity [24], an effect opposite to that reported for SST-14 at the same doses [28]. However, at a higher dose (10 μg) CST-14 infusion induces seizures and barrel rotation, a characteristic behavior of SST-14 injection, suggesting that, under saturating conditions CST-14 may act through SSTRs *in vivo*.

To determine the function of CST-14 *in vivo*, de Lecea *et al.* [1] infused rats with CST-14 into the brain ventricles and analyzed the electroencephalogram (EEG) as a measure of cortical excitability. Polygraphic monitoring of arousal states subsequent to the administration of CST-14 indicated that rats spent up to 75% of the 4-hour recording time in slow-wave sleep compared to 40% in saline-treated control animals. A significant reduction of paradoxical (REM) sleep with the highest dose of CST-14 was also detected. Again, the physiological responses induced by CST-14 are in clear contrast to the reported enhancement of REM sleep with a similar dose of somatostatin [29].

The slow wave sleep-inducing activity of CST-14 has been replicated by others, using synthetic human CST-17 [7]. Moreover, CST-14 actively induced sleep when infused during the dark period, in rats that have already accomplished their physiological demand of sleep. Consistent with the hypothesis that CST-14 is a sleep factor, sleep deprivation increases the steady state concentration of preprocortistatin mRNA, and its levels oscillate along the light/dark cycle [17].

How does CST-14 induce sleep? CST-14 may enhance the intrinsic activity of cortical neurons by its hyperpolarizing activity on principal cells. Also, regulated release of CST-14 at the appropriate circadian time may antagonize the excitatory effects of ACh and promote sleep. CST-14 may also enhance SW sleep by influencing the activity of neuronal networks by its activity on a hyperpolarization-activated cationic conductance known as the h-current. The h-current induces a small after-depolarization in thalamocortical cells that causes a refractory period between spindle waves [30]. Activation and subsequent deactivation of $I_h$ in thalamocortical cells thus appears to be a key mechanism through which synchronized oscillations are terminated and prevented within defined intervals. The h-current has also been shown to be important for the establishment of rhythmicity in other neuronal systems [30].

CST-14, but not SST-14, enhances the amplitude of $I_h$ on hippocampal slices (P. Schweitzer, personal communication). Activation of $I_h$ by CST-14 in cortical networks may thus have a prominent role in regulating the synchronous activity that characterizes slow-wave sleep.

## Additional Functions For CST-14

By analogy with SST, it is expected that interaction of CST-14 with either SSTRs or CST-specific receptors will mediate different functions. Rauca et al [31] have shown that cortistatin can protect against neuronal injury caused by ischemia. Several groups have shown that icv injection of CST impairs long-term memory in passive avoidance tests [32, 33]. Tallent et al [34] have shown that transgenic mice overexpressing CST-14 do not produce long-term potentiation in the dentate gyrus. This effect appears to be mediated by reduction of postsynaptic NMDA receptor function. Thus, in rats, inhibition of glutamatergic neurotransmission by CST may be the cellular basis for long-term memory impairment. However, the effects of CST-14 on cholinergic activity in the cortex may also account for its amnesic activity. The neurotransmitter ACh has been proposed to be a key mediator in the consolidation of memory [35]. Infusion of CST-14 may thus impair memory by blocking the cognitive enhancing effects of ACh. Future experiments with cortistatin knock-out animals may shed light into the possible role of this peptide in cognition.

Recent studies demonstrated the endocrine activities of CST. Indeed, CST has been shown to bind to Growth Hormone (GH) secretagogue receptor in human tissue [36]. Furthermore, Broglio et al.[37] showed that CST inhibited both basal and stimulated GH secretion in the same manner as SST. Increasing evidence also suggests that CST may play a role in the regulatory mechanisms between the neuroendocrine and immune systems. Dalm et al. [38] first reported the expression of CST mRNA in human lymphoid tissue, immune cells and bone marrow. They also observed high expression levels of CST mRNA in monocyte-derived macrophages and dendritic cells. In contrast, no SST mRNA was detectable in immune cell types. In addition, both CST and SSTR2 mRNAs have been shown to be up-regulated during differentiation of monocytes into both macrophages and dendritic cells. Therefore, it has been proposed that CST would be an endogenous ligand of SSTR2 rather than SST in the immune system, pointing to a regulatory role of CST in the immune system [39].

## CST-14 Binding to Receptors Other Than SSTRs

CST and SST are members of a diverse group of approximately 20 families of neuropeptides with varying degrees of sequence, structural and evolutionary relatedness. Neuropeptide receptors are also structurally and functionally related with some cross talk between neuropeptides and receptors for other neuropeptide families [40]. For example, SST-14 is able to activate the urotensin II receptor and bind opioid receptors [41]. As discussed above, *in vitro* assays show CST binds to and activates the known SSTRs. However, CST-14 does not bind the urotensin II receptor [42] or μ- and δ-opioid receptors [43].

Many of the physiological effects of CST suggest CST's activity is indeed mediated by non-SSTRs. Indeed, it has been shown recently that CST-14 and CST-17, but not SST-14, bind to the Growth Hormone (GH)-secretagogue receptors with similar affinity than ghrelin [36]. Furthermore, the human orphan G-protein coupled receptor MrgX2 has been described as the first human cortistatin selective receptor [44]. Indeed, both CST-14 and CST-17 bind MrgX2 with high affinity, and showed no affinity for SST-14, confirming our data on structure/activity relationships [24]. Moderate levels of MgrX2 are observed in subsets of neurons in the CA2, CA3 and CA4 of the hippocampus. MgrX2 is not detected in the cerebral cortex. However, the highest expression levels of this receptor were found in the lumbar dorsal root ganglion, suggesting a putative role of CST in nociception.

*Table III.* Similarities and differences between CST-14 and SST-14.

| CST-14 | SST-14 | Refs |
|---|---|---|
| **Similarities** | | |
| 14-residue cyclic peptides | | [46], [1] |
| Co-localize with GABA | | [47], [15] |
| Nanomolar affinity to SSTRs *in vitro* | | [7, 24, 25] |
| Inhibitory peptides | | [1, 48] |
| Enhance M-current | | [4], [48] |
| **Differences** | | |
| Primarily expressed in cerebral cortex and hippocampus | Widely distributed throughout the brain | [15], [49] |
| Decreases locomotor activity | Increases locomotor activity | [24], [28, 50] |
| Enhances Slow-wave sleep | Enhances REM sleep | [1, 7], [29] |
| Enhances h- current | Does not enhance h-current | [4] |
| Antagonizes ACh in cortex | Potentiates ACh responses | [1], [51] |
| Not responsive to kainate | Stimulated by kainate in the dentate gyrus | [16] |
| Binds to MgrX2 receptor | Does not bind to MgrX2 | [44] |

## Concluding Remarks

CST-14 shares homology with SST-14 in bioactive peptide amino acid sequence, gene structure, partial coexpression, activation of common receptors and signaling pathways, and neuronal inhibition via the M-current. These observations may suggest a duplication of function between these two related peptides. Indeed the lack of a significant phenotype in mice lacking the gene for SST [45] suggests CST at least partly duplicates SST function. However, several experiments *in vivo* reveal substantial differences between the two peptides and cumulatively lead to the conclusion that CST-14 is not an alternative, or "back-up" SST (Table III).

Its relatively restricted distribution in the CNS, compared with other neuropeptides expressed in the neocortex, makes cortistatin an interesting target for pharmacological intervention on diseases in which cortical neurotransmission is impaired. Also, cortistatin's unique effects on promoting slow-wave sleep may help to understand the mechanisms underlying the maintenance of synchronized activity in the cortex. The interactions of cortistatin with other transmitter systems may have implications in synaptic plasticity and cognitive function.

## REFERENCES

1.      de Lecea, L., Criado, J. R., Prospero-Garcia, O., Gautvik, K. M., Schweitzer, P., Danielson, P. E., Dunlop, C. L., Siggins, G. R., Henriksen, S. J.and Sutcliffe, J. G. (1996). A cortical neuropeptide with neuronal depressant and sleep-modulating properties. *Nature*. 381: 242-245.

2.      Veber, D. F., Holly, F. W., Nutt, R. F., Bergstrand, S. J., Brady, S. F., Hirschmann, R., Glitzer, M. S.and Saperstein, R. (1979). Highly active cyclic and bicyclic somatostatin analogues of reduced ring size. *Nature*. 280: 512-4.

3.      de Lecea, L., Ruiz-Lozano, P., Danielson, P., Foye, P., Peelle-Kirley, J., Frankel, W.and Sutcliffe, J. G. (1997). cDNA cloning, mRNA distribution and chromosomal mapping of mouse and human preprocortistatin. *Genomics*. 42: 499-506.

4.      Spier, A. D. and de Lecea, L. (2000). Cortistatin: a member of the somatostatin neuropeptide family with distinct physiological functions. *Brain Res Brain Res Rev*. 33: 228-241.

5.      Gluschankof, P., Morel, A., Gomez, S., Nicolas, P., Fahy, C.and Cohen, P. (1984). Enzymes processing somatostatin precursors: an Arg-Lys esteropeptidase from the rat brain cortex converting somatostatin-28 into somatostatin- 14. *Proc Natl Acad Sci U S A*. 81: 6662-6.

6.      Puebla, L., Mouchantaf, R., Sasi, R., Khare, S., Bennett, H. P., James, S.and Patel, Y. C. (1999). Processing of rat preprocortistatin in mouse AtT-20 cells. *J Neurochem*. 73: 1273-7.

7.      Fukusumi, S., Kitada, C., Takekawa, S., Kizawa, H., Sakamoto, J., Miyamoto, M., Hinuma, S., Kitano, K.and Fujino, M. (1997). Identification and characterization of a novel human cortistatin-like peptide. *Biochem Biophys Res Commun*. 232: 157-63.

8.      Lalley, P. A., Sakaguchi, A. Y., Eddy, R. L., Honey, N. H., Bell, G. I., Shen, L. P., Rutter, W. J., Jacobs, J. W., Heinrich, G., Chin, W. W.and et al. (1987). Mapping polypeptide hormone genes in the mouse: somatostatin, glucagon, calcitonin, and parathyroid hormone. *Cytogenet Cell Genet*. 44: 92-7.

9.    Ejeskar, K., Abel, F., Sjoberg, R., Backstrom, J., Kogner, P.and Martinsson, T. (2000). Fine mapping of the human preprocortistatin gene (CORT) to neuroblastoma consensus deletion region 1p36.3-->p36.2, but absence of mutations in primary tumors. *Cytogenet Cell Genet.* 89: 62-6.

10.   Seidah, N. G., Day, R., Marcinkiewicz, M.and Chretien, M. (1993). Mammalian paired basic amino acid convertases of prohormones and proproteins. *Ann N Y Acad Sci.* 680: 135-46.

11.   Galanopoulou, A. S., Kent, G., Rabbani, S. N., Seidah, N. G.and Patel, Y. C. (1993). Heterologous processing of prosomatostatin in constitutive and regulated secretory pathways. Putative role of the endoproteases furin, PC1, and PC2. *J Biol Chem.* 268: 6041-9.

12.   Schafer, M. K., Day, R., Cullinan, W. E., Chretien, M., Seidah, N. G.and Watson, S. J. (1993). Gene expression of prohormone and proprotein convertases in the rat CNS: a comparative in situ hybridization analysis. *J Neurosci.* 13: 1258-79.

13.   Patel, Y. C.and Galanopoulou, A. (1995). Processing and intracellular targeting of prosomatostatin-derived peptides: the role of mammalian endoproteases. *Ciba Found Symp.* 190: 26-40; discussion 40-50.

14.   Ramirez, J. L., Sasi, R., Kumar, U., Mouchantaf, R., Otero, V., Rubinstein, M., Low, M. J.and Patel, Y. C. (1999). Brain somatostatin receptors (sstrs) are upregulated in the somatostatin deficient mouse. *Soc. Neurosci. Abs.* 25: 175.4.

15.   de Lecea, L., del Rio, J. A., Criado, J. R., Alcantara, S., Morales, M., Henriksen, S. J., Soriano, E.and Sutcliffe, J. G. (1997). Cortistatin is expressed in a distinct subset of cortical interneurons. *J. Neurosci.* 17: 5868-5880.

16.   Calbet, M., Guadano-Ferraz, A., Spier, A. D., Maj, M., Sutcliffe, J. G., Przewlocki, R.and de Lecea, L. (1999). Cortistatin and somatostatin mRNAs are differentially regulated in response to kainate. *Brain Res Mol Brain Res.* 72: 55-64.

17.   Bourgin, P., Fabre, V., Huitron-Resendiz, S., Sanchez-Alavez, M., Calbet, M., Criado, J., Henriksen, S., Sutcliffe, J., Prospero-Garcia, O.and de Lecea, L. (2001). Cortistatin is a selective regulator of deep sleep. *J. Neurosci.* 20:7760-7765.

18.   Moore, S. D., Madamba, S. G., Joels, M.and Siggins, G. R. (1988). Somatostatin augments the M-current in hippocampal neurons. *Science.* 239: 278-80.

19.   Schweitzer, P., Madamba, S., Champagnat, J.and Siggins, G. R. (1993). Somatostatin inhibition of hippocampal CA1 pyramidal neurons: mediation by arachidonic acid and its metabolites. *J Neurosci.* 13: 2033-49.

20.   Tallent, M. K.and Siggins, G. R. (1997). Somatostatin depresses excitatory but not inhibitory neurotransmission in rat CA1 hippocampus. *J Neurophysiol.* 78: 3008-18.

21.   Tallent, M. K.and Siggins, G. R. (1999). Somatostatin acts in CA1 and CA3 to reduce hippocampal epileptiform activity. *J Neurophysiol.* 81: 1626-35.

22.   Vasilaki, A., Lanneau, C., Dournaud, P., De Lecea, L., Gardette, R.and Epelbaum, J. (1999). Cortistatin affects glutamate sensitivity in mouse hypothalamic neurons through activation of sst2 somatostatin receptor subtype. *Neuroscience.* 88: 359-64.

23.   Patel, Y. C. (1999). Somatostatin and its receptor family. *Front Neuroendocrinol.* 20: 157-98.

24.   Criado, J., Li, H., Liapakis, G., Spina, M., Henriksen, S., Koob, G., Reisine, T., Sutcliffe, J., Goodman, M.and de Lecea, L. (1999). Structural and compositional determinants of cortistatin activity. *J Neurosci Res.* 56: 611-19.

25.   Siehler, S., Seuwen, K.and Hoyer, D. (1998). [125I]Tyr10-cortistatin14 labels all five somatostatin receptors. *Naunyn Schmiedebergs Arch Pharmacol.* 357: 483-9.

26.   Leroux, P., Weissmann, D., Pujol, J. F.and Vaudry, H. (1993). Quantitative autoradiography of somatostatin receptors in the rat limbic system. *J Comp Neurol.* 331: 389-401.

27.   Spier, A. D.and de Lecea, L. (2000). Autoradiographic localization of cortistatin binding sites in the central nervous system: comparison with somatostatin. *Soc. Neurosci. Abs.* 27: 719.7.

28.   Rezek, M., Havlicek, V., Hughes, K. R.and Friesen, H. (1976). Cortical administration of somatostatin (SRIF): effect on sleep and motor behavior. *Pharmacol Biochem Behav.* 5: 73-77.

44

29.     Danguir, J. (1986). Intracerebroventricular infusion of somatostatin selectively increases paradoxical sleep in rats. *Brain Res.* 367: 26-30.

30.     Luthi, A.and McCormick, D. A. (1998). H-current: properties of a neuronal and network pacemaker. *Neuron.* 21: 9-12.

31.     Rauca, C., Schafer, K.and Hollt, V. (1999). Effects of somatostatin, octreotide and cortistatin on ischaemic neuronal damage following permanent middle cerebral artery occlusion in the rat [In Process Citation]. *Naunyn Schmiedebergs Arch Pharmacol.* 360: 633-8.

32.     Flood, J. F., Uezu, K.and J.E., M. (1997). The cortical neuropeptide, cortistatin-14, impairs post-training memory processing. *Brain Res.* 775: 250-252.

33.     Sanchez-Alavez, M., Gomez-Chavarin, M., Navarro, L., Jimenez-Anguiano, A., Murillo-Rodriguez, E., Prado-Alcala, R. A., Drucker-Colin, R.and Prospero-Garcia, O. (2000). Cortistatin modulates memory processes in rats. *Brain Res.* 858: 78-83.

34.     Tallent, M. K., Baratta, M. V., de Lecea, L.and Siggins,'G. R. (2000). Transgenic mice overexpressing cortistatin show a deficit in dentate gyrus long-term potentiation. *Soc. Neurosci Abs*432-1.

35.     Aigner, T. G. (1995). Pharmacology of memory: cholinergic-glutamatergic interactions. *Curr Opin Neurobiol.* 5: 155-60.

36.     Deghenghi, R., Papotti, M., Ghigo, E.and Muccioli, G. (2001). Cortistatin, but not somatostatin, binds to growth hormone secretagogue (GHS) receptors of human pituitary gland. *J Endocrinol Invest.* 24: RC1-3.

37.     Broglio, F., Koetsveld Pv, P., Benso, A., Gottero, C., Prodam, F., Papotti, M., Muccioli, G., Gauna, C., Hofland, L., Deghenghi, R., Arvat, E., Van Der Lely, A. J.and Ghigo, E. (2002). Ghrelin secretion is inhibited by either somatostatin or cortistatin in humans. *J Clin Endocrinol Metab.* 87: 4829-32.

38.     Dalm, V. A., van Hagen, P. M., van Koetsveld, P. M., Langerak, A. W., van der Lely, A. J., Lamberts, S. W.and Hofland, L. J. (2003). Cortistatin rather than somatostatin as a potential endogenous ligand for somatostatin receptors in the human immune system. *J Clin Endocrinol Metab.* 88: 270-6.

39.     Dalm, V. A., van Hagen, P. M., van Koetsveld, P. M., Achilefu, S., Houtsmuller, A. B., Pols, D. H., van der Lely, A. J., Lamberts, S. W.and Hofland, L. J. (2003). Expression of somatostatin, cortistatin, and somatostatin receptors in human monocytes, macrophages, and dendritic cells. *Am J Physiol Endocrinol Metab.* 285: E344-53.

40.     Civelli, O., Reinscheid, R. K.and Nothacker, H. P. (1999). Orphan receptors, novel neuropeptides and reverse pharmaceutical research. *Brain Res.* 848: 63-5.

41.     Pelton, J. T., Kazmierski, W., Gulya, K., Yamamura, H. I.and Hruby, V. J. (1986). Design and synthesis of conformationally constrained somatostatin analogues with high potency and specificity for mu opioid receptors. *J Med Chem.* 29: 2370-5.

42.     Nothacker, H. P., Wang, Z., McNeill, A. M., Saito, Y., Merten, S., O'Dowd, B., Duckles, S. P.and Civelli, O. (1999). Identification of the natural ligand of an orphan G-protein-coupled receptor involved in the regulation of vasoconstriction. *Nat Cell Biol.* 1: 383-5.

43.     Connor, M., Ingram, S. L.and Christie, M. J. (1997). Cortistatin increase of a potassium conductance in rat locus coeruleus in vitro. *Br J Pharmacol.* 122: 1567-72.

44.     Robas, N., Mead, E.and Fidock, M. (2003). MrgX2 is a high potency cortistatin receptor expressed in dorsal root ganglion. *J Biol Chem.* 78:44400-44404.

45.     Juarez, R. A., Rubenstein, M., Chan, E. C.and Low, M. J. (1997). Increased growth following normal development in middle-aged somatostatin-deficient mice. *Soc. Neurosci. Abs.* 23: 659.10.

46.     Brazeau, P., Vale, W., Burgus, R., Ling, N., Butcher, M., Rivier, J.and Guillemin, R. (1973). Hypothalamic polypeptide that inhibits the secretion of immunoreactive pituitary growth hormone. *Science.* 179: 77-9.

47.     Schmechel, D. E., Vickrey, B. G., Fitzpatrick, D.and Elde, R. P. (1984). GABAergic neurons of mammalian cerebral cortex: widespread subclass defined by somatostatin content. *Neurosci Lett.* 47: 227-32.

48. Jacquin, T., Champagnat, J., Madamba, S., Denavit-Saubie, M.and Siggins, G. R. (1988). Somatostatin depresses excitability in neurons of the solitary tract complex through hyperpolarization and augmentation of IM, a non- inactivating voltage-dependent outward current blocked by muscarinic agonists. *Proc Natl Acad Sci U S A.* 85: 948-52.

49. Morrison, J. H., Benoit, R., Magistretti, P. J.and Bloom, F. E. (1983). Immunohistochemical distribution of pro-somatostatin-related peptides in cerebral cortex. *Brain Res.* 262: 344-51.

50. Rezek, M., Havlicek, V., Hughes, K. R.and Friesen, H. (1977). Behavioural and motor excitation and inhibition induced by the administration of small and large doses of somatostatin into the amygdala. *Neuropharmacology.* 16: 157-162.

51. Mancillas, J. R., Siggins, G. R.and Bloom, F. E. (1986). Somatostatin selectively enhances acetylcholine-induced excitations in rat hippocampus and cortex. *Proc Natl Acad Sci U S A.* 83: 7518-21.

# 4
# MOLECULAR EVOLUTION OF SOMATOSTATIN GENES

Hervé Tostivint[1], Michele Trabucchi[1,2], Mauro Vallarino[2], J. Michael Conlon[3], Isabelle Lihrmann[1] and Hubert Vaudry[1]
[1]Laboratory of Cellular and Molecular Neuroendocrinology INSERM U413, UA CNRS, IFRMP 23, University of Rouen, France; [2]Department of Experimental Biology, University of Genova, Italy; [3]Department of Biochemistry, Faculty of Medicine and Health Sciences, United Arab Emirates University, 17666 Al-Ain, UAE

## INTRODUCTION

Polyploidy (one or more duplications of the whole genome) has been proposed as an important driving force in the evolution of vertebrates (Ohno, 1970) and it has been suggested that two such genome duplication events occurred in rapid succession, early in the vertebrate lineage, some 500 My ago: the first one, soon after the emergence of vertebrates and the second one during early gnathostome evolution (Furlong and Holland, 2002). The occurrence of these duplication events exactly coincides with the period of diversification of most neuropeptide families currently known, including the NPY family (Larhammar, 1996; Conlon, 2002), the GnRH family (Fernald and White, 1999), the vasopressin/oxytocin family (Acher, 1996) and the PACAP/GHRH/glucagon family (Sherwood et al., 2000; Vaudry et al., 2000). There is now evidence that somatostatin also belongs to a family of neuropeptides whose members arose from gene duplication early during vertebrate evolution.

The occurrence of immunoreactive somatostatin has been reported in all vertebrate groups (Conlon et al., 1997a), in several invertebrate species (Conlon et al., 1988a) and even in plants (LeRoith et al., 1985a) and bacteria (LeRoith et al., 1985b,c). Owing to the development of peptide isolation and DNA cloning techniques, it has been possible to characterize somatostatin-related molecules from over 50 vertebrate species. At the present time, different forms of somatostatin have been identified, which differ both in sequence and size. The biological actions of these somatostatin variants are mediated by a family of seven-transmembrane domain G-protein-coupled receptors that, in mammals, comprises five distinct subtypes, termed SSTR1-5 (Patel, 1999; see also chapter 6 of this book). The counterparts of mammalian

SSTR1, SSTR2, SSTR3 and SSTR5 have recently been cloned in teleost fish (Zupanc et al., 1999; Lin et al., 1999b; 2000; 2002; 2003).

The objective of the present chapter is to review the current knowledge on the existence of somatostatin isoforms in various vertebrate groups, to discuss their phylogenetical relationship and to sketch possible history of their corresponding genes.

## Peptides derived from pre-prosomatostatin 1

First identified in mammals, somatostatin 1 (SS1; also termed SST-14) has subsequently been found with the same amino acid sequence in almost all other vertebrate groups, including agnathans, cartilaginous fish, ray-finned fish, lungfish, amphibians, reptiles and birds (Table 1). It thus appears that SS1, whose sequence has remained unchanged during more than 500 My, is the most conserved neuropeptide identified so far in vertebrates. Not only the primary structure of SS1 but also the overall organization of prosomatostatin 1 (PSS1) has been strongly preserved during evolution, the sequence of SS1 being systematically located at the C-terminus of the precursor. It is noteworthy that the genome of the tetraodontiform *Takifugu rubripes* (pufferfish) contains two copies of the SS1 gene (personal observation). One of the two peptides (referred to as 1b; Figure1) exhibits a Lys→Arg substitution at position 4. In mammals, processing of PSS1 can generate an N-terminally extended form of SS1 called somatostatin-28 (SST-28) (Esch et al, 1980; Pradayrol et al., 1980). The primary structure of the N-terminal domain of SST-28 has also been conserved, although to a lesser extent, from fish to mammals (Figure 1).

The two peptides SS1 and SST-28 are synthesized in variable amounts by different PSS1 producing cells as a result of differential precursor processing. Thus, SS1 is most abundant in pancreatic islets, stomach and neural tissues, while SST-28 predominates in D-cells of the intestinal mucosa. SS1 and SST-28 are generated through processing at two cleavage sites, a dibasic Arg-Lys site and a monobasic Arg site, respectively. Co-transfection experiments have shown that SS1 and SST-28 are independently synthesized from the PSS1 molecule through the action of two different sets of prohormone convertases (PCs), *i.e.* PC1 and PC2 at the Arg-Lys site, and furin and PACE4 at the Arg site (Galanopoulou et al., 1993; Brakch et al., 1995). Interestingly, these two cleavage motifs have been fully conserved in all vertebrate PSS1s (Figure 1). In addition, several basic residues located upstream the Arg-Lys and Arg sites, have been preserved in the PSS1 sequences. These basic amino acids, occurring at P2, P4 and/or P6 are known to facilitate the hydrolytic activity of PCs (Seidah et al., 1998).

The size of the N-terminal extension of SS1 is more variable: it encompasses 12 residues in all tetrapods and some fish such as the anglerfish *Lophius americanus* (Hobart et al., 1980) and the American white sturgeon *Acipenser*

**Table 1.** Characterization of the sequence of SS1 in vertebrates

| Species | Peptide | cDNA | Gene | Reference |
|---|:---:|:---:|:---:|---|
| Human, *Homo sapiens* | ● | | | Conlon and McCarthy, 1984 |
| | | ● | | Shen et al., 1982 |
| | | | ● | Shen and Rutter, 1984 |
| Monkey, *Macaca fascicularis* | | ● | | Travis and Sutcliffe, 1988 |
| Pig, *Sus scrofa* | ● | | | Schally et al., 1976 |
| Cow, *Bos taurus* | | ● | | Su et al., 1988 |
| | | | ● | Furu et al., 1999 |
| Sheep, *Ovis aries* | ● | | | Brazeau et al., 1973 |
| | | ● | | Bruneau and Tillet, 1998 |
| Rat, *Rattus norvegicus* | ● | | | Benoit et al., 1980 |
| | | ● | | Funckes et al., 1983 |
| | | ● | | Goodmann et al., 1982 |
| | | | ● | Tavianini et al., 1984 |
| | | | ● | Montminy et al., 1984 |
| Mouse, *Mus musculus* | | ● | | Fuhrmann et al., 1990 |
| | | | ● | Fuhrmann et al., 1990 |
| Guinea pig, *Cavia porcellus* | ● | | | Conlon, 1984 |
| Chicken, *Gallus gallus* | | ● | | Nata et al., 1991 |
| Pigeon, *Colomba livia* | ● | | | Spiess al., 1979 |
| Python, *Python molurus* | ● | | | Conlon et al., 1997b |
| Alligator, *Alligator mississipiensis* | ● | | | Wang and Conlon, 1993a |
| Turtle, *Pseudemys scripta* | ● | | | Conlon and Hicks, 1990 |
| Tortoise, *Gopherus agassizii* | ● | | | Wang et al., 1999b |
| Leopard frog, *Rana pipens* | ● | | | Takami et al., 1985 |
| European green frog, *Rana ridibunda* | ● | | | Vaudry et al., 1992 |
| | | ● | | Tostivint et al., 1996 |
| African clawed toad, *Xenopus laevis* | | ● | | Tostivint et al., 2003a,b |
| Amphiuma, *Amphiuma tridactylum* | ● | | | Cavanaugh et al., 1996 |
| African lungfish, *Protopterus annectens* | | ● | | Trabucchi et al., 1999 |
| Anglerfish, *Lophius americanus* | ● | | | Noe et al., 1979 |
| | | ● | | Hobart et al., 1980 |
| | | ● | | Goodman et al.,, 1980 |
| Trout, *Oncorhynchus mykiss* | | ● | | Kittilson et al., 1999 |
| Coho salmon, *Oncorhynchus kisutch* | ● | | | Plisetskaya et al., 1986 |
| Catfish, *Ictalurus punctatus* | | ● | | Minth et al., 1982 |
| Pacu, *Piaractus mesopotamicus* | ● | | | Ferraz de Lima et al., 1999 |
| Goldfish, *Carassius auratus* | | ● | | Lin et al., 1999a |
| Zebrafish, *Danio rerio* | | ● | | Devos et al., 2002 |
| Pufferfish, *Takifugu rubripes* | | | ● | Ensembl Annotation Project, 2003a |
| Siver erawana, *Osteogossum bicirrhosum* | | ● | | Al-Mahrouki et al., 2000a |
| Freshwater butterflyfish, *Pantodon buchholzi* | | ● | | Al-Mahrouki et al., 2000b |
| Weakly electric fish, *Gnathonemus petersii* | | ● | | Al-Mahrouki et al., 2000c |
| Chitala, *Chitala chitala* | | ● | | Al-Mahrouki et al., 2000d |
| White sucker, *Catostomus commersoni* | | ● | | Al-Mahrouki et al., 2000e |
| Eel, *Anguilla anguilla* | ● | | | Conlon et al., 1988c |
| Flounder, *Platichtythes flesus* | ● | | | Conlon et al.,1987 |
| Stone flounder, *Kareius bicoloratus* | ● | | | Andoh and Nagasawa, 1998 |
| *Daddy sculpin,* Cottus scorpius | ● | | | Conlon et al., 1987 |
| Tilapia, *Oreochromis nilotica* | ● | | | Nguyen et al., 1995 |
| Bowfin, *Amia calva* | ● | | | Wang et al., 1993b |
| American paddlefish, *Polyodon spathula* | ● | | | Conlon, unpublished data |
| Pallid sturgeon, *Scaphirhynchus albus* | ● | | | Kim et al., 2000 |
| White sturgeon, *Acipenser transmontanus* | | ● | | Trabucchi et al., 2002 |
| Ray, *Torpedo marmorata* | ● | | | Conlon et al., 1985 |
| Sea lamprey, *Petromyzon marinus* | ● | | | Sower et al., 1994 |
| River lamprey, *Lampetra fluviatilis* | ● | | | Conlon et al., 1995a |
| Southern-hemisphere lamprey, *Geotria australis* | ● | | | Wang et al., 1999a |
| Hagfish, *Myxine glutinosa* | ● | | | Conlon et al., 1988b |

*transmontanus* (Trabucchi et al., 2002) but only 11 residues in the lungfish *Protopterus annectens* (Trabucchi et al., 1999), and 10 residues in several other fish such as the rainbow trout *Oncorhynchus mykiss* (Kittilson et al., 1999), the channel catfish *Ictalurus punctatus* (Minth et al., 1982), the goldfish *Carassius auratus* (Lin et al., 1999a) and the bowfin *Amia calva* (Wang et al., 1993b).

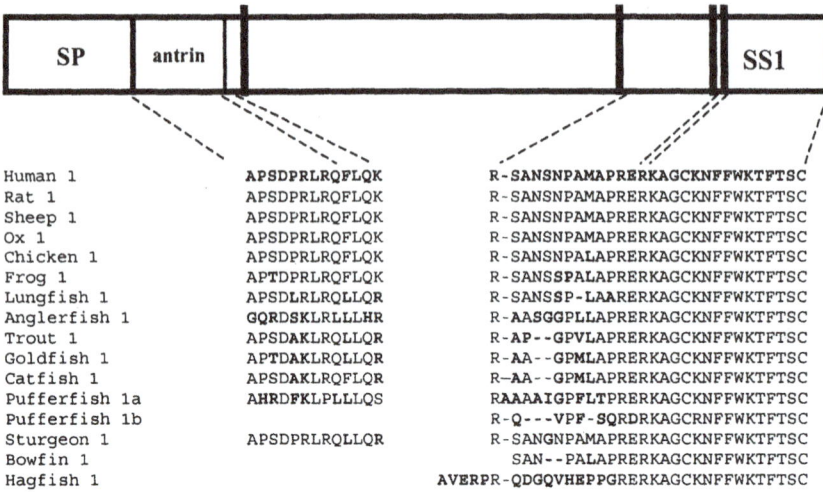

```
Human 1        APSDPRLRQFLQK    R-SANSNPAMAPRERKAGCKNFFWKTFTSC
Rat 1          APSDPRLRQFLQK    R-SANSNPAMAPRERKAGCKNFFWKTFTSC
Sheep 1        APSDPRLRQFLQK    R-SANSNPAMAPRERKAGCKNFFWKTFTSC
Ox 1           APSDPRLRQFLQK    R-SANSNPAMAPRERKAGCKNFFWKTFTSC
Chicken 1      APSDPRLRQFLQK    R-SANSNPALAPRERKAGCKNFFWKTFTSC
Frog 1         APTDPRLRQFLQK    R-SANSSPALAPRERKAGCKNFFWKTFTSC
Lungfish 1     APSDLRLRQLLQR    R-SANSSP-LAARERKAGCKNFFWKTFTSC
Anglerfish 1   GQRDSKLRLLLHR    R-AASGGPLLAPRERKAGCKNFFWKTFTSC
Trout 1        APSDAKLRQLLQR    R-AP--GPVLAPRERKAGCKNFFWKTFTSC
Goldfish 1     APTDAKLRQLLQR    R-AA--GPMLAPRERKAGCKNFFWKTFTSC
Catfish 1      APSDAKLRQFLQR    R-AA--GPMLAPRERKAGCKNFFWKTFTSC
Pufferfish 1a  AHRDFKLPLLLQS    RAAAAIGPFLTPRERKAGCKNFFWKTFTSC
Pufferfish 1b                   R-Q---VPF-SQRDRKAGCRNFFWKTFTSC
Sturgeon 1     APSDPRLRQLLQR    R-SANGNPAMAPRERKAGCKNFFWKTFTSC
Bowfin 1                        SAN--PALAPRERKAGCKNFFWKTFTSC
Hagfish 1                       AVERPR-QDGQVHEPPGRERKAGCKNFFWKTFTSC
```

Figure 1. *Alignment of the amino acid sequences of PSS1-derived peptides. Residues that are identical to those of the human sequence are indicated in thin letters. SP, signal peptide.*

As a matter of fact, in non-mammalian species, the monobasic Arg site seems to be seldom used and pancreatic bowfin SST-26 is the only N-terminally extended form of SS1 identified to date (Wang et al., 1993b). It is interesting to note that, in the islet organ and gut of the Atlantic hagfish *Myxine glutinosa*, cleavage of PSS1 occurs predominantly 6 residues upstream the monobasic Arg site to generate an SST-34 peptide (Conlon et al., 1988b). Nevertheless, the strong conservation of the amino acid sequence of SST-28, at least in tetrapods, supports the view that this polypeptide could play an important role. Consistent with this notion, it has been found that SST-28 exhibits greater affinity than SS1 for the SSTR5 receptor subtype (O'Carroll et al., 1992; Panetta et al., 1994; Patel et al., 1994) and that SST-28 but not SS1 causes internalization of SSTR5 (Roth et al., 1997; Roosterman et al., 1997).

In mammals, processing of PSS1 also occurs at a single Lys residue located in the N-terminal region of the precursor to yield a decapeptide called antrin (Benoit et al., 1987; Rabbani and Patel, 1990). The sequence of antrin has also been strongly conserved across vertebrates, particularly in tetrapods, and in all species investigated to date, the peptide is flanked by a monobasic amino acid residue at its C-terminus (Figure 1). It has been postulated that antrin may serve as a sorting signal for targeting PSS1 products to the regulated secretory pathway (Sevarino and Stork, 1991; Mouchantaf et al., 2001). However, the significance of its cleavage at the putative monobasic site is currently unknown (Patel et al., 1997).

## Peptides derived from pre-prosomatostatin II

In addition to PSS1, most teleost fish possess a second somatostatin precursor, containing the $[Tyr^7, Gly^{10}]$SST-14 sequence at its C-terminus (Table 2). In spite of the existence of a putative dibasic Arg-Lys cleavage site at its N-terminal extremity, the peptide $[Tyr^7, Gly^{10}]$SST-14 is generally not produced. Rather, this precursor appears to be cleaved at a monobasic Arg site located upstream to generate an N-terminally extended peptide (named SSII) encompassing 25 to 28 amino acid residues (Figure 2).

The degree of conservation of the prepro-somatostatin II (PSSII) sequence is relatively low. Even in the $[Tyr^7, Gly^{10}]$SST-14 region, amino acid substitutions have been reported in the eel *Anguilla anguilla* (Conlon et al., 1988c), the tilapia *Oreochromis nilotica* (Nguyen et al., 1995) and the goldfish *Carassius auratus* (Lin et al., 1999a).

The existence of two distinct PSSII has been recently demonstrated in the rainbow trout (Moore et al., 1999), and is strongly suggested in the goldfish (Uesaka et al., 1995; Lin et al., 1999a). Duplication of the corresponding genes, as well as those encoding the two SS1 variants in fugu, likely occurred after genome tetraploidization in the teleost ancestor (Taylor et al., 2001).

***Table 2.*** Characterization of the sequences of SSII in teleosts

| Species | Sequences determined from | | | Reference |
| --- | --- | --- | --- | --- |
| | Peptide | CDNA | Gene | |
| Anglerfish, *Lophius americanus* | | ● | | Hobart et al., 1980 |
| | ● | | | Andrews et al.,, 1984b |
| Trout, *Oncorhynchus mykiss* | | ● | | Moore et al., 1995; 1999 |
| Coho salmon, *Oncorhynchus kisutch* | ● | | | Plisetskaya et al., 1986 |
| Goldfish, *Carassius auratus* | ● | | | Uesaka et al., 1995 |
| | | ● | | Lin et al., 1999a |
| Medaka, *Oryzias latipes* | | ● | | Mita et al., 2001 |
| Eel, *Anguilla anguilla* | ● | | | Conlon et al., 1988c |
| Flounder, *Platichtythes flesus* | ● | | | Conlon et al., 1987 |
| Daddy Sculpin, *Cottus scorpius* | ● | | | Conlon et al., 1987 |
| Tilapia, *Oreochromis nilotica* | ● | | | Nguyen et al., 1995 |
| Catfish, *Ictalurus punctatus* | ● | | | Andrews et al., 1984a |
| | | ● | | Magazin et al., 1982 |
| Zebrafish, *Danio rerio* | | ● | | Argenton et al., 1999 |

The channel catfish *Ictalurus punctatus* also possesses, in addition to PSS1, a second somatostatin precursor (Magazin et al., 1982; Andrews et al., 1984a). This precursor isoform generates a peptide of 22 residues that shows only very limited sequence identity with other teleost SSIIs, but that appears very similar to the zebrafish SSII variant (Argenton et al., 1999; Figure 3).

PSSII-derived peptides have long been thought to occur exclusively in the islet organ and/or gut but recent studies indicate that the PSSII gene can also be expressed in the brain (Moore et al., 1999; Lin et al., 1999a). The functional

52

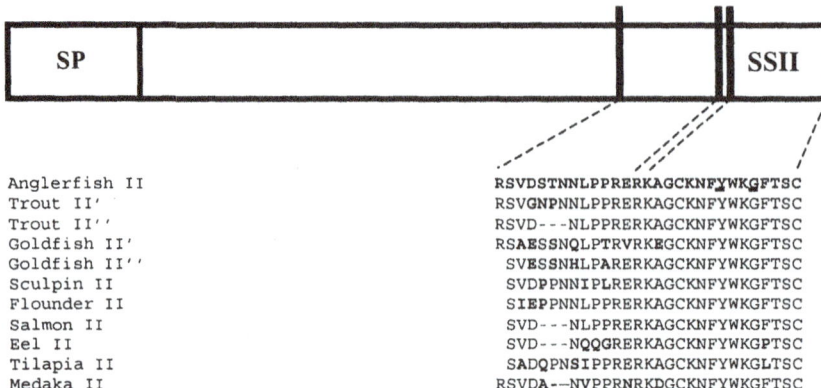

**Figure 2.** *Alignment of the amino acid sequences of PSSII-derived peptides. Residues of the anglerfish sequence that differ from the SS1 sequence are underlined. Residues that are conserved are indicated in thin letters. SP, signal peptide.*

significance of PSSII-related peptides is poorly understood. Intravenous injections of SS1 and SSII in rainbow trout have been reported to produce differential effects on carbohydrate and lipid metabolism (Eilertson and Sheridan, 1993) but, the receptor binding characteristics of PSSII-derived peptides for fish SSTRs are still unknown.

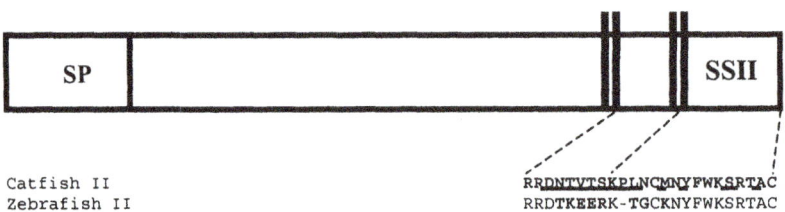

**Figure 3.** *Alignment of the amino acid sequences of atypical teleost SSIIs. Residues of the catfish sequence that differ from the SS1 sequence are underlined. Residues that are conserved are indicated in thin letters. SP, signal peptide.*

# Peptides derived from pre-prosomatostatin 2 and prepro-cortistatin

For two decades after the discovery of SS1, it has been assumed that a single gene encoding a somatostatin precursor was present in the genome of tetrapods, and it was generally accepted that only fish possessed a second gene encoding a somatostatin precursor variant which was exclusively expressed in peripheral organs (Shen and Rutter, 1984; Conlon, 1990a). This view was invalidated when two forms of somatostatin, namely SS1 and the novel variant [Pro$^2$, Met$^{13}$]SST-14 (SS2) were isolated from the brain of a single species, the European green frog *Rana ridibunda* (Vaudry et al., 1992). In 1995, Nishii et al. have reported the existence of the somatostatin variant [Pro$^2$]SST-14 in the pituitary of the Russian sturgeon *Acipenser gueldenstaedti* but, at that time, it was not clear whether two somatostatin isoforms existed in this species since the occurrence of SS1 had not been demonstrated.

The cDNAs encoding the two precursors, *i.e.* PSS1 and PSS2, have subsequently been cloned in frog (Tostivint et al., 1996). While the general organization of frog PSS2 is similar to that of PSS1s (Figure 4), the two precursors do not exhibit appreciable sequence identity in their N-terminal domains (Tostivint et al., 1996). In addition, PSS2 mRNA are almost exclusively detected in a few discrete areas of the brain whereas PSS1 mRNA are more widely distributed in the brain, pancreas and gut (Tostivint et al., 1996). Conversely, the PSS2 gene, but not the PSS1 gene, is actively expressed in the intermediate lobe of frog pituitary (Tostivint et al., 2002).

*Table 3.* Characterization of the sequences of PSS2 / PCST-derived peptides

| Species | Sequences determined from | | | Reference |
| --- | --- | --- | --- | --- |
| | Peptide | CDNA | Gene | |
| Human, *Homo sapiens* | | ● | | Fukusumi et al., 1997 |
| | | ● | | De Lecea et al., 1997 |
| | | | ● | *NCBI Annotation Project, 2002* |
| Rat, *Rattus norvegicus* | | ● | | De Lecea et al., 1996 |
| Mouse, *Mus musculus* | | ● | | De Lecea et al., 1997 |
| | | | ● | Calbet et al., 1999 |
| Chicken, *Gallus gallus* | | ● | | Trabucchi et al., 2003 |
| European green frog, *Rana ridibunda* | ● | | | Vaudry et al., 1992 |
| | | ● | | Tostivint et al., 1996 |
| African lungshish, *Protopterus annectens* | | ● | | Trabucchi et al., 1999 |
| Goldfish, *Carassius auratus* | | ● | | Lin et al., 1999a |
| Zebrafish, *danio rerio* | | ● | | *Clark et al., 1998* |
| Pufferfish, *Takifugu rubripes* | | | ● | Ensembl Annotation Project, 2003b |
| Russian sturgeon, *Acipenser gueldenstaedti* | ● | | | Nishii et al., 1995 |
| White sturgeon, *Acipenser transmontanus* | | ● | | Trabucchi et al., 2002 |

The occurrence of both PSS1 and PSS2 genes has subsequently been demonstrated in the lungfish (Trabucchi et al., 1999), the goldfish (also named PSSIII; Lin et al., 1999a), the zebrafish (in which it has been named PSS3; Devos et al., 2002), the American white sturgeon (Trabucchi et al., 2002) and the chicken (Trabucchi et al., 2003; Table 3). In these five species, the sequence of the SS2 peptide is [Pro$^2$]SST-14 (Figure 4). A comparison of the structure of the corresponding precursors indicates an overall sequence identity of 52 to 67%. Surprisingly, the chicken PSS2 appears more related to the fish PSS2 (about 60% identity) than to the frog PSS2 (only 35% identity). The existence of PSS1 and PSS2 in the American sturgeon indicates that the [Pro$^2$]SST-14 isoform isolated by Nishii et al. (1995) in the Russian sturgeon *Acipenser gueldenstaedti* actually corresponds to the SS2 variant.

In addition to the dibasic processing site Arg-Lys yielding SS2, sturgeon, goldfish, lungfish and chicken PSS2 also possess an Arg residue that can potentially generate an SST-24-like molecule (Figure 4). In frog, the fact that the Arg is substituted by a Lys residue and that the amino acids located at P2, P4 and P6 are not basic residues, suggests that PCs cannot operate at this site.

The precursor of an SS2-related peptide, called cortistatin (CST), has been characterized in the rat (de Lecea et al., 1996), the mouse (de Lecea et al., 1997) and human (de Lecea et al., 1997; Fukusumi et al., 1997; Table 3). Several lines of evidence suggest that PSS2 and procortistatin (PCST) actually derive from orthologous genes: *i)* their processing products, SS2 and CST, both contain the Gly→Pro substitution at position 2 (Figure 4); *ii)* they are mainly expressed in the brain but not in the pancreas or gut; and *iii)* the mouse and human genomes do not appear to encompass any sequence that would be more related to PSS2 than PCST. It should be noted however that, apart their C-terminal region, PCST and PSS2 possess very little sequence similarity. In addition, the structure of CST is itself globally more variable, both in size and in sequence, than the SS2 one. For instance, the human CST is likely 17 residues in length while the rodent CST is only 14 (Figure 4). Taken together, all these data indicate that the SS2 / CST genes must have evolved at vastly different rates, depending on both the species and the periods: the rate of evolution of the gene was globally low in fish and in most tetrapods, but more rapid in the amphibian lineage and even faster in the mammalian lineage.

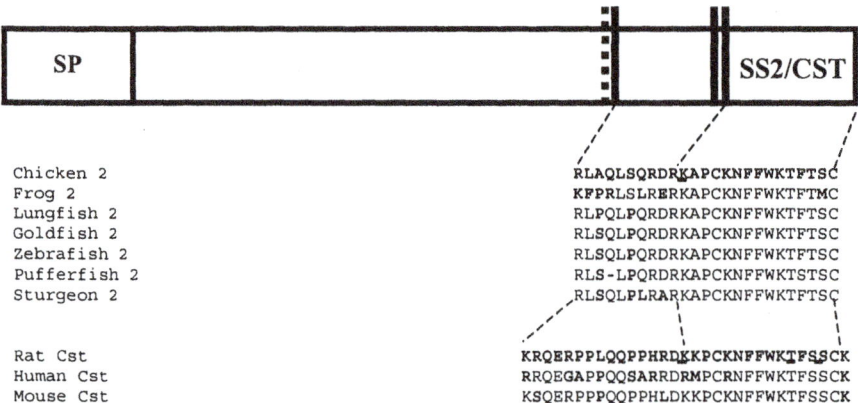

**Figure 4.** *Alignment of the amino acid sequences of PSS2- and PCST-derived peptides. Residues of the chicken and rat sequences that differ from the SS1 sequence are underlined. Residues that are conserved are indicated in thin letters. SP, signal peptide.*

Like all other somatostatin-related peptides, CST is flanked at its N-terminal extremity by a dibasic cleavage site. The presence of additional dibasic residues in the rat and human, but not in the mouse PCST sequence suggests that N-terminally extended forms of CST may also be produced. Consistent with this view, it has been shown that PCST is efficiency processed to both CST-14 and CST-29 in AtT-20 cells (Puebla et al., 1999).

As expected from their structural similarities, CST and SS2 can mimic several of the effects of SS1 (de Lecea et al., 1996; Jeandel et al., 1997). However, CST and SS1 have also been shown to induce differential responses, especially on cortical electrical activity, locomotor behavior and paradoxical sleep in rat (de Lecea et al., 1996). Whether the effects of CST are mediated via a specific receptor or only via SSTRs is currently a matter of speculation (Spier and de Lecea, 2000; see also chapter 2 of this book). Nevertheless, it has been recently reported that the orphan receptor MrgX2 is selectively activated by CST (Robas et al., 2003).

## Other somatostatin-related peptides

Two somatostatin variants have been characterized in the lampreys *Geotria australis*, *Lampetra fluviatilis* and *Petromyzon marinus* (Table 4). One of them is the tetradecapeptide SS1 and the second one is a 33- to 37-residue peptide which exhibits a conserved [Ser$^{12}$]SST-14 sequence at its C-terminus (Figure 5). In both *Lampetra fluviatilis* and *Petromyzon marinus*, SS1 predominates in the brain while the [Ser$^{12}$]SST-14-containing variants are primarily found in the pancreas (Andrews et al., 1988; Sower et al., 1994; Conlon et al., 1995a). Recent studies have shown that, in *Geotria australis*, the processing of the [Ser$^{12}$]SST-14 precursor is tissue-specific: in the pancreas, [Ser$^{12}$]SST-14 is the major molecular form while, in the intestine, the [Ser$^{12}$]SST-14 precursor is processed almost exclusively to the 33-amino acid form (Conlon et al., 1995b; Wang et al., 1999a). Analysis of the sequence of these N-terminally extended forms of [Ser$^{12}$]SST-14 does not reveal any similarity with either the SS2 or SSII precursors, so that they cannot be grouped objectively with any of them.

Another somatostatin variant, [Ser$^{5}$]SST-14, has been characterized from the pancreas of a holocephalan fish, the Pacific ratfish *Hydrolagus colliei* (Conlon, 1990b; Table 4; Figure 5). Since the peptide SS1 has never been characterized in this species, it is currently not known if this [Ser$^{5}$]SST-14 variant derives from the PSS1 gene or from a paralogous gene.

*Table 4.* Characterization of the sequences of other somatostatin-related peptides

| Species | Sequences determined from | | References |
| --- | --- | --- | --- |
| | Peptide | CDNA | |
| Sea lamprey, *Petromyzon marinus* | ● | | Andrews et al., 1988 |
| River lamprey, *Lampetra fluviatilis* | ● | | Conlon et al., 1995a |
| Southern-hemisphere lamprey, *Geotria australis* | ● | | Conlon et al., 1995b |
| Pacific ratfish, *Hydrolagus colliei* | ● | | Conlon, 1990b |

Finally, the occurrence of somatostatin-like immunoreactivity, that does not derive from the PCST gene, in the proximal midgut of somatostatin-deficient mice strongly suggests the existence of a novel somatostatin-related gene in

56

mammals (Ramirez et al., 2002). In support of this hypothesis, a novel peptide, termed thritten, whose sequence corresponds to S-28$_{(1-13)}$ but does not derive from PSS1 or PCST, has been recently characterized in the rat and monkey intestinal tract (Ensinck et al., 2002). Further studies will be necessary to identify the mysterious gene encoding this novel somatostatin-related peptide.

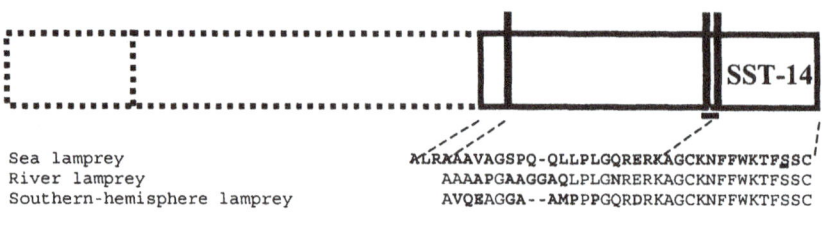

```
Sea lamprey                  ALRKAAVAGSPQ-QLLPLGQRERKAGCKNFFWKTFSSC
River lamprey                AAAAPGAAGGAQLPLGNRERKAGCKNFFWKTFSSC
Southern-hemisphere lamprey  AVQEAGGA--AMPPPGQRDRKAGCKNFFWKTFSSC

Ratfish                                    AGCKSFFWKTFTSC
```

*Figure 5.* *Alignment of the amino acid sequences of other somatostatin-related peptides. Residues of the sea lamprey and ratfish sequences that differ from the SS1 sequence are underlined. Residues that are conserved are indicated in thin letters. SP, signal peptide.*

## Evolution of the somatostatin gene family

The data on somatostatin-related peptides currently available support the existence of at least three distinct somatostatin genes in vertebrates. The presence of the SS1 gene in the hagfish indicates that this is a very ancient gene and likely the ancestor of all other somatostatin genes (Figure 6).

The SSII gene, which apparently occurs only in teleosts, has probably arisen from a duplication event that took place in a common ancestor of teleosts, after the divergence from the tetrapod lineage (Su et al., 1988; Figure 6). The SS2 / CST gene, which is present in the sturgeon (Figure 6) clearly appeared before the transition from Chondrichthyes to Osteichthyes. The exact time of this duplication event may have coincided with the whole-genome duplication that occurred during early vertebrate evolution and which has affected several other neuropeptide-encoding gene families. The lamprey [Ser$^{12}$]SST-14 variant may thus represent an SS2 gene product, just as the ratfish [Ser$^{5}$]SST-14 one. Cloning of the lamprey [Ser$^{12}$]SST-14 gene and the ratfish [Ser$^{5}$]SST-14 gene should provide crucial information to test this hypothesis.

During its evolutionary history, the SS2 sequence has been affected by two important mutational events: the acquisition and preservation of a Pro residue at position 2 and, in mammals, the addition of a Lys residue at the C-terminus. Recent structure-activity relationship studies have demonstrated the importance of these two amino acids for the specific effects of CST on the locomotor activity, sleep/wake cycle and the excitability of hippocampal CA1 neurons in rat (Criado et al., 1999).

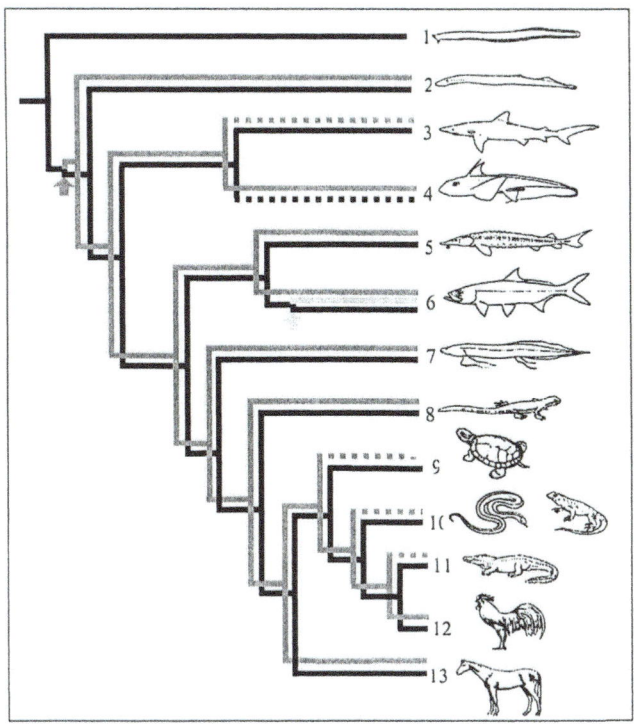

***Figure 6.*** *Possible gene duplication events in the somatostatin family shown in a schematic vertebrate evolution tree. This figure illustrates the hypothesis of the most parsimonious scenario of the somatostatin family evolution, in which the lamprey [Ser¹²]SST-14 and the ratfish [Ser⁵]SST-14 variants are both considered as PSS2 products. Duplication generating the SS2 / CST gene;* ⬆ *Duplication generating the SSII gene;* ▬▬ *SS1 gene;* ▬▬ *SS2 / CST gene;* ▬▬ *SSII gene. Broken lines, uncharacterized genes. 1, Myxiniformes; 2, Petromizoniformes; 3, Elasmobranchs; 4, Holocephalans; 5, Chondrosteans; 6, Telosts; 7, Lungfish; 8, Amphibians; 9, Chelonians; 10, Ophidians and Lacertilians; 11, Crocodilians; 12, Birds; 13, Mammals.*

Interestingly, the evolution of the somatostatin gene family exhibits striking similarities with that of the GnRH gene family (Fernald and White, 1999). Both families are composed of at least three genes that are differentially expressed in the brain and in peripheral tissues. In both families, the sequence of one of the biologically active peptides (GnRH-II or SS1) has been highly conserved, whereas the sequence of the second member (GnRH-I or SS2) is more variable. The presence of the third peptide (GnRH-III or SSII) is apparently restricted to the teleosts.

The occurrence of GnRH in prochordates is now well established (Powell et al., 1996). In contrast, although immunoreactive somatostatin has been detected in numerous invertebrate species (Conlon et al., 1988a), so far, no somatostatin-related peptide has been characterized in invertebrates. Inspection of the recently characterized *Ciona intestinalis* genome (Dehal et al., 2002) reveals the existence of a somatostatin receptor-like sequence but not any somatostatin peptide-like sequence. The endogeneous ligand of this receptor thus remains unknown. A previous study has shown that the *Drosophila* genome also encompasses somatostatin-like receptors whose natural ligands are allatostatins (Birgül et al., 1999; Kreienkamp et al., 2002), a family of peptides that do not possess structural similarity to somatostatin (Bendena et al., 1999). The exploration of the amphioxus genome should provide crucial information regarding the origin of somatostatin and the possible evolutionary relationship of the somatostatin ancestor with other structurally-related peptides, such as urotensin II (Conlon et al., 1997a; Coulouarn et al, 1998).

# REFERENCES

1. Acher R. Molecular evolution of fish neurohypophysial hormones: neutral and selective evolutionary mechanisms. Gen Comp Endocrinol 1996; 102:157-72
2. Al-Mahrouki A.A., Irwin D.M., Youson J.H. Characterization of variant somatostatin cDNAs from several osteoglossomorphs: molecular identification and comparative analysis. Genbank 2000a; AF292650. 2000b; AF292651. 2000c; AF292652. 2000d; AF292653. 2000e; AF292654 (*Direct submission*)
3. Andoh T., Nagasawa H. Purification and structural determination of insulins, glucagons and somatostatin from stone flounder, *Kareius bicoloratus*. Zool Sci 1998; 15:939-43
4. Andrews P.C., Pubols M.H., Hermodson M.A., Shears B.T., Dixon J.E. Structure of the 22-residue somatostatin from catfish. An O-glycosylated peptide having multiple forms. J Biol Chem 1984a; 259:13267-72
5. Andrews P.C., Hawke D., Shively J.E., Dixon J.E. Anglerfish preprosomatostatin II is processed to somatostatin-28 and contains hydroxylysine at residue 23. J Biol Chem 1984b; 259:15021-4
6. Andrews P.C., Pollock H.G., Elliott W.M., Youson J.H., Plisetskaya E.M. Isolation and characterization of a variant somatostatin-14 and two related somatostatins of 34 and 37 residues from lamprey (*Petromyzon marinus*). J Biol Chem 1988; 268:15809-14
7. Argenton F., Zecchin E., Bortolussi, M. Early appearance of pancreatic hormone-expressing cells in the zebrafish embryo. Mech Dev 1999; 87:217-21
8. Bendena W.G., Donly B.C., Tobe S.S. Allatostatins: a growing family of neuropeptides with structural and functional diversity. Ann N Y Acad Sci 1999;897:311-29
9. Benoit R., Bohlen P., Brazeau P., Ling N., Guillemin R. Isolation and characterization of rat pancreatic somatostatin. Endocrinology 1980; 107:2127-9
10. Benoit R., Ling N., Esch F. A new prosomatostatin-derived peptide reveals a pattern for prohormone cleavage at monobasic sites. Science 1987; 238:1126-9
11. Birgül N., Weise C., Kreienkamp H.J., Richter D. Reverse physiology in drosophila: identification of a novel allatostatin-like neuropeptide and its cognate receptor structurally related to the mammalian somatostatin/galanin/opioid receptor family. EMBO J 1999; 18:5892-900.
12. Brakch N., Galanopoulou A.S., Patel Y.C., Boileau G., Seidah N.G. Comparative proteolytic processing of rat prosomatostatin by convertases PC1, PC2, furin, PACE4 and PC5 in constitutive and regulated secretory pathways. FEBS Lett 1995; 362:143-6

13. Brazeau P., Vale W., Burgus R., Ling N., Butcher M., Rivier J., Guillemin R. Hypothalamic polypeptide that inhibits the secretion of immunoreactive pituitary growth hormone. Science 1973; 179:77-9

14. Bruneau G., Tillet Y. Localization of the preprosomatostatin-mRNA by *in situ* hybridization in the ewe hypothalamus. Peptides 1998; 19:1749-58

15. Calbet M., Guadano-Ferraz A., Spier A.D., Maj M., Sutcliffe J.G., Przewlocki R., de Lecea L. Cortistatin and somatostatin mRNAs are differentially regulated in response to kainate. Mol Brain Res 1999; 72:55-64

16. Cavanaugh E.S., Nielsen P.F., Conlon J.M. Isolation and structural characterization of proglucagon-derived peptides, pancreatic polypeptide, and somatostatin from the urodele *Amphiuma tridactylum*. Gen Comp Endocrinol 1996; 101:12-20

17. Clark M., Johnson S.L., Lehrach H., Lee R., Li F., Marra M., Eddy S., Hillier L., Kucaba T., Martin J., Beck C., Wylie T. , Underwood K., Steptoe M., Theising B., Allen M., Bowers Y. , Person B., Swaller T., Gibbons M., Pape D., Harvey N., Schurk R., Ritter E., Kohn S., Shin T., Jackson Y., Cardenas M., McCann R., Waterston R., Wilson R. WashU Zebrafish EST Project. Genbank 1998; BG307388 *(Direct submission)*

18. Conlon J.M. Isolation and structure of guinea pig gastric and pancreatic somatostatin. Life Sci 1984; 35:213-20

19. Conlon J. M. "Somatostatin: aspects of molecular evolution." In *ProgreSSTin Comparative Endocrinology*. A. Epple, C.G. Scanes, M.H. Stetson, M.H, eds. New York: Wiley-Liss, 1990a; 10-5

20. Conlon J.M. [Ser$^5$]somatostatin-14: isolation from the pancreas of a holocephalan fish, the Pacific ratfish, *Hydrolagus colliei*. Gen Comp Endocrinol 1990b; 80:314-20

21. Conlon J.M. The origin and evolution of peptide YY (PYY) and pancreatic polypeptide (PP). Peptides 2002; 23:269-78

22. Conlon J.M., Mc Carthy D.M. Fragments of prosomatostatin from a human pancreatic tumour. Mol Cell Endocrinol 1984; 38:81-6

23. Conlon J.M., Agoston D.V., Thim L. An elasmobranchian somatostatin: primary structure and tissue distribution in *Torpedo marmorata*. Gen Comp Endocrinol 1985; 60:406-13

24. Conlon J.M., Davis M.S., Falkmer S., Thim L. Structural characterization of peptides derived from prosomatostatins I and II isolated from the pancreatic islets of two species of teleostean fish: the daddy sculpin and the flounder. Eur J Biochem 1987; 168:647-52

25. Conlon J.M., Reinecke M., Thorndyke M.C., Falkmer S. Insulin and other islet hormones (somatostatin, glucagon and PP) in the neuroendocrine system of some lower vertebrates and that of invertebrates - a minireview. Horm Metab Res 1988a; 20:406-10

26. Conlon J.M., Askensten U., Falkmer S., Thim L. Primary structures of somatostatins from the islet organ of the hagfish suggest an anomalous pathway of post-translational processing of prosomatostatin-1. Endocrinology 1988b; 122:1855-9

27. Conlon J.M., Deacon C.F., Hazon N., Henderson I.W., Thim L. Somatostatin-related and glucagon-related peptides with unusual structural features from the European eel *(Anguilla anguilla)*. Gen Comp Endocrinol 1988c; 72:181-9

28. Conlon J.M., Hicks J.W. Isolation and structural characterization of insulin, glucagon and somatostatin from the turtle, *Pseudemys scripta*. Peptides 1990; 11:461-6

29. Conlon J.M., Bondereva V., Rusakov Y., Plisetskaya E.M., Mynarcik D.C., Whittaker J. Characterization of insulin, glucagon and somatostatin from the river lamprey, *Lampetra fluviatilis*. Gen Comp Endocrinol 1995a; 100:96-105

30. Conlon J.M., Nielsen P.F., Youson J.H., Potter I.C. Proinsulin and somatostatin from the islet organ of the southern-hemisphere lamprey *Geotria australis*. Gen Comp Endocrinol 1995b; 100:413-22

31. Conlon J.M., Tostivint H., Vaudry H. Somatostatin- and urotensin II-related peptides: molecular diversity and evolutionary perspectives. Regul Pept 1997a; 69:95-103

32. Conlon J.M., Secor S.M., Adrian T.E., Mynarcik D.C., Whittaker J. Purification and characterization of islet hormones (insulin, glucagon, pancreatic, polypeptide and somatostatin) from the Burmese python, *Python molurus*. Regul Pept 1997b; 71:191-8

60

33. Coulouarn Y., Lihrmann I., Jégou S., Anouar Y., Tostivint H., Beauvillain J.C., Conlon J.M., Bern H.A., Vaudry H. Cloning of the cDNA encoding the urotensin II precursor in frog and human reveals intense expression of the urotensin II gene in motoneurons of the spinal cord. Proc Natl Acad Sci USA 1998; 95:15803-8

34. Criado J.R., Li H., Jiang X., Spina M., Huitron-Resendiz S., Liapakis G., Calbet M., Siehler S., Henriksen S.J., Koob G., Hoyer D., Sutcliffe J.G., Goodman M., de Lecea L. Structural and compositional determinants of cortisatin activity. J Neurosci Res 1999; 56:611-9

35. Dehal P., Satou Y., Campbell R.K., Chapman J., Degnan B., De Tomaso A., Davidson B., Di Gregorio A., Gelpke M., Goodstein D.M., Harafuji N., Hastings K.E., Ho I., Hotta K., Huang W., Kawashima T., Lemaire P., Martinez D., Meinertzhagen I.A., Necula S., Nonaka M., Putnam N., Rash S., Saiga H., Satake M., Terry A., Yamada L., Wang H.G., Awazu S., Azumi K., Boore J., Branno M., Chin-Bow S., DeSantis R., Doyle S., Francino P., Keys D.N., Haga S., Hayashi H., Hino K., Imai K.S., Inaba K., Kano S., Kobayashi K., Kobayashi M., Lee B.I., Makabe K.W., Manohar C., Matassi G., Medina M., Mochizuki Y., Mount S., Morishita T., Miura S., Nakayama A., Nishizaka S., Nomoto H., Ohta F., Oishi K., Rigoutsos I., Sano M., Sasaki A., Sasakura Y., Shoguchi E., Shin-i T., Spagnuolo A., Stainier D., Suzuki M.M., Tassy O., Takatori N., Tokuoka M., Yagi K., Yoshizaki F., Wada S., Zhang C., Hyatt P.D., Larimer F., Detter C., Doggett N., Glavina T., Hawkins T., Richardson P., Lucas S., Kohara Y., Levine M., Satoh N., Rokhsar D.S. The draft genome of *Ciona intestinalis*: insights into chordate and vertebrate origins. Science 2002;298:2157-67

36. de Lecea L., Criado J.R., Prospero-Garcia O., Gautvik K.M., Schweitzer P., Danielson P.E., Dunlop C.L., Siggins G.R., Henriksen S.J., Sutcliffe, J.G. A cortical neuropeptide with neuronal depressant and sleep-modulating properties. Nature 1996; 381:242-5

37. de Lecea L., Ruiz-Lozano P., Danielson P.E., Peelle-Kirley J., Foye P.E., Frankel W.N., Sutcliffe J.G. Cloning, mRNA expression, and chromosomal mapping of mouse and human preprocortistatin. Genomics 1997; 42:499-506

38. Devos N., Deflorian G., Biemar F., Bortolussi M., Martial J.A., Peers B., Argenton F. Differential expression of two somatostatin genes during zebrafish embryonic development. Mech Dev 2000;115:133-37

39. Eilertson C.D., Sheridan M.A. Differential effects of somatostatin-14 and somatostatin-25 on carbohydrate and lipid metabolism in rainbow trout Oncorhynchus mykiss. Gen Comp Endocrinol 1993; 92:62-70

40. Ensembl Annotation Project. 2003a; scaffold_1451 (SS1a), scaffold_18 (SS1b). 2003b; scaffold_2390

41. Ensinck J.W., Baskin D.G., Vahl T.P., Vogel R.E., Laschansky E.C., Francis B.H., Hoffman R.C., Krakover J.D., Stamm M.R., Low M.J., Rubinstein M., Otero-Corchon V., D'Alessio D.A. Thrittene, homologous with somatostatin-28 $_{(1-13)}$, is a novel peptide in mammalian gut and circulation. Endocrinology 2002;143:2599-609

42. Esch F., Bohlen P., Ling N., Benoit R., Brazeau P., Guillemin R. Primary structure of ovine hypothalamic somatostatin-28 and somatostatin-25. Proc Natl Acad Sci USA 1980; 77:6827-31

43. Fernald R.D., White R.B. Gonadotropin-releasing hormone genes: phylogeny, structure, and functions. Front Neuroendocrinol 1999; 20:224-40

44. Ferraz de Lima J.A., Oliveira B., Conlon J.M. Purification and characterization of insulin and peptides derived from proglucagon and prosomatostatin from the fruit-eating fish, the pacu *Piaractus mesopotamicus*. Comp Biochem Physiol 1999; 122:127-35

45. Fuhrmann G., Heilig R., Kempf J., Ebel A. Nucleotide sequence of the mouse preprosomatostatin gene. Nucleic Acids Res 1990; 18:1287

46. Fukusumi S., Kitada C., Takekawa S., Kizawa H., Sakamoto J., Miyamoto M., Hinuma S., Kitano K., Fujino M. Identification and characterization of a novel human cortistatin-like peptide. Biochem Biophys Res Commun 1997; 232:157-63

47. Funckes C.L., Minth C.D., Deschenes R., Magazin M., Tavianini M.A., Sheets M., Collier K., Weith H.L., Aron D.C., Roos B.A., Dixon J.E. Cloning and characterization of a mRNA-encoding rat preprosomatostatin. J Biol Chem 1983; 238:8781-7

48. Furlong R.F., Holland P.W. Were vertebrates octoploid? Philos Trans R Soc Lond B Biol Sci 2002; 357:531-44

49. Furu L.M., Kazmer G.W., Strausbaugh L., Zinn S.A. Cloning and characterization of the bovine somatostatin gene. J Anim Sci 1999; 77:492-3

50. Galanopoulou A.S., Kent G., Rabbani S.N., Seidah N.G., Patel Y.C. Heterologous processing of prosomatostatin in constitutive and regulated secretory pathways: putative role of the endoproteases furin, PC1n and PC2. J Biol Chem 1993; 268:6041-9

51. Goodman R.H., Jacobs J.W., Chin W.W., Lund P.K., Dee P.C., Habener J.F. Nucleotide sequence of a cloned structural gene coding for a precursor of pancreatic somatostatin. Proc Natl Acad Sci USA 1980; 77:5869-73

52. Goodman R.H., Jacobs J.W., Dee P.C., Habener J.F. Somatostatin-28 encoded in a cloned cDNA obtained from a rat medullary thyroid carcinoma. J Biol Chem 1982; 257:1156-9

53. Hobart P., Crawford R., Shen L.P., Pictet R., Rutter W.J. Cloning and sequence analysis of cDNAs encoding two distinct somatostatin precursors found in the endocrine pancreas of anglerfish. Nature 1980; 288:137-41

54. Jeandel L., Okuno A., Kobayashi T., Kikuyama S., Tostivint H., Lihrmann I., Chartrel N., Conlon J.M., Fournier A., Tonon M.C., Vaudry H. Effects of the two somatostatin variants somatostatin-14 and [Pro$^2$, Met$^{13}$]somatostatin-14 on receptor binding, adenylyl cyclase activity and growth hormone release from the frog pituitary. J Neuroendocrinol 1998; 10:187-92

55. Kim J.B., Gadsboll V., Whittaker J., Barton B.A., Conlon J.M. Gastroenteropancreatic hormones (insulin, glucagon, somatostatin, and multiple forms of PYY) from the pallid sturgeon, *Scaphirhynchus albus* (Acipenseriformes). Gen Comp Endocrinol 2000; 120:353-63

56. Kittilson J.D., Moore C.A., Sheridan M.A. Polygenic expression of somatostatin in rainbow trout, *Oncorhynchus mykiss*: evidence of a preprosomatostatin encoding somatostatin-14. Gen Comp Endocrinol 1999; 114:88-96

57. Kizer J.S., Tropsha A. A motif found in propeptides and prohormones that may target them to secretory vesicles. Biochem Biophys Res Commun 1991; 174:586-92

58. Kreienkamp H.J., Larusson, H.J., Witte I., Roeder T., Birgül N, Hönck H.H., Harder S., Ellinghausen G., Buck F., Richter D. Functional annotation of two orphan G-coupled receptors, Drostar1 and –2, from *Drosophola melanogaster* and their ligands by reverse pharmacology. J Biol Chem 2002; 277:39937-43

59. Larhammar D. Evolution of neuropeptide Y, peptide YY and pancreatic polypeptide. Regul Pept 1996; 62:1-11

60. LeRoith D., Pickens W., Wilson G.L., Miller B., Berelowitz M., Vinik A.I., Collier E., Cleland C.F. Somatostatin-like material is present in flowering plants. Endocrinology 1985a; 117:2093-7

61. LeRoith D., Pickens W., Vinik A.I., Shiloach J. Bacillus subtilis contains multiple forms of somatostatin-like material. Biochem Biophys Res Commun 1985b; 127:713-9

62. LeRoith D., Pickens W., Crosby L.K., Berelowitz M., Holtgrefe M., Shiloach J. Evidence for multiple molecular weight forms of somatostatin-like material in Escherichia coli. *Biochim Biophys Acta* 1985c; 838:335-42

63. Lin X., Otto C.J., Peter R.E. Expression of three distinct somatostatin messenger ribonucleic acids (mRNAs) in goldfish brain: characterization of the complementary deoxyribonucleic acids, distribution and seasonal variation of the mRNAs, and action of a somatostatin-14 variant. Endocrinology 1999a; 140:2089-99

64. Lin X., Janovick J.A., Brothers S., Conn P.M., Peter R.E. Molecular cloning and expression of two type one somatostatin receptors in goldfish brain. Endocrinology 1999b; 140:5211-9

65. Lin X., Janovick J.A., Cardenas R., Conn P.M., Peter R.E. Molecular cloning and expression of a type-two somatostatin receptor in goldfish brain and pituitary. Mol Cell Endocrinol 2000; 166:75-87

66. Lin X., Nunn C., Hoyer D., Rivier J., Peter R.E. Identification and characterization of a type five-like somatostatin receptor in goldfish pituitary. Mol Cell Endocrinol 2002; 189:105-16

67. Lin X., Peter R.E. Somatostatin-like receptors in goldfish: cloning of four new receptors. Peptides 2003 ; 24:53-63

68. Magazin M., Minth C.D., Funckes C.L., Deschenes R., Tavianini M.A., Dixon J.E. Sequence of a cDNA encoding pancreatic preprosomatostatin-22. Proc Natl Acad Sci USA 1982; 79:5152-6

69. Minth C.D., Taylor W.L., Magazin M., Tavianini M.A., Collier K., Weith H.L., Dixon J.E. The structure of cloned DNA complementary to catfish pancreatic somatostatin-14 messenger RNA. J Biol Chem 1982; 257:10372-7

70. Mita, K., Ishikawa, Y., Yamauchi, M. Establishment of cDNA database of medaka, *Oryzias latipes*. Genbank 2001; AU168379 *(Direct submission)*

71. Montminy M.R., Goodman R.H., Horovitch S.J., Habener J.F. Primary structure of the gene encoding rat preprosomatostatin. Proc Natl Acad Sci USA 1984; 81:3337-40

72. Moore C.A., Kittilson J.D., Dahl S.K., Sheridan M.A. Isolation and characterization of a cDNA encoding for preprosomatostatin containing [Tyr$^7$, Gly$^{10}$]-somatostatin-14 from the endocrine pancreas of rainbow trout, *Oncorhynchus mykiss*. Gen Comp Endocrinol 1995; 98:253-61

73. Moore C.A., Kittilson J.D., Ehrman M.M., Sheridan M.A. Rainbow trout *(Oncorhynchus mykiss)* posseSSTwo somatostatin mRNAs that are differentially expressed. Am J Physiol 1999; 277:R1553-61

74. Mouchantaf R., Kumar U., Sulea T., Patel Y.C. A conserved alpha-helix at the amino terminus of prosomatostatin serves as a sorting signal for the regulated secretory pathway. J Biol Chem 2001; 276:26308-16

75. Nata K., Kobayashi T., Karahashi K., Kato S., Yamamoto H., Yonekura H., Okamoto, H. Nucleotide sequence determination of chicken somatostatin precursor cDNA. Genbank 1991; X60191 *(Direct submission)*

76. NCBI Annotation Project. Genbank 2002; NT021937 *(Direct submission)*

77. Nguyen T.M., Wright J.R., Nielsen P.F., Conlon J.M. Characterization of the pancreatic hormones from the Brockmann body of the tilapia: implications for islet xenograft studies. Comp Biochem Physiol 1995; 111C:33-44

78. Nishii, M., Moverus B., Bukovskaya O.S., Takahashi A., Kawauchi H. Isolation and characterization of [Pro$^2$]somatostatin-14 and melanotropins from Russian sturgeon, *Acipenser gueldenstaedti* Brandt. Gen Comp Endocrinol 1995; 99:6-12

79. Noe B.D., SpieSSTJ., Rivier J.E., Vale W. Isolation and characterization of somatostatin from anglerfish pancreatic islet. Endocrinology 1979; 105:1410-5

80. O'Carroll A.M., Lolait S.J., Konig M., Mahan L.C. Molecular cloning and expression of a pituitary somatostatin receptor with preferential affinity for somatostatin-28. Mol Pharmacol 1992; 42:939-46

81. Ohno, Susumu, *Evolution by Gene Duplication*. New York: Springer-Verlag, 1970.

82. Panetta R., Greenwood M.T., Warszynska A., Demchyshyn L.L., Day R., Niznik H.B., Srikant C.B., Patel Y.C. Molecular cloning, functional characterization, and chromosomal localization of a human somatostatin receptor (somatostatin receptor type 5) with preferential affinity for somatostatin-28. Mol Pharmacol 1994; 45:417-27

83. Patel Y.C . Somatostatin and its receptor family. Front Neuroendocrinol 1999; 20:157-98

84. Patel Y.C., Panetta R., Escher E., Greenwood M., Srikant C.B. Expression of multiple somatostatin receptor genes in AtT-20 cells. Evidence for a novel somatostatin-28 selective receptor subtype. J Biol Chem 1994; 269:1506-9

85. Patel Y.C., Galanopoulou A.S., Rabbani S.N., Liu J.L., Ravazzola M., Amherdt M. Somatostatin-14, somatostatin-28, and prosomatostatin[1-10] are independently and

efficiently processed from prosomatostatin in the constitutive secretory pathway in islet somatostatin tumor cells (1027B2). Mol Cell Endocrinol 1997; 131:183-94

86. Plisetskaya E.M., Pollock H.G., Rouse J.B., Hamilton J.W., Kimmel J.R., Andrews P.C., Gorbman A. Characterization of Coho salmon (*Oncorhynchus kisutch*) islet somatostatins. Gen Comp Endocrinol 1986; 63:252-63

87. Powell J.F., Reska-Skinner S.M., Prakash M.O., Fischer W.H., Park M., Rivier J.E., Craig A.G., Mackie G.O., Sherwood N.M. Two new forms of gonadotropin-releasing hormone in a protochordate and the evolutionary implications. Proc Natl Acad Sci USA 1996; 93:10461-4

88. Pradayrol L., Jornvall H., Mutt V., Ribet A. N-terminally extended somatostatin: the primary structure of somatostatin-28. FEBS Lett 1980; 109:55-8

89. Puebla L., Mouchantaf R., Sasi R., Khare S., Bennett H.P., James S., Patel Y.C. Processing of rat preprocortistatin in mouse AtT-20 cells. J Neurochem 1999; 73:1273-7

90. Rabbani S.N., Patel Y.C. Peptides derived by processing of rat prosomatostatin near the amino-terminus: characterization, tissue distribution, and release. Endocrinology. 1990; 126:2054-61

91. Ramirez J.L., Mouchantaf R., Kumar U., Otero-Corchon V., Rubinstein M., Low M.J., Patel Y.C. Brain somatostatin receptors are up-regulated in somatostatin-deficient mice. Mol Endocrinol 2002;16:1951-63

92. Roostermann D., Roth A., Kreienkamp H.J., Richter D., Meyerhof W. Distinct agonist-mediated endocytosis of cloned rat somatostatin receptor subtypes expressed in insulinoma cells. J Neuroendocrinol 1997; 9:741-51

93. Roth A., Kreienkamp H.J., Nehring R.B., Roosermann D., Meyerhof W., Richter D. Endocytosis of the rat somatostatin receptors: subtype discrimination, ligand specificity, and delineation of carboxy terminal positive and negative sequence motifs. DNA Cell Biol 1997; 16:111-9

94. Schally A.V., Dupont A., Arimura A., Redding T.W., Nishi N., Lithicum G.L., Schlesinger, D.H. Isolation and structure of somatostatin from porcine hypothalami. Biochemistry 1976; 15:509-14

95. Seidah N.G., Day R., Marcinkiewisz M., Chrétien M. Precursor convertases: an evolutionary ancient, cell-specific, combinatorial mechanism yielding diverse bioactive peptides and proteins. Ann NY Acad Sci 1998; 839:9-24

96. Sevarino K.A., Stork P. Multiple preprosomatostatin sorting signals mediate secretion via discrete cAMP- and tetradecanoylphorbolacetate-responsive pathways. J Biol Chem 1991; 266:18507-13

97. Shen L.P., Pictet R.L., Rutter W.J. Human somatostatin I: sequence of the cDNA. Proc Natl Acad Sci USA 1982; 79:4575-9

98. Shen L.P., Rutter W.J. Sequence of the human somatostatin gene. Science 1984; 224:168-71

99. Sherwood N.M., Krueckl S.L., McRory J.E. The origin and function of the pituitary adenylate cyclase-activating polypeptide (PACAP)/glucagon superfamily. Endocr Rev 2000 ; 21:619-70

100. Sower S.A., Chiang Y.C., Conlon J.M. Polygenic expression of somatostatin in lamprey. Peptides 1994; 15:151-4

101. Spier A.D., de Lecea L. Cortistatin: a member of the somatostatin neuropeptide family with distinct physiological functions. Brain Res Rev 2000; 33:228-241

102. SpieSSTJ., Rivier J.E., Rodkey J.A., Bennett C.D., Vale W. Isolation and characterization of somatostatin from pigeon pancreas. Proc Natl Acad Sci USA 1979; 76:2974-8

103. Su C.J., White J.W., Li W.H., Luo C.C., Frazier M.L., Saunders G.F., Chan L. Structure and evolution of somatostatin genes. Mol Endocrinol 1988; 2:209-16

104. Takami M., Reeve J.R., Hawke D., Shively J.E., Basinger S., Yamada T. Purification of somatostatin from frog brain: coisolation with retinal somatostatin-like immunoreactivity. J Neurochem 1985; 45:1869-74

105. Tavianini M.A., Hayes T.E., Magazin M.D., Minth C.D., Dixon J.E. Isolation,

characterization, and cDNA sequence of the rat somatostatin gene. J Biol Chem 1984; 259:11798-803

106. Taylor J.S., Van de Peer Y., Braasch I., Meyer A. Comparative genomics provides evidence for an ancient genome duplication event in fish. Philos Trans R Soc Lond B Biol Sci. 2001; 356:1661-79

107. Tostivint H., Lihrmann I., Bucharles C., Vieau D., Coulouarn Y., Fournier A., Conlon J.M., Vaudry H. Occurrence of two somatostatin variants in the frog brain: characterization of the cDNAs, distribution of the mRNAs, and receptor-binding affinities of the peptides. Proc Natl Acad Sci USA 1996; 93:12605-10

108. Tostivint H., Vieau D., Chartrel N., Boutelet I., Galas L., Fournier A., Lihrmann I., Vaudry H. Expression and processing of the [Pro$^2$,Met$^{13}$]somatostatin-14 precursor in the intermediate lobe of the frog pituitary. Endocrinology 2002;143:3472-81

109. Trabucchi M., Tostivint H., Lihrmann I., Jegou S., Vallarino M., Vaudry H. Molecular cloning of the cDNAs and distribution of the mRNAs encoding two somatostatin precursors in the African lungfish *Protopterus annectens*. J Comp Neurol 1999; 410:643-52

110. Trabucchi M., Tostivint H., Lihrmann I., Sollars C., Vallarino M., Dores R.M., Vaudry H. Polygenic expression of somatostatin in the sturgeon *Acipenser transmontanus*: Molecular cloning and distribution of the mRNAs encoding two somatostatin precursors. *J. Comp. Neurol.* 2002;443: 332-45

111. Trabucchi M., Tostivint H., Lihrmann I., Blähser S., Vallarino M., Vaudry H. Characterization of the cDNA encoding a somatostatin variant in the chicken brain: Comparison of the distribution of the two somatostatin precursor mRNAs. J Comp Neurol. 2003;461:441-51

112. Travis G.H., Sutcliffe J.G. Phenol emulsion-enhanced DNA-driven subtractive cDNA cloning: isolation of low-abundance monkey cortex-specific mRNAs. Proc Natl Acad Sci USA 1988; 85:1696-700

113. Uesaka T., Yano K., Yamasaki M., Ando M. Somatostatin-, vasoactive intestinal peptide-, and granulin-like peptides isolated from intestinal extracts of goldfish, *Carassius auratus*. Gen Comp Endocrinol 1995; 99:298-306

114. Vaudry D., Gonzalez B.J., Basille M., Yon L., Fournier A., Vaudry H. Pituitary adenylate cyclase-activating polypeptide and its receptors: from structure to functions. Pharmacol Rev 2000; 52:269-324

115. Vaudry H., Chartrel N., Conlon J.M. Isolation of [Pro$^2$,Met$^{13}$]somatostatin-14 and somatostatin-14 from the frog brain reveals the existence of a somatostatin gene family in a tetrapod. Biochem Biophys Res Commun 1992; 188:477-82

116. Wang Y., Conlon J.M. Neuroendocrine peptides (NPY, GRP, VIP, somatostatin) from the brain and stomach of the alligator. Peptides 1993a; 13:573-9

117. Wang Y., Youson J.H., Conlon J.M. Prosomatostatin-I is processed to somatostatin-26 and somatostatin-14 in the pancreas of the bowfin, *Amia calva*. Regul Pept 1993b; 47:33-9

118. Wang Y., Nielsen P.F., Youson J.H., Potter I.C., Conlon J.M. Multiple forms of glucagon and somatostatin isolated from the intestine of the southern-hemisphere lamprey *Geotria australis*. Gen Comp Endocrinol 1999a; 113:274-82

119. Wang Y., Lance V.A., Nielsen P.F., Conlon J.M. Neuroendocrine peptides (insulin, pancreatic polypeptide, neuropeptide Y, galanin, somatostatin, substance P, and neuropeptide γ) from the desert tortoise, *Gopherus agassizii*. Peptides 1999b; 20:713-22

120. Zupanc G.K., Siehler S., Jones E.M., Seuwen K., Furuta H., Hoyer D., Yano H. Molecular cloning and pharmacological characterization of a somatostatin receptor subtype in the gymnotiform fish *Apteronotus albifrons*. Gen Comp Endocrinol 1999; 115:333-45

# 5
# ALTERED PATTERNS OF GROWTH HORMONE SECRETION IN SOMATOSTATIN KNOCKOUT MICE

Malcolm J. Low, Marcelo Rubinstein,* and Veronica Otero-Corchon
*Vollum Institute, Oregon Health Sciences University, Portland OR 97201, U.S.A. and *Instituto de Investigaciones en Ingeniería Genética y Biología Molecular, Consejo Nacional de Investigaciones Científicas y Técnicas and Departamento de Ciencias Biológicas, Facultad de Ciencias Exactas y Naturales, Universidad de Buenos Aires, Buenos Aires 1428, Argentina*

## 1. INTRODUCTION

Three decades have passed since the original isolation and characterization of somatostatin (SST) as a hypophysiotropic inhibitor of growth hormone (GH) secretion from the anterior pituitary gland (1). Five distinct high-affinity SST receptor (SSTR) subtypes have been cloned and they are widely expressed, although with unique cell-type specificity and subcellular distribution, in the central nervous system, endocrine glands, gastrointestinal tract, cardiopulmonary system, and immune system (2). Despite the advances in characterization of SST gene regulation, SST prohormone processing, SSTR distribution and intracellular signaling pathways mediating SST action, many questions remain unanswered concerning the functional role of SST peptides in each of these organ systems, and particularly in extrahypothalamic brain. Immunological, antisense, and pharmacological approaches to the abrogation of SST expression or activity all have been used to determine SST function *in vivo*, but each technique also has important limitations in its selectivity, specificity, and efficacy. Therefore, we have addressed the problem of SST function by the generation of gene knockout mice lacking all forms of proSST derived peptide ligands (3). An initial description of the phenotypic effects of this gene mutation on GH secretion, GH-regulated hepatic mRNAs, and sexually dimorphic somatic growth forms the basis of this chapter. We also discuss the implications of our data in relationship to the published work on SSTR subtype knockout mice and the future prospects of utilizing SST-deficient mice to infer physiological roles of the peptide in the regulation of additional endocrine axes and modulation of emotional and cognitive processes.

## 1.1 Sexual Dimorphism of Neuroendocrine Function

Exposure of the developing brain to gonadal steroids in perinatal life results in permanent morphological changes in specific brain areas (4-6). Furthermore, the expression of many genes controlling neurotransmitter biosynthesis and receptor interactions is sexually diergic [functionally dimorphic) (reviewed in (7)]. This imprinting may be mediated by testosterone and androgen receptors or by the aromatization of testosterone to estradiol and subsequent interaction with estrogen receptors. Sexual differentiation of the brain is known to influence diverse aspects of reproduction in vertebrates. For example, sexual differentiation of the anteroventral periventricular nucleus of the hypothalamus modulates GnRH secretion and lordosis behavior in rats (8), sexual differentiation of hypothalamic nuclei and oxytocin receptors is species-specific in voles and has powerful effects on mating choice (9,10), and sexual differentiation of peptidergic neural circuits is responsible for the elaboration of courtship songs in seasonally breeding songbirds (11-13) and gender-specific vocalizations in fish morphs (14). More controversial have been the reported sex differences in the volumes of the interstitial nuclei of the anterior hypothalamus in humans (15) and their possible associations with sexual preference or behavior.

Ultradian endocrine rhythms generated by the hypothalamus and pituitary gland, which are not directly linked to gonadal function, are also sexually differentiated. A prime example is the sexual diergism of pulsatile growth hormone (GH) secretion [reviewed in (16)] (Fig. 1). Male rats exhibit relatively narrow GH pulses with a frequency of about 3-4 hr$^{-1}$ and prolonged nadir values below 1-2 ng/ml (17). In contrast, female rats exhibit relatively broader pulses with an irregular frequency and nadir values between 5-20 ng/ml (18). Similarly, there is sexual diergism of GH secretion between men and women indicating phylogenetic conservation of neuroendocrine hormone rhythms (19).

## 1.2 Sexually Diergic GH Secretion in Rats and Mice

The episodic release of GH from pituitary somatotrophs results from a complex interplay primarily between two hypothalamic peptides, the stimulatory GH releasing hormone (GHRH) and the inhibitory SST (20,21). Direct measurement of GHRH and SST in hypophyseal-portal blood of male rats indicates a rhythmic pattern of release of the two peptides at regular 3-4 hr intervals about 180° out of phase (22). In contrast, female rats appear to have a different release pattern of the two neuropeptides characterized by continuous secretion of SST and episodic, rather than rhythmic bursts of GHRH. This results in a very different pattern of GH secretion in females compared to males, which may account for the striking sex difference in somatic growth (23).

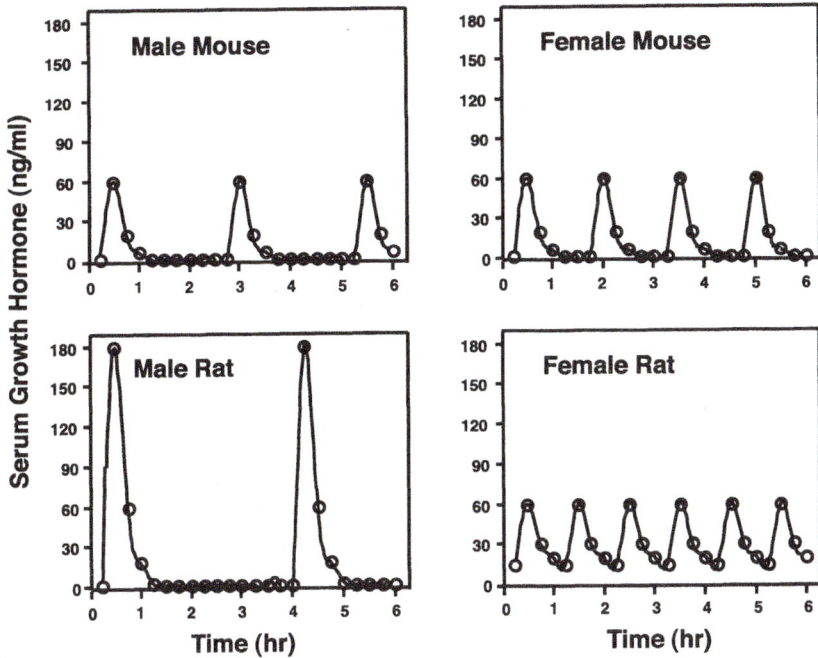

*Figure 1. A comparison of sexually diergic GH secretion in mice and rats.. Idealized profiles of pulsatile GH secretion in males and females of both species based on published data in references (24-26).*

Unlike the rat, direct measurement of pulsatile GH in the mouse has been hampered by technical difficulties of chronic blood sampling in unrestrained animals. In a single reported study male and female mice showed equivalent pulse amplitude, but greater pulse frequency in females, resulting in greater integrated levels of GH over time in the females (26). The ultradian patterns of GH secretion in both rats and mice are illustrated schematically in Figure 1. The common, defining characteristic of masculinized GH secretion in the two species is the relatively protracted interpulse interval of nadir GH levels compared to females. This interval is approximately 3 hr in male rats, but closer to 2 hr in male mice.

The underlying mechanism for such sharp gender-specific differences in GH secretion of rodents remains unidentified. A first clue derives from a lesioning study in male rats that demonstrated feminization of prolactin receptor expression in the liver following electrolytic destruction of the anterior periventricular hypothalamus (27). As discussed below, the sexually dimorphic expression of hepatic prolactin-receptor gene expression is determined by the pattern of GH secretion. Hypothalamic SST peptide and mRNA levels are normally higher in male rats compared to females (28,29) suggesting that gender differences in hypothalamic SST signaling to pituitary somatotrophs play a key role in the sexually diergic GH responses (20). Furthermore, passive immunization of rats with SST antisera causes an increase in circulating GH levels consistent with a direct inhibitory action of SST on pituitary somatotrophs (25). However, SST plays another important role in GH secretion by more indirect means. SST neurons within the periventricular nucleus of the hypothalamus express GH receptors and reciprocally innervate GHRH neurons in the arcuate nucleus (30-35). GH can therefore feedback on GHRH secretion utilizing SST interneurons. This pathway has been shown to be important in mutant mice lacking the SSTR2 (sst2) (36), and by the infusion of a SSTR antagonist that crosses the blood brain barrier in rats (37). Immunolocalization studies of the SSTR1 in hypothalamus suggest that it is an autoreceptor on SST neurons and may therefore play a role in inhibition of SST release with a net effect opposite to that of SSTR2 (38).

The number of periventricular SST neurons and quantity of SST mRNA are upregulated two- to three-fold by the combined organizational and activational effects of testosterone in male rodents compared to female (29,39,40). Additionally, treatment of adult castrated male rats with 17-ß-estradiol, or female rats with testosterone, feminizes or masculinizes the GH secretory profile, respectively (41,42). Of the five known SSTR subtypes, characterized by distinct cellular distributions and subtle differences in intracellular signaling (2), sexual diergism has been noted for expression of SSTR1, 2, 3, and 5 in the anterior pituitary and SSTR1 but not SSTR2 in the rat hypothalamus (43-45). Together, these data all support the hypothesis that SST is a determinant of gender differences in GH secretion.

Sex differences in GH pulsatility are responsible for sexual differentiation of the liver, a major target organ of GH (16). In particular, many members of the cytochrome P450 (CYP) and the major urinary protein (MUP) gene families are differentially regulated by the sex-specific pattern of GH secretion (46,47). It should be noted, however, that the specific CYPs exhibiting GH-dependent sexual diergism differ between rats and mice, and even among different mouse strains. Regardless of this caveat, a key factor determining the masculinized hepatic phenotype is the long interpulse nadir of GH levels, which permits the resensitization of GH receptor signaling by Janus kinase/signal transducer and activation of transcription (JAK-STAT) cascades (24) specifically involving STAT 5a and 5b proteins (48-52). The hepatic microsomal enzymes regulated by GH have important physiological roles in the hydroxylation and metabolism of both endogenous steroid hormones and xenotoxins, thereby providing a possible link between sexual differentiation of the brain and sex differences in susceptibility to chemical toxicity and carcinogenicity (53).

# 2.  SOMATOSTATIN DEFICIENCY: THE PRESENT

Our gene targeting strategy in embryonic stem (ES) cells replaced critical SST promoter elements and exon 1, which contains the translational start site for preproSST, with a *neo* selection cassette (3). $F_2$ hybrid mice (129P2/OlaHsd, C57BL/6J) homozygous for the null allele ($SST^{-/-}$) were generated after two generations of breeding from ES cell derived chimeric founders. $F_1$ heterozygotes were also backcrossed for five successive generations to produce $N_5$ incipient congenic C57BL/6J mice. The SST null allele was transmitted from mating pairs of $SST^{+/-}$ mice in a Mendelian ratio of 1:2:1 for $SST^{+/+}$, $SST^{+/-}$ and $SST^{-/-}$ genotypes, respectively. $SST^{-/-}$ mice of both sexes were healthy, fertile and raised normal litters.

*In situ* hybridization histochemistry utilizing a $[^{35}S]$-radiolabeled antisense SST oligonucleotide revealed abundant RNA signal in neurons within the cortex, hippocampus, amygdala, bed nucleus of the stria terminalis, reticular nucleus of the thalamus, and hypothalamus of wild type mice, but no signal above background in the mutant mice (Fig. 2). The major hypothalamic nuclei expressing the SST gene are the periventricular, suprachiasmatic, and arcuate. To verify that the gene mutation also resulted in the loss of SST peptides, we analyzed tissue extracts by a radioimmunoassay that detects both the 14 and 28 amino acid forms of SST. SST-like immunoreactivity was reduced by 50% in stomach and brain from $SST^{+/-}$ mice and was below the assay sensitivity in $SST^{-/-}$ mice. Similarly, no SST or SST prohormone was detected by immunohistochemistry in brain, stomach, or pancreatic islets of the $SST^{-/-}$ mice, utilizing specific antisera directed against epitopes in either the mid portion of SST-14 or a "cryptic peptide" sequence derived from the amino terminal prohormone (54).

## 2.1.  Somatic Growth of SST Knockout Mice

Direct observation of the first litters of $F_2$ mice indicated that the absence of SST did not result in either gigantism or a loss of sexual dimorphism in body size. Growth curves obtained from both $F_2$ and $N_5$ mice (Fig. 3) confirmed this impression, although a gradual increase in body weight due to obesity did emerge for male $SST^{-/-}$ mice, after age 18 weeks [repeated measures ANOVA; F (10, 140) = 3.5, $p<0.001$ genotype x time interaction]. Body length of $SST^{-/-}$ mice, measured from nose to anus, was not significantly different from sibling $SST^{+/+}$ mice, and the sexual dimorphism in length was not altered (data not shown). Pituitary size was unchanged between the $SST^{+/+}$ and $SST^{-/-}$ male and female mice. However, pituitary GH content was reduced by 50% in both sexes of $SST^{-/-}$ mice compared to gender matched controls [F (3,14)=9.12, $p=0.01$ for each comparison], possibly reflecting an altered balance between GH synthesis and release. To assess the integrated levels of secreted GH bioactivity over time, we measured the levels of two stable circulating proteins

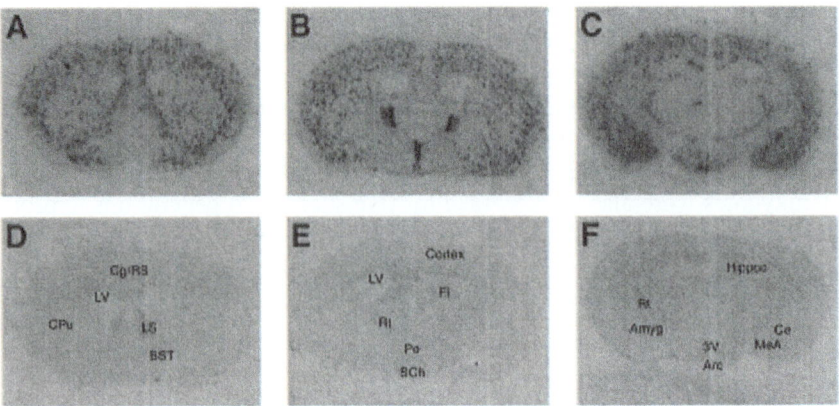

**Figure 2.** *In situ hybridization of coronal brain sections from a SST$^{+/+}$ (A-C) and a SST$^{-/-}$ (D-F) mouse demonstrating the loss of all SST gene expression in the mutant strain. Amygdala, Amyg; Arcuate nucleus, Arc; Bed nucleus of the stria terminalis, BST; Central nucleus of the amygdala, Ce; Caudate-Putamen, CPu; Cingulate/ Retrosplenial cortex, Cg/RS; fimbria, Fi; Hippocampus, Hippoc; Lateral ventricle, LV; Lateral septal nuclei, LS; Medial nucleus of the amygdala, MeA; Periventricular nucleus of the hypothalamus, Pe; Reticular nucleus of the thalamus, Rt; Suprachiasmatic nucleus, SCh; and Third ventricle, 3V.*

produced by the liver in response to GH, insulin-like growth factor-1 (IGF-1) and IGF-binding protein 3 (IGF-BP3). SST$^{+/+}$ females had higher IGF-1 levels than SST$^{+/+}$ males [F(3,19)=3.11, $p<0.01$] consistent with the higher mean GH levels reported in females. In contrast, IGF-1 levels of SST$^{-/-}$ males and females did not differ from each other and were intermediate to both sexes of SST$^{+/+}$ mice. IGF-BP3 levels, measured by a radiolabeled Western ligand binding assay (55), were also elevated in SST$^{+/+}$ females compared to SST$^{+/+}$ males [F(3,16)=6.98, $p<0.01$], but there was no comparable gender difference in IGF-BP3 levels for SST$^{-/-}$ mice (data not shown).

To indirectly determine whether SST-deficiency was associated with altered GH pulsatility independently of integrated levels, we took advantage of the firmly established role of pulsatile GH in regulating sexual differentiation of the liver (24,47). Northern blots were used to quantify mRNA levels of several GH-regulated genes and the data were normalized to 18S RNA content (3). The male-predominant gene transcripts for both MUPS group I/II and III exhibited levels that were approximately five-fold higher in SST$^{+/+}$ male compared to SST$^{+/+}$ female mice [F(3, 20)=48.40, $p<0.001$]. However, there was a dramatic reduction in MUPS expression in SST$^{-/-}$ male mice to levels indistinguishable from females. Identical results were found in cohorts of both F$_2$ hybrid and N$_5$ incipient congenic mice indicating that this phenotype is

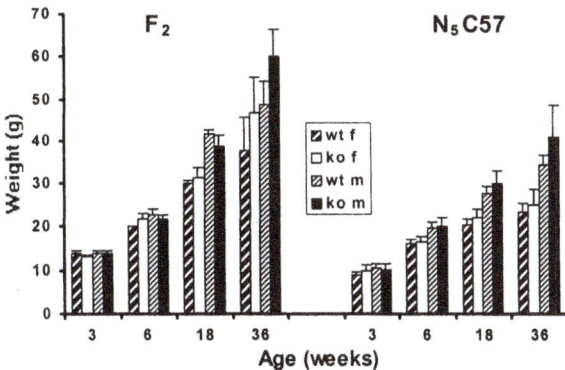

*Figure 3. Body weights of wild-type (wt) and SST-deficient (ko) mice carried on either the F₂ hybrid (129P2/Ola, C57BL/6) (left) or N₅ incipient congenic C57BL/6 (right) genetic backgrounds at ages 3, 6, 18, and 36 weeks. Male, m; female, f.*

is most likely not affected by epistatic interactions of the targeted allele with genetic background gene alleles. In contrast, prolactin receptor mRNA levels were more than three-fold higher in SST[+/+] females compared to SST[+/+] males ($p<0.02$, one-tailed T-test) consistent with previously published results in mice for this typical female-predominant gene (56). The absence of SST resulted in an increase of prolactin receptor expression in SST[-/-] male compared to SST[+/+] male mice [$F(3,20)=4.91$, $p<0.05$] but no significant change in SST[-/-] female mice. Finally, we analyzed expression of the female predominant *Cyp450-15α2A4* mRNA in liver (57). Both SST[-/-] male and female mice exhibited a significant increase in hybridization signal compared to SST[+/+] male mice [$F(3,20)=6.13$, $p<0.05$ and $<0.01$, respectively]. When all four of the sexually diergic hepatic mRNAs are considered together, there is a compelling pattern of feminization in the SST[-/-] male mice compared to wildtypes, without reciprocal changes in the SST[-/-] female mice.

## 2.2. GH Secretion in SST Knockout Mice

As a surrogate for the direct serial measurement of pulsatile GH in the mice, we obtained randomly-timed blood samples by rapid tail-bleeds between the hours of 05:00 and 23:00 from adult male and female mice of both genotypes over the course of 6 days. A total of 408 plasma samples from 107 mice (26-28 per group) were assayed for GH using a homologous mouse GH radioimmunoassay and the data are plotted as frequency histograms in Figure 4. Male and female SST[+/+] mice had identical median GH levels of 3 ng/ml. Furthermore, the overall distribution of values was similar with the exception that no female value was greater than 36 ng/ml while 5 of 107 male values ranged from 50-122 ng/ml. In marked contrast, the distribution of GH values was shifted higher for both male and female SST[-/-] mice, which had medians of 21 and 17 ng/ml, respectively ($P < 0.0001$, Mann-Whitney U test for each paired comparison between SST$^{+/+}$ and SST$^{-/-}$ groups). Less than 10% of the random GH values from SST$^{-/-}$ mice were below 6 ng/ml compared to 75% for both sexes of SST$^{+/+}$ mice (Chi square = 166.1, $P < 0.0001$).

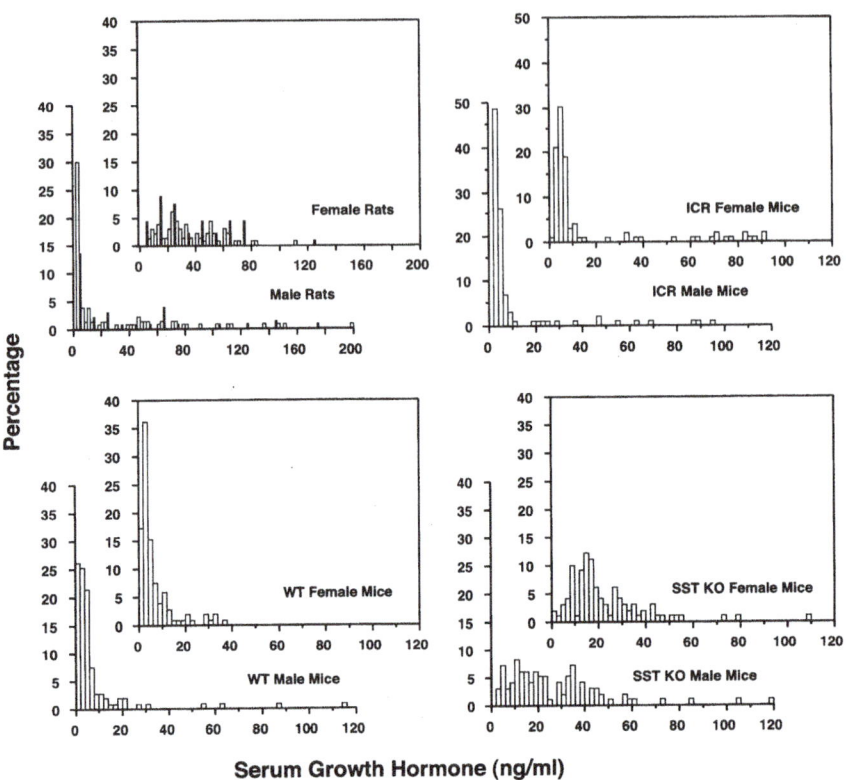

*Figure 4. Histograms of randomly timed plasma GH levels obtained from SST wild type and deficient mice on the $N_5$ incipient congenic C57BL/6J genetic background. Data are plotted as the percentage of all values within a group using a bin size of 2 ng/ml. The comparative data for male and female rats were combined from published representative GH secretory profiles (24,25); the comparative data for male and female ICR mice were from published representative GH secretory profiles (26); and the data from WT and SST KO mice are from our study (3)*

To facilitate a comparison between the present data and that of MacLeod, et al. (26), who directly measured pulsatile GH secretion in cannulated mice at 15 min intervals over 8 hr periods, we plotted the values from MacLeod's representative pulsatile GH profiles [3 male and 3 female outbred CD-1 (ICR) mice; n = 99 data points for each sex; see original data in (26)] as frequency histograms in Figure 4. Similarly, we plotted pulsatile GH data obtained from male and female rats and published by the Waxman (24) and Tannenbaum (25) laboratories as frequency histograms in Figure 4. The distributions of GH values are remarkably similar among male rats, male and female ICR mice, and male and female N5 C57BL/6J SST$^{+/+}$ mice, despite the different methods by which the samples were obtained. The highly similar distributions between the C57BL/6J SST$^{+/+}$ and MacLeod's ICR mice suggest that the C57BL/6J SST$^{+/+}$ mice secrete GH in a pulsatile rhythm that is similar to the ICR mice and characterized by large amplitude pulses and low interpulse nadir levels in both sexes. In contrast, male and female SST$^{-/-}$ mice had very few random GH levels < 3 ng/ml and clearly elevated median GH levels, but the maximal amplitude of GH was not changed. The distribution of GH values in both sexes of SST$^{-/-}$ mice was most similar to female rats. These data suggest that SST$^{-/-}$ mice continue to secrete GH in a pulsatile rhythm, although SST is apparently required to inhibit GH secretion between GHRH pulses and thereby maintain low nadir interpulse GH levels. The decreased pituitary content of GH is also consistent with increased secretion of GH into the circulation. However, it is not possible from this analysis to make firm conclusions concerning the frequency or regularity of GH pulses in the SST$^{-/-}$ mice.

## 2.3. Models to Explain the Dysregulation of GH Secretion

How does SST modulate GH secretion in a sexually diergic pattern? Based on previous studies and the present experiments, we propose that SST tone is normally higher in the hypothalamus of males compared to females. The higher tone is a product of increased SST gene expression, synaptic release of SST, and expression or signaling of SSTRs in the hypothalamus and pituitary gland. Although we did not observe basal sex differences in steady-state levels of mRNA for SSTR1 in mice as described in rats (45) we did measure significant changes in SSTR1, 2, and 5 in the SST-deficient mice (3). Furthermore, the increases in SSTR2 and 5 in hypothalamus and pituitary, respectively, occurred only in male SST$^{-/-}$ mice, consistent with the premise that SST receptor-mediated signaling is different between male and female mice in these tissues. There is strong evidence that organizational effects of gonadal steroids on the developing male brain together with continued testosterone exposure are responsible for the increased SST tone.

Intrinsic pulsatility of GH is determined by the pulsatile release of GHRH, possibly associated with effects of GH secretagogues, which in turn are linked directly to a circadian pacemaker (20) (Fig. 5). The higher SST tone of the male hypothalamus is required for short-loop GH feedback inhibition of the hypothalamic circuitry and produces more attenuation of GHRH secretion or GHRH action on GH secretion, resulting in less frequent pulses with longer interpulse intervals (mouse) or more regular pulses with low interpulse GH levels (rat). Based on this model, the absence of SST would be predicted to convert the male-specific pattern of GH pulsatility to a default female pattern as was observed experimentally. Despite our data indicating a key role for

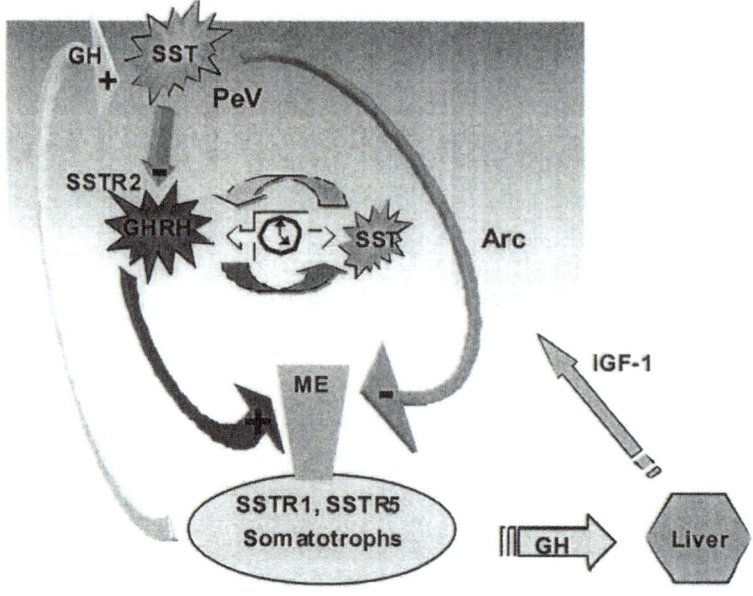

*Figure 5. A model of the proposed hypothalamic circuitry involving SST and GHRH responsible for GH secretion. An intrinsic hypothalamic clock determines the default ultradian pulsatile release of GHRH from the arcuate nucleus. The frequency of GHRH release is negatively regulated by the short-loop feedback of pituitary GH on periventricular SST neurons. Greater SST tone in the male hypothalamus mediates GH secretion through two mechanisms: the frequency of GHRH pulses, and subsequently GH pulses, is reduced by the activation of SSTR2 on GHRH neurons; and the interpulse secretion of GH is suppressed by the direct activation of SSTR1 and SSTR5 on pituitary somatotrophs.*

SST in maintaining the sexually diergic pattern of GH secretion, this model will need further development because other hypothalamic peptides such as galanin, CCK and NPY are also sexually diergic and likely interact within the SST-GHRH neuronal circuitry to regulate GH secretion (58,59).

## 3. SOMATOSTATIN DEFICIENCY: THE FUTURE

### 3.1. Regulation of Endocrine Systems

#### 3.1.1. Pituitary Hormone Secretion

Like somatotrophs, pituitary corticotrophs express SST receptors, but their functional importance is unclear. SST does not directly suppress normal corticotroph secretion, however it can inhibit the augmented ACTH secretion that occurs in glucocorticoid deficiency suggesting that expression of SSTRs in corticotrophs may be inhibited by glucocorticoids (60,61). The *in vivo* administration of SST inhibits ACTH and cortisol responses to insulin-induced hypoglycemia probably through inhibition of hypothalamic CRH secretion (62). Interestingly, SST has been implicated in the hippocampal mediation of negative feedback of the HPA axis by dexamethasone (63,64) but these results have been inconsistent due in part to the use of cysteamine, a nonspecific pharmacological depletor of SST. Preliminary experiments suggest that the HPA axis of SST$^{-/-}$ mice is dysregulated. There is a loss of sexual dimorphism in ACTH/corticosterone secretion and the stress-activation of corticosterone secretion is exaggerated in both sexes of mice (unpublished data). The anatomic loci and endocrine mechanisms responsible for these changes are under investigation and may expand the known physiological role of SST in neuroendocrine modulation.

### 3.1.2. Pancreatic Islet Hormones and Glucose Homeostasis

Utilizing *in vitro* preparations from SSTR2 KO mice, the SSTR2 has been recently demonstrated to play a critical role in secretion of glucagon from alpha cells, but not insulin from beta cells in the pancreatic islets (65). It might be expected therefore that SST-deficient mice would exhibit an alteration in glucose homeostasis. Our preliminary experiments, however, suggest that SST-deficient mice have normal pancreatic islet anatomy and normal basal glucose and insulin levels (unpublished data). Further studies are necessary to determine whether insulin and/or glucagon secretion from the pancreas will be altered in SST$^{-/-}$ mice following challenges to glucose homeostasis.

## 3.2. SST and Modulation of Higher Brain Functions

### 3.2.1. Altered SST Expression in Neuropsychiatric Disorders: Correlation or Causality?

An extensive literature has associated changes in SST or SSTR expression with neuropsychiatric disease including dementia, epilepsy, and major affective disorders (66,67). Furthermore, SST (see Figure 2) and SST binding sites are widely distributed in the brain. Experimental depletion of SST by cysteamine disrupts avoidance learning and decreases locomotor activity (68) while SST agonists have been reported to have opposite effects (69). More recently, SSTR2 KO mice were shown to have a marked increase in anxiety-related behavior, together with decreased locomotor and exploratory behavior in novel, stress-provoking situations (70). Our preliminary studies show that SST-deficient mice are slightly more seizure-prone than controls and have altered acoustic startle responses (unpublished data). Therefore, together with the mouse models of individual SSTR deletions described elsewhere in this volume, SST KO mice may help to unravel the multiple functions of the SST system in brain function that are distinct from the control of hormone secretion.

# 4.    SUMMARY

In summary, we have examined the physiological role of SST in the regulation of GH secretion by a gene knockout approach. The absence of SST results in elevated nadir and median GH levels in both sexes and a feminized pattern of GH-dependent gene expression in the liver of male mice. Despite the magnitude of these changes, SST$^{-/-}$ mice do not exhibit increased IGF-1 levels or gigantism and they retain a normal sexual dimorphism in postnatal somatic growth. A comparison of the distribution of a large set of randomly obtained GH values from mice in our study with GH data from formal pulse-analysis studies suggests that the pattern of GH secretion in both sexes of SST$^{-/-}$ mice is most similar to female rats. Measurement of pulsatile GH from cannulated SST-deficient mice will be necessary to confirm this hypothesis. We conclude that SST has an essential physiological role to determine the normal secretory profile of GH in both sexes of mice and specifically to masculinize GH secretion, however our data challenge the concept that a typical masculinized profile of GH pulsatility or masculinized liver is necessary for the greater somatic growth characteristics of male compared to female animals after puberty.

## ACKNOWLEDGEMENTS

The authors are greatly indebted to our colleagues and collaborators who have contributed to the ideas and experiments discussed in this chapter. In particular we want to acknowledge Dr. A. Parlow who assayed mouse GH and Dr. Y. Patel who quantified SSTR mRNA by RT-PCR.

## REFERENCES

1.    Brazeau, P., Vale, W., Burgus, R., Ling, N., Butcher, M., Rivier, J., and Guillemin, R. 1973. Hypothalamic polypeptide that inhibits the secretion of immunoreactive pituitary growth hormone. *Science.* 179:77-79.

2.    Patel, Y.C. 1999. Somatostatin and its receptor family. *Front Neuroendocrinol.* 20:157-198.

3.    Low, M.J., Otero-Corchon, V., Parlow, A.F., Ramirez, J.L., Kumar, U., Patel, Y.C., and Rubinstein, M. 2001. Somatostatin is required for masculinization of growth hormone-regulated hepatic gene expression but not of somatic growth. *J. Clin. Invest.* 107:1571-1580.

4.    MacLusky, N.J., and Naftolin, F. 1981. Sexual differentiation of the central nervous system. *Science.* 211:1294-1302.

5.    Leal, S., Andrade, J.P., Paula-Barbosa, M.M., and Madeira, M.D. 1998. Arcuate nucleus of the hypothalamus: effects of age and sex. *J Comp Neurol.* 401:65-88.

6.    Bleier, R., Byne, W., and Siggelkow, I. 1982. Cytoarchitectonic sexual dimorphisms of the medial preoptic and anterior hypothalamic areas in guinea pig, rat, hamster, and mouse. *J Comp Neurol.* 212:118-130.

7.    Cooke, B., Hegstrom, C.D., Villeneuve, L.S., and Breedlove, S.M. 1998. Sexual differentiation of the vertebrate brain: principles and mechanisms. *Front Neuroendocrinol.* 19:323-362.

8.    Simerly, R.B. 1989. Hormonal control of the development and regulation of tyrosine hydroxylase expression within a sexually dimorphic population of dopaminergic cells

in the hypothalamus. *Brain Res Mol Brain Res.* 6:297-310.

9. Shapiro, L.E., Leonard, C.M., Sessions, C.E., Dewsbury, D.A., and Insel, T.R. 1991. Comparative neuroanatomy of the sexually dimorphic hypothalamus in monogamous and polygamous voles. *Brain Res.* 541:232-240.

10. Insel, T.R., Gelhard, R., and Shapiro, L.E. 1991. The comparative distribution of forebrain receptors for neurohypophyseal peptides in monogamous and polygamous mice. *Neuroscience.* 43:623-630.

11. Gahr, M., and Metzdorf, R. 1999. The sexually dimorphic expression of androgen receptors in the song nucleus hyperstriatalis ventrale pars caudale of the zebra finch develops independently of gonadal steroids. *J Neurosci.* 19:2628-2636.

12. Bottjer, S.W., Roselinsky, H., and Tran, N.B. 1997. Sex differences in neuropeptide staining of song-control nuclei in zebra finch brains. *Brain Behav Evol.* 50:284-303.

13. Arnold, A.P., and Schlinger, B.A. 1993. Sexual differentiation of brain and behavior: the zebra finch is not just a flying rat. *Brain Behav Evol.* 42:231-241.

14. Goodson, J.L., and Bass, A.H. 2000. Forebrain peptides modulate sexually polymorphic vocal circuitry. *Nature.* 403:769-772.

15. Byne, W., Lasco, M.S., Kemether, E., Shinwari, A., Edgar, M.A., Morgello, S., Jones, L.B., and Tobet, S. 2000. The interstitial nuclei of the human anterior hypothalamus: an investigation of sexual variation in volume and cell size, number and density. *Brain Res.* 856:254-258.

16. Jansson, J.O., Eden, S., and Isaksson, O. 1985. Sexual dimorphism in the control of growth hormone secretion. *Endocr Rev.* 6:128-150.

17. Tannenbaum, G.S., and Martin, J.B. 1976. Evidence for an endogenous ultradian rhythm governing growth hormone secretion in the rat. *Endocrinology.* 98:562-570.

18. Eden, S. 1979. Age- and sex-related differences in episodic growth hormone secretion in the rat. *Endocrinology.* 105:555-560.

19. Jaffe, C.A., Ocampo-Lim, B., Guo, W., Krueger, K., Sugahara, I., DeMott-Friberg, R., Bermann, M., and Barkan, A.L. 1998. Regulatory mechanisms of growth hormone secretion are sexually dimorphic. *J Clin Invest.* 102:153-164.

20. Wagner, C., Caplan, S.R., and Tannenbaum, G.S. 1998. Genesis of the ultradian rhythm of GH secretion: a new model unifying experimental observations in rats. *Am J Physiol.* 275:E1046-1054.

21. Tannenbaum, G. 1994. Multiple levels of cross-talk between somatostatin (SRIF) and growth hormone (GH)- releasing factor in genesis of pulsatile GH secretion. *Clinics in Pediatric Endocrinology.* 3:97-110.

22. Plotsky, P.M., and Vale, W. 1985. Patterns of growth hormone-releasing factor and somatostatin secretion into the hypophysial-portal circulation of the rat. *Science.* 230:461-463.

23. Jansson, J.O., Ekberg, S., Isaksson, O., Mode, A., and Gustafsson, J.A. 1985. Imprinting of growth hormone secretion, body growth, and hepatic steroid metabolism by neonatal testosterone. *Endocrinology.* 117:1881-1889.

24. Waxman, D.J., Pampori, N.A., Ram, P.A., Agrawal, A.K., and Shapiro, B.H. 1991. Interpulse interval in circulating growth hormone patterns regulates sexually dimorphic expression of hepatic cytochrome P450. *Proc Natl Acad Sci U S A.* 88:6868-6872.

25. Painson, J.C., and Tannenbaum, G.S. 1991. Sexual dimorphism of somatostatin and growth hormone-releasing factor signaling in the control of pulsatile growth hormone secretion in the rat. *Endocrinology.* 128:2858-2866.

26. MacLeod, J.N., Pampori, N.A., and Shapiro, B.H. 1991. Sex differences in the ultradian pattern of plasma growth hormone concentrations in mice. *J Endocrinol.* 131:395-399.

27. Norstedt, G., Mode, A., Hokfelt, T., Eneroth, P., Elde, R., Ferland, L., Labrie, F., and Gustafsson, J.A. 1983. Possible role of somatostatin in the regulation of the sexually differentiated steroid metabolism and prolactin receptor in rat liver. *Endocrinology.* 112:1076-1090.

28. Argente, J., Chowen, J.A., Zeitler, P., Clifton, D.K., and Steiner, R.A. 1991. Sexual dimorphism of growth hormone-releasing hormone and somatostatin gene expression in the hypothalamus of the rat during development. *Endocrinology.* 128:2369-2375.

29. Chowen-Breed, J.A., Steiner, R.A., and Clifton, D.K. 1989. Sexual dimorphism and testosterone-dependent regulation of somatostatin gene expression in the periventricular nucleus of the rat brain. *Endocrinology.* 125:357-362.

30. Horvath, S., Palkovits, M., Gorcs, T., and Arimura, A. 1989. Electron microscopic immunocytochemical evidence for the existence of bidirectional synaptic connections between growth hormone-releasing hormone- and somatostatin-containing neurons in the hypothalamus of the rat. *Brain Res.* 481:8-15.

31. Burton, K.A., Kabigting, E.B., Clifton, D.K., and Steiner, R.A. 1992. Growth hormone receptor messenger ribonucleic acid distribution in the adult male rat brain and its colocalization in hypothalamic somatostatin neurons. *Endocrinology.* 131:958-963.

32. Bertherat, J., Dournaud, P., Berod, A., Normand, E., Bloch, B., Rostene, W., Kordon, C., and Epelbaum, J. 1992. Growth hormone-releasing hormone-synthesizing neurons are a subpopulation of somatostatin receptor-labelled cells in the rat arcuate nucleus: a combined in situ hybridization and receptor light-microscopic radioautographic study. *Neuroendocrinology.* 56:25-31.

33. Pellegrini, E., Bluet-Pajot, M.T., Mounier, F., Bennett, P., Kordon, C., and Epelbaum, J. 1996. Central administration of a growth hormone (GH) receptor mRNA antisense increases GH pulsatility and decreases hypothalamic somatostatin expression in rats. *J Neurosci.* 16:8140-8148.

34. Rogers, K.V., Vician, L., Steiner, R.A., and Clifton, D.K. 1988. The effect of hypophysectomy and growth hormone administration on pre-prosomatostatin messenger ribonucleic acid in the periventricular nucleus of the rat hypothalamus. *Endocrinology.* 122:586-591.

35. Tannenbaum, G.S., Zhang, W.H., Lapointe, M., Zeitler, P., and Beaudet, A. 1998. Growth hormone-releasing hormone neurons in the arcuate nucleus express both Sst1 and Sst2 somatostatin receptor genes. *Endocrinology.* 139:1450-1453.

36. Zheng, H., Bailey, A., Jiang, M.H., Honda, K., Chen, H.Y., Trumbauer, M.E., Van der Ploeg, L.H., Schaeffer, J.M., Leng, G., and Smith, R.G. 1997. Somatostatin receptor subtype 2 knockout mice are refractory to growth hormone-negative feedback on arcuate neurons. *Mol Endocrinol.* 11:1709-1717.

37. Baumbach, W.R., Carrick, T.A., Pausch, M.H., Bingham, B., Carmignac, D., Robinson, I.C., Houghten, R., Eppler, C.M., Price, L.A., and Zysk, J.R. 1998. A linear hexapeptide somatostatin antagonist blocks somatostatin activity in vitro and influences growth hormone release in rats. *Mol Pharmacol.* 54:864-873.

38. Helboe, L., Stidsen, C.E., and Moller, M. 1998. Immunohistochemical and cytochemical localization of the somatostatin receptor subtype sst1 in the somatostatinergic parvocellular neuronal system of the rat hypothalamus. *J Neurosci.* 18:4938-4945.

39. Herbison, A.E. 1995. Sexually dimorphic expression of androgen receptor immunoreactivity by somatostatin neurones in rat hypothalamic periventricular nucleus and bed nucleus of the stria terminalis. *J Neuroendocrinol.* 7:543-553.

40. Nurhidayat, Tsukamoto, Y., Sigit, K., and Sasaki, F. 1999. Sex differentiation of growth hormone-releasing hormone and somatostatin neurons in the mouse hypothalamus: an immunohistochemical and morphological study. *Brain Res.* 821:309-321.

41. Painson, J.C., Veldhuis, J.D., and Tannenbaum, G.S. 2000. Single exposure to testosterone in adulthood rapidly induces regularity in the growth hormone release process. *Am J Physiol.* 278:E933-940.

42. Painson, J.C., Thorner, M.O., Krieg, R.J., and Tannenbaum, G.S. 1992. Short-term adult exposure to estradiol feminizes the male pattern of spontaneous and growth hormone-releasing factor-stimulated growth hormone secretion in the rat. *Endocrinology.* 130:511-519.

43. Kimura, N., Tomizawa, S., and Arai, K.N. 1998. Chronic treatment with estrogen up-regulates expression of sst2 messenger ribonucleic acid (mRNA) but down-regulates expression of sst5 mRNA in rat pituitaries. *Endocrinology.* 139:1573-1580.

44. Senaris, R.M., Lago, F., and Dieguez, C. 1996. Gonadal regulation of somatostatin receptor 1, 2 and 3 mRNA levels in the rat anterior pituitary. *Brain Res Mol Brain Res.* 38:171-175.

45. Zhang, W.H., Beaudet, A., and Tannenbaum, G.S. 1999. Sexually dimorphic expression of sst1 and sst2 somatostatin receptor subtypes in the arcuate nucleus and anterior pituitary of adult rats. *J Neuroendocrinol.* 11:129-136.

46. McIntosh, I., and Bishop, J.O. 1989. Differential expression in male and female mouse liver of very similar mRNAs specified by two group 1 major urinary protein genes. *Mol Cell Biol.* 9:2202-2207.

47.     Sundseth, S.S., Alberta, J.A., and Waxman, D.J. 1992. Sex-specific, growth hormone-regulated transcription of the cytochrome P450 2C11 and 2C12 genes. *J Biol Chem.* 267:3907-3914.

48.     Teglund, S., McKay, C., Schuetz, E., van Deursen, J.M., Stravopodis, D., Wang, D., Brown, M., Bodner, S., Grosveld, G., and Ihle, J.N. 1998. Stat5a and Stat5b proteins have essential and nonessential, or redundant, roles in cytokine responses. *Cell.* 93:841-850.

49.     Udy, G.B., Towers, R.P., Snell, R.G., Wilkins, R.J., Park, S.H., Ram, P.A., Waxman, D.J., and Davey, H.W. 1997. Requirement of STAT5b for sexual dimorphism of body growth rates and liver gene expression. *Proc Natl Acad Sci U S A.* 94:7239-7244.

50.     Park, S.H., Liu, X., Hennighausen, L., Davey, H.W., and Waxman, D.J. 1999. Distinctive roles of STAT5a and STAT5b in sexual dimorphism of hepatic P450 gene expression. Impact of STAT5a gene disruption. *J Biol Chem.* 274:7421-7430.

51.     Davey, H.W., Wilkins, R.J., and Waxman, D.J. 1999. STAT5 Signaling in Sexually Dimorphic Gene Expression and Growth Patterns. *Am J Hum Genet.* 65:959-965.

52.     Davey, H.W., Park, S.H., Grattan, D.R., McLachlan, M.J., and Waxman, D.J. 1999. STAT5b-deficient mice are growth hormone pulse-resistant. Role of STAT5b in sex-specific liver p450 expression. *J Biol Chem.* 274:35331-35336.

53.     Waxman, D.J., and Chang, T.K.H. Hormonal regulation of liver cytochrome P450 enzymes. In: Oritiz de Montellano PR, editor. *Cytochrome P450, Structure, Mechanism, and Biochemistry.* 2'nd ed. New York and London: Plenum Press; 1995. p 391-417.

54.     Lechan, R.M., Goodman, R.H., Rosenblatt, M., Reichlin, S., and Habener, J.F. 1983. Prosomatostatin-specific antigen in rat brain: localization by immunocytochemical staining with an antiserum to a synthetic sequence of preprosomatostatin. *Proc Natl Acad Sci U S A.* 80:2780-2784.

55.     Kato, Y., Hu, H.Y., and Sohmiya, M. 1996. Short-term treatment with different doses of human growth hormone in adult patients with growth hormone deficiency. *Endocr J.* 43:177-183.

56.     Norstedt, G., and Palmiter, R. 1984. Secretory rhythm of growth hormone regulates sexual differentiation of mouse liver. *Cell.* 36:805-812.

57.     Sueyoshi, T., Yokomori, N., Korach, K.S., and Negishi, M. 1999. Developmental action of estrogen receptor-alpha feminizes the growth hormone-Stat5b pathway and expression of Cyp2a4 and Cyp2d9 genes in mouse liver. *Mol Pharmacol.* 56:473-477.

58.     Rajendren, G., Levenkova, N., and Gibson, M.J. 2000. Galanin immunoreactivity in mouse basal forebrain: sex differences and discrete projections of galanin-containing cells beyond the blood-brain barrier. *Neuroendocrinology.* 71:27-33.

59.     Urban, J.H., Bauer-Dantoin, A.C., and Levine, J.E. 1993. Neuropeptide Y gene expression in the arcuate nucleus: sexual dimorphism and modulation by testosterone. *Endocrinology.* 132:139-145.

60.     Lamberts, S.W., Zuyderwijk, J., den Holder, F., van Koetsveld, P., and Hofland, L. 1989. Studies on the conditions determining the inhibitory effect of somatostatin on adrenocorticotropin, prolactin and thyrotropin release by cultured rat pituitary cells. *Neuroendocrinology.* 50:44-50.

61.     Fehm, H.L., Voigt, K.H., Lang, R., Beinert, K.E., Raptis, S., and Pfeiffer, E.F. 1976. Somatostatin: a potent inhibitor of ACTH-hypersecretion in adrenal insufficiency. *Klin Wochenschr.* 54:173-175.

62.     Petraglia, F., Facchinetti, F., D'Ambrogio, G., Volpe, A., and Genazzani, A.R. 1986. Somatostatin and oxytocin infusion inhibits the rise of plasma beta-endorphin, beta-lipotrophin and cortisol induced by insulin hypoglycaemia. *Clin Endocrinol (Oxf).* 24:609-616.

63.     Ferrara, C., Cocchi, D., and Muller, E.E. 1991. Somatostatin in the hippocampus mediates dexamethasone-induced suppression of corticosterone secretion in the rat. *Neuroendocrinology.* 53:428-431.

64.     Radke, J.M., and Vincent, S.R. 1988. Effects of systemic and intracerebroventricular cysteamine on dexamethasone-induced suppression of corticosterone levels in the rat. *Neuroendocrinology.* 48:258-263.

65.     Strowski, M.Z., Parmar, R.M., Blake, A.D., and Schaeffer, J.M. 2000. Somatostatin inhibits insulin and glucagon secretion via two receptors subtypes: an in vitro study of pancreatic islets from somatostatin receptor 2 knockout mice. *Endocrinology.* 141:111-117.

66.     Vecsei, L., and Widerlov, E. 1988. Brain and CSF somatostatin concentrations in patients with psychiatric or neurological illness. An overview. *Acta Psychiatr Scand.* 78:657-667.
67.     Robbins, R.J., Brines, M.L., Kim, J.H., Adrian, T., de Lanerolle, N., Welsh, S., and Spencer, D.D. 1991. A selective loss of somatostatin in the hippocampus of patients with temporal lobe epilepsy. *Ann Neurol.* 29:325-332.
68.     Haroutunian, V., Mantin, R., Campbell, G.A., Tsuboyama, G.K., and Davis, K.L. 1987. Cysteamine-induced depletion of central somatostatin-like immunoactivity: effects on behavior, learning, memory and brain neurochemistry. *Brain Res.* 403:234-242.
69.     Vecsei, L., and Widerlov, E. 1988. Effects of intracerebroventricularly administered somatostatin on passive avoidance, shuttle-box behaviour and open-field activity in rats. *Neuropeptides.* 12:237-242.
70.     Viollet, C., Vaillend, C., Videau, C., Bluet-Pajot, M.T., Ungerer, A., L'Heritier, A., Kopp, C., Potier, B., Billard, J., Schaeffer, J.et al. 2000. Involvement of sst2 somatostatin receptor in locomotor, exploratory activity and emotional reactivity in mice. *Eur J Neurosci.* 12:3761-3770.

# 6

# SOMATOSTATIN RECEPTOR GENE FAMILY - SUBTYPE SELECTIVITY FOR LIGAND BINDING

Jason P. Hannon[¥], Christian Bruns, Gisbert Weckbecker & Daniel Hoyer[¥].
[¥]*Neuroscience and Transplantation Research, Novartis Institutes for Biomedical Research,,*
*CH-4002 Basel, Switzerland*

## INTRODUCTION

Somatostatin (SST; SRIF - somatotrophin release inhibiting factor) produced by neuronal, neuroendocrine, inflammatory, immune cells and tumours, regulates its many (patho)-physiological functions through specific cell membrane receptors. Five different SST receptors have been identified and classified $sst_1$-$sst_5$, respectively. Based on pharmacological features and molecular cloning, they are separated into two families, $SRIF_1$ and $SRIF_2$. They all belong to the superfamily of G-protein coupled receptors with seven α-helical transmembrane spanning domains (Hoyer *et al.*, 1994). These receptors have high affinity for the naturally occurring peptides, SST-14 and SST-28, but also for the related cortistatin (CST) peptides (rCST-14, hCST-17 and CST-29) which are specifically produced in the brain. Whether there are specific CST receptors is presently unknown, therefore this chapter will generally refer to SST receptors. Activation of SST receptors inhibits adenylyl cyclase and regulates a number of signalling pathways including ion channels ($K^+$, $Ca^{2+}$), protein serine/threonine- and tyrosine phosphatases as well as phospholipases (PL) C and $A_2$ and the protein kinase C/mitogen activated protein kinase pathways.

Whilst the SST receptor subtypes have been characterized by molecular cloning and pharmacology, the availability of 'selective' ligands is still limited, and thus their development is essential in order to better understand the (patho)-physiological functions of SST receptor activation. SST receptors play

**Abbreviations**: Calcitonin gene-related peptide (CGRP); Chinese hamster lung fibroblast cells (CCL-39); Cortistatin (CST); Gamma aminobutyric acid (GABA); Growth hormone (GH); Knock-out (KO); Phospholipase (PL); PSD95/SAP90 Discs Large ZO-1 homologous (PDZ-binding domain); Receptor activity modifying proteins (RAMPS); SST Somatostatin; SRIF (Somatotropin release inhibiting factor).

an important inhibitory role in the secretion of a number of gastroenterological and pancreatic peptides. However, the SST receptor sub-type specifically involved in any given effect, is not always defined. SST receptor agonists that bind to more than one SST receptor subtype, such as octreotide (Sandostatin®), have been available for more than a decade. They are indicated for the treatment of acromegaly, gastroenterological and pancreatic tumours and other functional gastrointestinal disorders. Recently, selective SST receptor agonists have been synthesized and shown to inhibit growth hormone (GH), glucagon and insulin secretion; in addition to pain, epileptic seizures, vascular remodeling after angioplasty/transplantation, and tumour cell growth. This review summarizes the major aspects regarding SST receptor subtypes by focusing on the molecular biology, pharmacology and structure-activity relationship of their ligands.

## Somatostatin and Cortistatin

SST was first identified, in 1973, as a low molecular weight peptide from hypothalamic extracts capable of inhibiting GH secretion from cultured anterior pituitary cells (Brazeau et al., 1973). First characterized as a fourteen amino acid peptide, produced from a single gene encoding preprosomatostatin (116 amino acids) which is processed to prosomatostatin (96 amino acids) and enzymatically cleaved to produce two bioactive forms of SST – SST-14 and SST-28 that has an $NH_2$-terminal extension of fourteen amino acids (Pradayrol et al., 1980; Patel & O'Neil, 1988; Patel & Galanopoulou, 1995). Both peptides possess a disulphide bridge between $Cys^3$ and $Cys^{14}$ (see Figure 1 & Chapter 1), and act as neurotransmitters or hormones, depending on the site of action and/or the target cell type.

More recently, a prepropeptide cDNA has been cloned from rat brain tissue with structural similarity to that of preprosomatostatin (see Chapter 3). This preprohormone gives rise to two products, CST-14 and CST-29. CST-14 has eleven amino acids identical to SST-14 and found to displace iodine-labelled SST-14 as potently as SST itself (De Lecea et al., 1996). A human form of prepro-CST has also been discovered from which a putative seventeen amino acid peptide (hCST-17) is produced (Fukusumi et al., 1997). CST mRNA has only been detected in the brain, and possibly blood cells, yet the corresponding peptide remains elusive since no selective antibodies are presently available. However, processing and secretion of rat cortistatins have been demonstrated in mouse AtT-20 cells (Puebla et al., 1999). Thus, cortistatins can be produced in pituitary cells, and release studies should clarify the proportions of CST and SST peptides produced in the brain. It remains to be determined whether the effects produced by CST are mediated via SST receptor activation only, or whether there are specific CST receptors awaiting cloning and/or identification. For convenience, we shall refer to SST receptors throughout this chapter. However, it should be understood that thus far, the cloned and native SST receptors display high affinity for both families of peptides, namely the somatostatins and cortistatins.

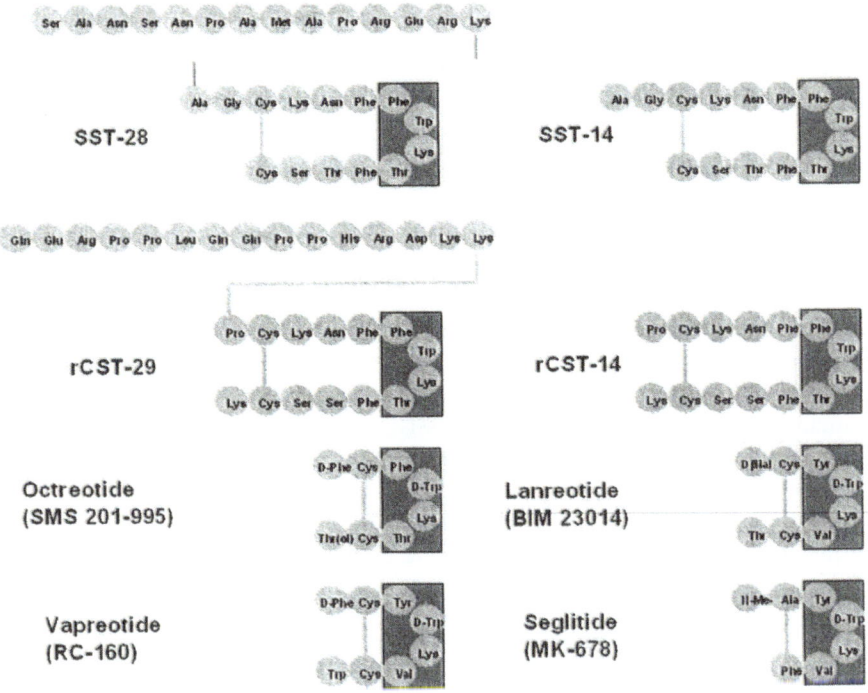

*Figure 1: Amino acid sequences of human somatostatin and its analogue peptides. The boxed area indicates a β-turn and the conserved amino acids Trp and Lys, essential for biological activity.*

## SST Receptor Subtypes

SST produces its diverse effects by binding to several membrane bound receptors (Bell & Reisine, 1993). These receptors were initially proposed to be subdivided into two main classes, $SRIF_1$ and $SRIF_2$, based on binding studies with iodinated synthetic analogues of SST. The $SRIF_1$ receptor demonstrated differential structural and pharmacological activity with high (nM) affinity for the short chain synthetic SST analogues octreotide (SMS 201-955, Sandostatin[®]) and MK-678 (seglitide) (Reubi, 1984; Tran *et al.*, 1985; Martin *et al.*, 1991; Raynor & Reisine, 1992; Raynor *et al.*, 1992). Conversely, the $SRIF_2$ receptor displayed low (μM) affinity for octreotide and MK-678, but high affinity for CGP 23996, SST-14 and SST-28 (Reubi, 1984; Tran *et al.*, 1985; Martin *et al.*, 1991; Hoyer *et al.*, 1995a; 1995b).

Conclusive evidence for the existence of SST receptor subtypes has been provided by the cloning of five SST receptors, termed $sst_1$-$sst_5$; with the $sst_2$ receptor demonstrating a splice variant in certain species ($sst_{2A}$ and $sst_{2B}$; Vanetti *et al.*, 1992; Patel *et al.*, 1993). SST receptor subtypes have been cloned from human, rat, mouse, porcine, bovine and fish tissues (Bruno *et al.*, 1992; Kluxen *et al.*, 1992; Meyerhof *et al.*, 1992; O'Carroll *et al.*, 1992; Yasuda *et al.*, 1992; Yamada *et al.*, 1992a; Yamada *et al.*, 1992b; Demchyshyn *et al.*, 1993; Rohrer *et al.*, 1993; Xu *et al.*, 1993; Yamada *et al.*, 1993a; Panetta *et al.*, 1994; Schwabe *et al.*, 1996; Baumeister *et al.*, 1998; Moldovan *et al.*, 1998; Siehler *et al.*, 1999b; Feuerbach *et al.*, 2000). All five SST receptor subtypes have seven α-helical transmembrane domains and are coupled to guanine nucleotide binding proteins (Meyerhof, 1998). In humans, $sst_1$-$sst_5$ receptors are encoded by five non-allelic genes, which have been identified on chromosomes 14, 17, 22, 20 and 16, respectively (Corness *et al.*, 1993; Demchyshyn *et al.*, 1993; Yamada *et al.*, 1993b; Panetta *et al.*, 1994). Genes coding for $sst_{1, 3-5}$ are intronless, whereas the gene for mouse, rat (and possibly human) $sst_2$ produces spliced variants, $sst_{2A}$ and $sst_{2B}$ (Figure 2), which differ in the length and amino acid composition of their respective C-termini (Vanetti *et al.*, 1992; Patel *et al.*, 1993). The amino acid sequence of the cloned SST receptors range in size from 364 ($sst_5$) to 418 ($sst_2$) amino acids and their sequence homology varies from 39-57% between receptor subtypes (Patel & Srikant, 1997; see also Table 1).

Greater sequence homology is seen in the sequences of the transmembrane domains than in the extracellular N- and intracellular C-terminal domains. Between the different subtypes, the highest sequence homology is evident between $sst_1$ and $sst_4$, and $sst_2$, $sst_3$ and $sst_5$, respectively (Hoyer *et al.*, 1994). The individual receptor subtypes show high sequence homology between species demonstrating between 2 and 19% sequence divergence of the respective mouse, human and rat homologues (Meyerhof, 1998). This is illustrated in the phylogenetic tree of the SST receptor family (Figure 3), which clearly demonstrates the subdivision between the two main groups.

## Pharmacology of SST Receptor Subtypes

The pharmacology of the cloned SST receptors has been studied using radioligand-binding techniques, which demonstrate that all five SST receptor subtypes bind SST-14 and SST-28 with equally high affinity (Table 2). The $hsst_5$ receptor was claimed to have some selectivity for SST-28 over SST-14 depending on the binding conditions used (O'Carroll *et al.*, 1992). We have compared the two peptides at $hsst_5$ receptors using 4 different radioligands and 3 second messenger readouts, and there was very limited, if any, selectivity for SST-28 (Siehler & Hoyer, 1999a; 1999b; 1999c). The putative cleavage products of CST, hCST-17 and rCST-14, and iodinated CST analogues, display nM affinity for the five (human) SST receptors (Siehler *et al.*, 1998a; 1999a).

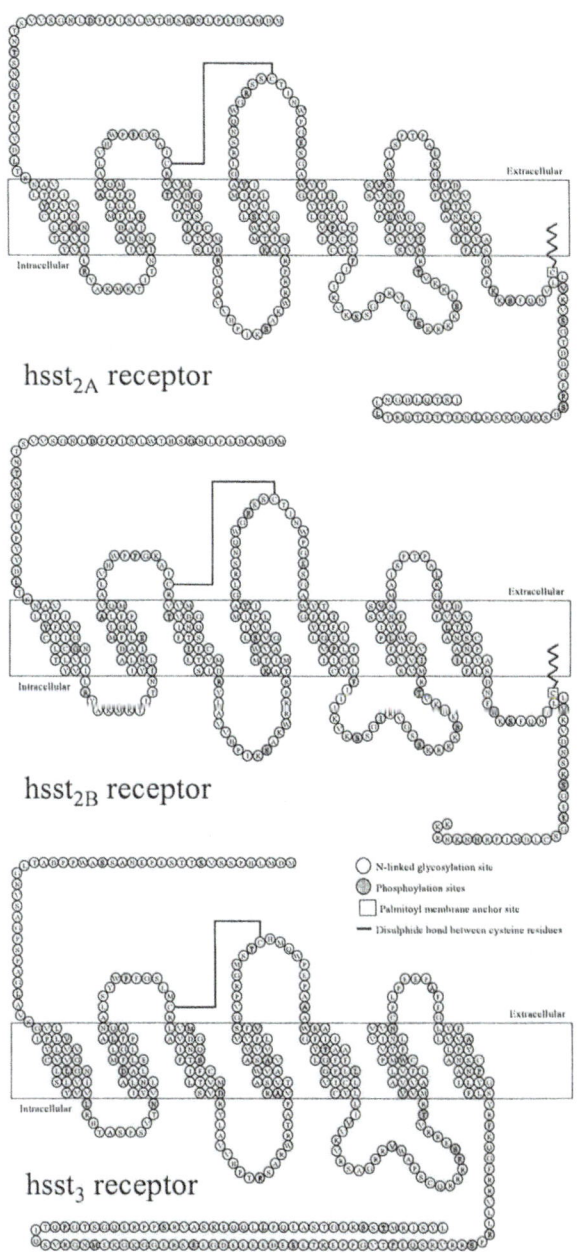

*Figure 2: Schematic representation of the putative human sst₂ splice variants and the sst₃ receptor (Yamada et al., 1992a; 1992b; Vanetti et al., 1992; Patel et al., 1993).*

Table 1: *Summary of the characteristics of human somatostatin receptor subtypes.*

| Subtype | sst$_1$ | sst$_2$ | sst$_3$ | sst$_4$ | sst$_5$ |
|---|---|---|---|---|---|
| Alternate names | SRIF$_2$, SSTR$_1$ SFIF$_{2A}$ | SRIF$_1$, SSTR$_2$ SFIF$_{1A}$ | SSTR$_3$ SFIF$_{1C}$ | SSTR$_4$, SSTR$_5$ SFIF$_{2B}$ | SSTR$_5$, SSTR$_4$ SFIF$_{1B}$ |
| G-protein coupling | Yes | Yes | Yes | Yes | Yes |
| Adenylyl cyclase activity | ↓ | ↓ | ↓ | ↓ | ↓ |
| Tyrosine phosphatase activity | ↑ | ↑ | ↑ | ↑ | ↑ |
| MAP kinase activity | ↑ | ↓ | ↑↓ | ↑ | ↓ |
| K$^+$ channels (GIRK) | | ↑ | ↓ | ↑ | ↑ |
| Ca$^{2+}$ channels | ↓ | ↓ | | | |
| Na$^+$ / H$^+$ exchanger | ↑ | | | | |
| AMPA/Kainate glutamate channels | ↑ | ↓ | | | |
| Phospholipase C / IP$_3$ activity | ↑ | ↑ | ↑ | ↑ | ↑↓ |
| Phospholipase A$_2$ activity | | | | ↑ | |
| Gene / Chromosome | SSR$_1$ / 14q13 | SSR$_{2A,2B}$ / 17q24 | SSR$_3$ / 22q13.1 | SSR$_4$ / 20p11.2 | SSR$_5$ / 16p13.3 |
| Amino Acids; Swiss Prot. # | 391; P30872 | 369/356; P30874 | 418; P32745 | 388; P31391 | 364; P35346 |
| Molecular weight (kDa) | 42.7 | 41.3 | 45.9 | 41.9 | 39.2 |
| Transcript size (kb) | 4.8 | 8.5 | 5.0 | 4.0 | 4.0 |
| Glycosylation sites | 3 | 4 | 2 | 1 | 3 |
| Phosphorylation sites | 6 | 8 | 5 | 3 | 5 |
| Tissue distribution | Brain, pituitary, islets, stomach, liver, kidneys | Brain, pituitary, islets, stomach, kidneys | Brain, pituitary, islets, stomach | Brain, islets, stomach, lungs, placenta | Pituitary, islets, stomach |

**Adapted from Hoyer et al., 2000**

Table 2: *Affinity of SST, CST and analogues for the five cloned human somatostatin receptors*

| Ligand (IC$_{50}$ or K$_d$ nM) | sst$_1$ | sst$_2$ | sst$_3$ | sst$_4$ | sst$_5$ |
|---|---|---|---|---|---|
| **Endogenous peptides** | | | | | |
| SST-14 | 0.1 - 2.3 | 0.1 - 1.3 | 0.2 - 2.1 | 0.3 - 4.1 | 0.1 - 2.2 |
| SST-28 | 0.1 - 2.4 | 0.1 - 4.1 | 0.1 - 7.9 | 0.3 - 2.2 | 0.05 - 0.4 |
| RCST$_{14}$ | 0.9 - 5.0 | 0.1 - 4.5 | 0.3 - 3.8 | 0.2 - 18.2 | 0.3 - 1.9 |
| rCST$_{29}$ | 2.8 | 7.1 | 0.2 | 3.0 | 13.7 |
| hCST$_{17}$ | 0.2 - 7.0 | 0.5 - 1.4 | 0.1 - 0.6 | 0.3 - 0.6 | 0.1 - 0.4 |
| **Peptide analogues** | | | | | |
| Octreotide (SMS 201-995) | 223 - >1000 | 0.1 - 2.1 | 2.5 - 34.5 | 398 - >1000 | 0.3 - 67 |
| Lanreotide (BIM 23014) | 177 - >1000 | 0.2 - 1.8 | 9.5 - 107 | 66 - >1000 | 0.5 - >1000 |
| Vapreotide (RC-160) | 83 - >1000 | 0.1 - 5.4 | 12.3 - 31 | 45 - 351 | 0.7 - 53.7 |
| Seglitide (MK-678) | >1000 | 0.1 - 1.5 | 12.9 - 36 | 6.5 - >1000 | 0.1 - >1000 |
| BIM 23268 | 18.4 | 15.1 | 61.6 | 16.3 | 0.37 |
| CH 275 | 1.8 - 4.9 | 740 - >1000 | 12 - >1000 | 4.3 - 874 | 980 - >1000 |
| BIM 23052 | 2.4 - 100 | 1.7 - 13.5 | 0.2 - 5.6 | 2.3 - 141 | 0.6 - 35 |
| BIM 23056 | 100 - >1000 | 132 - >1000 | 10.8 - 177 | 17 - 339 | 4.8 - 209 |
| L-362,855 | 501 - >1000 | 0.8 - 4.4 | 5.1 - 24 | 49 - >1000 | 0.02 - 68 |
| Sst3 ODN3 | >5 | >5 | 8.17 | >5 | >5 |
| Sst4 Alpha-Beta peptide | >5 | >5 | >5 | 7.08 | >5 |
| Cyanamid 154806 | 1200 - 3890 | 2.6 - 3.6 | 150 - 851 | 650-1738 | 2.0 - 331 |
| SOM230 | 7.70 | 9.00 | 9.00 | 5.83 | 10.22 |
| KE108 | 8.58 | 9.05 | 8.83 | 8.80 | 9.19 |
| **Non-peptide ligands** | | | | | |
| NNC 26-9100 | 1800 - 5000 | 621 - 3300 | 1400 - 6800 | 6 - 100 | 1900 - 4100 |
| SRA880 | 8.01 | 4.44 | 5.06 | 4.82 | 5.84 |
| BN 81674 | >5 | >5 | 9.04 | >5 | >5 |
| L-797,519 | 1.4 | 1.75 | 2240 | 170 | 3600 |
| L-779,976 | 2760 | 0.05 | 729 | 310 | 4260 |
| L-796,778 | 1255 | >10000 | 24 | 8650 | 1200 |
| L-803,087 | 199 | 4720 | 1280 | 0.7 | 3880 |
| L-817,818 | 3.3 | 52 | 64 | 82 | 0.4 |

Corness *et al.*, 1993; Demchyshyn *et al.*, 1993; Raynor *et al.*, 1993a; 1993b; Hoyer *et al.*, 1994; Patel & Srikant, 1994; Reisine & Bell, 1995; Bruns *et al.*, 1996; Liapakis *et al.*, 1996; Fukusumi *et al.*, 1997; Patel, 1997; Rohrer *et al.*, 1998; Ankersen *et al.*, 1998; Siehler *et al.*, 1998b; 1999a; Spier & De Lecea, 2000; Feniuk *et al.*, 2000; Bruns *et al.*, 2001; Hoyer *et al.*, 2002; Poitout *et al.*, 2001; Reubi *et al.*, 2000; 2002; Nunn *et al.*, 2003a; Gademann *et al.*, 2001

The cyclooctapeptide SST analogues, octreotide (SMS 201-995), BIM 23014, RC 160 and the hexapeptide MK-678 bind to $sst_2$ and $sst_5$ with high affinity, $sst_3$ with moderate affinity and show little or no affinity for $sst_1$ and $sst_4$ receptors (Raynor et al., 1993a; 1993b; Hoyer et al., 1994).

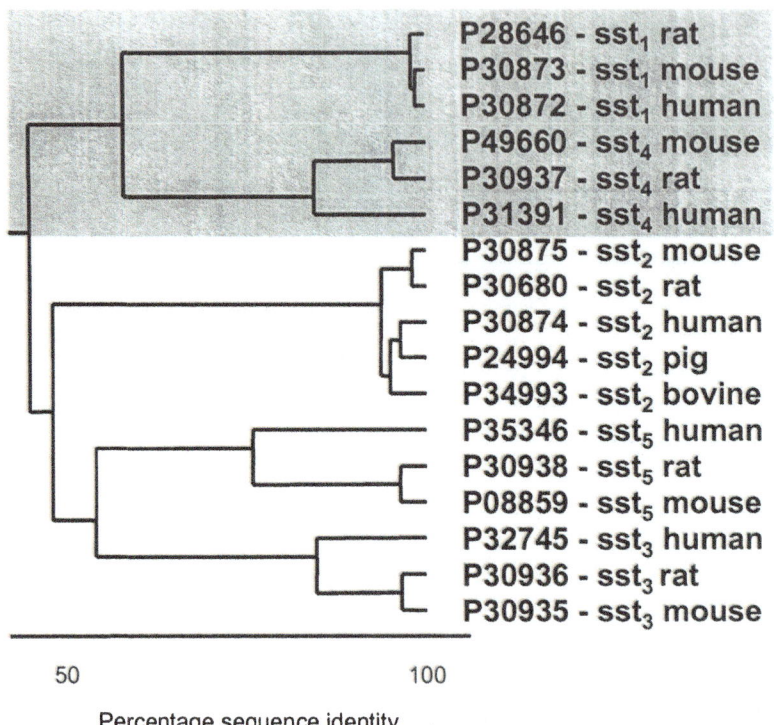

$$50 \qquad\qquad\qquad\qquad 100$$

Percentage sequence identity

*Figure 3: Phylogenetic tree of somatostatin receptors, demonstrating sequence homology between the respective receptor subtypes. SRIF$_2$ receptors are highlighted in dark grey, whilst those belonging to the SRIF$_1$ receptor family are highlighted in light grey. Swiss Prot accession numbers are shown on the right-hand side.*

Modification of these analogues can have a pronounced impact on their affinity, selectivity and stability. Octreotide, when coupled to chelators that tightly bind e.g. [111]Indium or [90]Yttrium, can be utilized for the diagnosis and treatment of SST receptor-expressing tumours (Stolz et al., 1998; Krenning et al., 2000). Binding to $sst_2$ is largely unaffected by such modifications, whereas their affinity for $sst_5$ was decreased up to 100-fold; suggesting that certain modifications to these ligands markedly alters their selectivity. BIM 23014 (lanreotide) and RC 160 (vapreotide) were subjected to similar modifications, yet retained high affinity binding to $sst_2$. RC 160 has been lipophilized by attaching a myristoyl moiety; this myristoylation led to an increased *in vitro* stability of the analogue without changing its affinity for the receptor (Dasgupta et al., 2000).

Considerable effort has been directed towards the identification of SST analogues, both peptidomimetics and non-peptide ligands, with increased selectivity for a distinct SST receptor subtype. However, reconciliation of data from binding analyses is difficult, especially with respect to subtype selectivity, as marked difference in affinity values may occur between SST receptors from different species.

*Peptide agonists at $sst_1$ receptors:* CH-275 (Des-AA$^{1,2,5}$[DTrp$^8$, IAMP$^9$]-SST and Des-AA$^{1,5}$[Tyr$^2$, DTrp$^8$, IAMP$^9$]-SST have been described as $sst_1$ selective ligands; accordingly, Des-AA$^{1,5}$[Tyr$^2$, DTrp$^8$, IAMP$^9$]-SST was shown to have high affinity for $sst_1$ expressing prostate tumours (Liapakis *et al.*, 1996; Reubi *et al.*, 1998). However, it has also been reported that CH-275 binds to $sst_4$ receptors with high affinity and would thus appear to be a prototypic agonist for the SST$_2$ receptor subclass (Patel, 1997). *$sst_2$ receptors:* BIM 23027 (c[N-Me-Ala-Tyr-DTrp-Lys-Abu-Phe]) and L-363,301 (c[Pro-Phe-DTrp-Lys-Thr-Phe]), have been demonstrated to be $sst_2$ receptor selective ligand (Raynor *et al.*, 1993a; 1993b; Bruns *et al.*, 1996; McKeen *et al.*, 1996; Koenig *et al.*, 1997). Whereas compounds such as DC2360 (DNal-c[Cys-Tyr-DTrp-Lys-Val-Cys]-Thr-OH) and EC5-21 (DPhe-c[Cys-Phe-DTrp-Lys-Thr-Cys]-Nal-NH$_2$), although exhibiting 19- and 35-fold selectivity for $sst_2$, also bind well to $sst_5$ receptors. *$sst_3$ receptors:* BIM 23056 (DPhe-Phe-Tyr-DTrp-Lys-Val-Phe-DNal-NH$_2$), has been shown to possess high affinity and selectivity at $sst_3$ receptors, but its relative affinity would appear to be species-dependent (Reisine & Bell, 1995; Bruns *et al.*, 1996). Moreover, in CHO cells expressing hsst$_5$ receptors (see below) this compound has been shown to be a potent full agonist for inhibition of cAMP formation (Carruthers *et al.*, 1999). *$sst_4$ receptors:* No selective peptide agonists were so far available for the $sst_4$ receptor subtype; however short beta- or alpha-beta peptides show promise, one agonist reported with an affinity in the 80 nM range and selectivity for the stt4 receptor (Gademann et al, 2001; Nunn et al, 2003a). *$sst_5$ receptors:* BIM 23268 (H-Cys-Phe-Phe-DTrp-Lys-Thr-Phe-Cys-NH$_2$) and BIM 23313 (H-Cys-Phe-Tyr(I)-DTrp-Lys-Thr-Phe-Cys-NH$_2$) appear to demonstrate specificity and high affinity for the $sst_5$ receptor (Shimon *et al.*, 1997a; Coy & Taylor, 1999); whereas L-362,855 (c[Aha-Phe-Trp-DTrp-Lys-Thr-Phe]), demonstrates modest selectivity for $sst_5$ *vis-à-vis* $sst_2$ but high affinity, and has been shown to act as a partial agonist in recombinant systems (Wilkinson *et al.*, 1997). *Peptides acting at all SST receptors*: The cyclohexapeptide SOM230 (Lewis et al 2001), shows high affinity to $sst_1$, $sst_2$, $sst_3$ and $sst_5$ receptors with good stability. SOM230 is a highly potent inhibitor of the GH/IGF-1 axis in rats, dogs and monkeys as well as in healthy human subjects and acromegalic patients (Bruns et al, 2001; 2002; Weckbecker et al, 2001; 2002; Lamberts et al, 2002; Van der Hoek et al, 2003). Similarly, the nonpeptide agonist KE108 binds with nanomolar affinity to $sst_{1-5}$ and behaves as an agonist, inhibiting cAMP production (Reubi et al, 2002). It also seems that the approach with beta or even gamma peptides has great value (see Gademann et al, 2001; Seebach et al, 2003).

*Non-peptide agonists:* With respect to non-peptide agonists, L-054,522 (L-Lysine, ($\beta$S)-N-[[4- (2,3-dihydro-2-oxo-1H-benzimidazol-1-yl) -1-piperidinyl] carbonyl]-$\beta$-methyl-D-tryptophyl-,1,1-dimethylethyl ester) has high affinity and selectivity for $sst_2$ receptors, acting as a full agonist it differentially mediates the inhibition of hormone release (Yang *et al.*, 1998). NNC 26-9100 (1-[3-[N-(5-Bromopyridin-2-yl)-N-(3,4-dichlorobenzyl) amino] propyl]-3-[3-(1H-imidazol-4-yl) propyl] thiourea) has moderate selectivity for the $sst_4$ receptor and was for some time the sole ligand with selectivity for that receptor (Ankersen *et al.*, 1998). These discoveries prompted further work and more recently, a series of high-affinity subtype-selective non-peptide agonists have been reported for each of the five human SST receptor subtypes. For example, L-797,591 ((R)-N-(6-Amino-2,2,4-trimethyl-hexyl)-3-naphthalen-1-yl-2-[3-phenethyl-3-(2-pyridin-2-yl-ethyl)-ureido]-propionamide), L-779,976 (4-(2-Oxo-2,3-dihydro-benzoimidazol-1-yl)-piperidine-1-carboxylic acid [(1R,2S)-1-[(((1R,3S) -3-aminomethyl- cyclohexyl methyl)-carbamoyl]-2-(1H-indol-3-yl)-propyl]-amide) and L-803,087 (2-[4-(5,7-Difluoro-2-phenyl-1 H -indol-3-yl) -butyrylamino] -5-guanidino- pentanoic acid methyl ester) demonstrate nM affinity and selectively for the $hsst_{1,2,4}$ receptor subtypes, respectively (Rohrer *et al.*, 1998; Rohrer & Berk, 1999; See Table 2 & Chapter 5 for more details). The availability of these high-affinity subtype-selective receptor agonists represents a major break-through in the field, and will undoubtedly enhance our understanding of the subtype-selective physiological functions of SST receptor activation.

*Antagonists:* In classical pharmacology, final proof for receptor-mediated effects comes from the combined use of selective agonists and antagonists. Evidently, with respect to antagonists, the SST receptor field is lagging behind that of most other peptide receptors for which antagonists have been 'easier' to find. *$sst_1$ receptors:* The non-peptide sst1 receptor antagonist SRA-880 (Hoyer et al, 2002), is based on a octahydrobenzo[$\gamma$]quinoline backbone. It is a selective high-affinity antagonist at the $sst_1$ receptor as demonstrated in various second messenger tests, with good bioavailability and is active in behavioural tests. *$sst_2$ receptors:* The first potential SST peptide antagonist, Cyanamid 154806 ([AC-4-NO$_2$-Phe-c(DCys-Tyr-DTrp-Lys-Thr-Cys)-DTyr-NH$_2$]), was initially described by Bass and colleagues (1996). The ligand binds to both $hsst_2$ and $hsst_5$ receptors with nM affinity and antagonizes the receptor effector coupling to adenylyl cyclase and a number of other functional responses mediated by SST receptors; both the L and DTyr analogues of the molecule possess antagonism at $sst_2$ receptors (Feniuk *et al.*, 2000). However, L- and D-Tyr[8]-CYN-154806 displayed close to full agonism in a cAMP accumulation assay ($pEC_{50}$ = 7.7), as well as in reported gene assays, but acted as apparently silent antagonists on [$^{35}$S]GTP$\gamma$S binding. The L type, showed also partial agonism in the guinea-pig ileum, whereas the D type was devoid of agonism; thus this compounds may well have agonism or antagonism in a system dependent way (Nunn et al, 2003b). Similar data have been reported with other small putative sst2 receptor antagonists (see Nunn et al, 2003c). Additional antagonists for the $sst_2$ receptor have been synthesized based on the cyclic hexapeptide structure containing DCys and DTrp, as in Cyanamid 154806, which demonstrate affinities in the nM range, at least 8-fold selectivity over $sst_3$ and $sst_5$ receptors and a lack of agonist activity in the μM range (Hocart *et al.*, 1999). Moreover, AC-178,335 (acetyl-DHis-DPhe-DIle-DArg-DTrp-DPhe-NH$_2$), has been shown, in anaesthetized rats, to induce a small transient antagonistic effect on SST with respect to GH release, presumably via an action on $sst_2$ receptors (Baumbach *et al.*, 1998). *$sst_3$ receptors:* The first $sst_3$-selective non-peptide antagonist BN-81674 (Poitout et al, 2001) has a Ki value of 0.92 nM at the hsst$_3$ receptor and about 10,000 fold selectivity over the other subtypes. It acts as a competitive antagonist against SST-14-mediated inhibition of cAMP accumulation in hsst$_3$-expressing CHO-K1 cells with an IC$_{50}$ value of 0.84 nM. Carbamoyl-des-AA$^{1,2,4,5,12,13}$ [DCys$^3$, Tyr$^7$, DAgl$^8$ (Me, 2-naphthoyl)]-SST-14 (sst3-ODN-8), binds selectively and with an affinity similar to that of SST-28 to $sst_3$ receptors and potently antagonizes $sst_3$-mediated events including inhibition of cAMP generation and PLC activation (Reubi *et al.*, 2000). *$sst_4$ receptors:* To date, no $sst_4$ selective antagonist has been reported. *$sst_5$ receptors:* As mentioned earlier, in addition to an agonist action at $sst_3$ and $sst_5$ receptors, BIM 23056, has also been reported to block hsst$_5$ signalling. Thus, antagonism for this subtype appears dependent on tissue type, receptor coupling or receptor density when tested in recombinant systems (Wilkinson *et al.*, 1997). Moreover, in a number of second messenger assays it was shown that the compound has (partial) agonist activity (Siehler & Hoyer, 1999a; 1999b; 1999c).

*SST receptor mutants:* The selective interaction of SST receptors and their ligands (as with any other receptor systems) is based on very specific structural elements both within SST, and its analogues (see above), and within the five SST receptor subtypes. The structural basis for the binding selectivity of SST receptors is currently only partially understood. Receptor mutants have been generated to map the binding sites responsible for high affinity binding of SST receptor specific ligands. For example, by systematically replacing single amino acids of sst$_1$ (the WT sst$_1$ receptor fails to bind octreotide), Kaupmann and colleagues (1995) identified two amino acids, Ser$^{305}$ and Gln$^{291}$, in transmembrane domains VI and VII that resulted in high affinity binding of the sst$_1$ mutant receptor for octreotide. Glycosylation of SST receptors is apparently another, still understudied, determinant of ligand binding. Nehring and co-workers (2000) found that a double sst$_3$ mutant, lacking both glycosylation sites, exhibited reduced affinity and impaired signalling whilst ligand selectivity was unaffected. The knowledge of the interactions between SST receptors and their ligands may enhance the identification of new SST analogues suitable as tools and therapeutics.

## Predictive Value of Radioligand Binding And Other *In Vitro* Data

Radioligand binding is one of the techniques which is most commonly used to determine the affinity of ligands for receptors, and incidentally to define the pharmacological profile of receptors. However, there has been quite some variation in the literature on the affinity values, and the claimed selectivity profiles, for a number of ligands derived from such studies (see Table 2). This is particularly true, although not exclusive, for the SST field. There are a number of reasons for these discrepancies, i.e. differences in the experimental conditions utilized (e.g. ions, buffers, concentrations, temperatures, kinetics, proteases, protease inhibitors) to perform binding studies. Moreover, the use of different cell lines, membrane preparations, SST receptor cDNAs from different species, variations in receptor density and differences in radioligands utilized only facilitates the, sometimes apparent, lack of correlation between data from different laboratories. In addition, the use of peptides as the almost sole source of ligands obviously does not alleviate the problem.

We have undertaken a systematic study of the pharmacological profiles of the five human SST receptors stably expressed in CCL-39 cells (Chinese hamster lung fibroblast) using [$^{125}$I]LTT-SST-28, [$^{125}$I][Tyr$^{10}$]CST-14, [$^{125}$I]CGP 23996 and [$^{125}$I][Tyr$^{3}$]octreotide when feasible (Siehler *et al.*, 1998a; 1998b; 1999a). The rank order of affinity was largely radioligand-independent at hsst$_{1-4}$ receptors, in contrast to hsst$_5$ receptors, where it was radioligand-dependent. Thus, the pharmacological profile of [$^{125}$I]LTT-SST-28- and [$^{125}$I][Tyr$^{10}$]CST-14-labelled hsst$_5$ sites correlated highly significantly, but did not correlate with the affinity profiles defined with [$^{125}$I]CGP 23996 and [$^{125}$I][Tyr$^{3}$]octreotide binding to hsst$_5$ receptors. Further, depending on the radiolabelled agonist used, and the receptor subtype studied, it would appear that binding can be essentially to a guanine nucleotide-sensitive (sst$_2$ or sst$_3$), -insensitive (sst$_1$ or sst$_4$), or a mixture of both states (sst$_5$); in the latter case, each radioligand defining a different rank order of affinity at the same receptor. These studies were extended to investigate various second messenger readouts, in the same cells, upon receptor stimulation. Again, marked differences were observed depending on ligand, receptor and second messenger studied (Siehler *et al.*, 1999a; Siehler & Hoyer, 1999a; 1999b; 1999c).

Nevertheless, there is obviously a correlation between what can be measured at recombinant receptors expressed in various cell types and native receptors. Indeed, we have shown previously that both autoradiography and membrane radioligand binding allow clear identification of sst$_1$ and sst$_2$ receptors in brain of various species (rat, mouse, monkey, human; Hoyer *et al.*, 1995b; Schoeffter *et al.*, 1995; Piwko *et al.*, 1995; 1996; 1997; Thoss *et al.*, 1995; 1996; 1997). Furthermore, in the rat lung most of the SST receptor binding is to sst$_4$

receptors (Schloos *et al.*, 1997), whereas in rat pancreas or guinea pig ileum, $sst_2$ receptors appear to be predominantly expressed (Fehlmann *et al.*, 2000).

Formerly, it was assumed that G-protein coupled receptors act by forming simple molecular complexes with the trimeric G proteins; this view is changing as more and more heptahelical receptors are reported to be capable of forming homo- or heterodimers which, in turn, activate G-proteins upon stimulation. Dimerization may then alter the functional properties of the (SST) receptor subtype in terms of ligand affinity, agonist-induced receptor internalization, and/or receptor upregulation. There is further evidence that some heptahelical receptors may also be subject to protein-protein interaction with, for example, RAMPS or a number of other accessory proteins/chaperones with PDZ-binding domains. These additional combinations may, or may not, affect the 'pharmacology' of a given receptor, as reported for an increasing number of heptahelical receptors ($b_2$ adrenergic, dopamine $D_2$, opiate, $GABA_B$, amylin and CGRP-like receptors (White *et al.*, 1998; Christopoulos *et al.*, 1999; Marshall *et al.*, 1999; Foord & Marshall, 1999; Tilakaratne *et al.*, 2000; Zumpe *et al.*, 2000). The same is also true for SST receptors, which form dimers within the SST receptor family, and with other families such as dopamine receptors (Rocheville *et al.*, 2000a; 2000b). Thus, $hsst_5$ shows agonist-induced homodimerization, which is associated with an increase in ligand-binding affinity due to conformational changes. In addition, $sst_5$ forms heterodimers with $sst_1$, but not with $sst_4$. Furthermore, it has been shown that dopamine and $sst_5$ receptors form heterodimers, creating a novel receptor that is pharmacologically distinct from its receptor homodimers (Rocheville *et al.*, 2000a). Although not analyzed in detail, it is conceivable that if such phenomena occur *in situ*, i.e. in native tissues, it may be difficult to predict how the pharmacological profile will be affected. One would assume that the homodimerization of two $sst_2$ receptors might retain high affinity for octreotide, although this has not been established. For instance, $GABA_{B1}$ and $GABA_{B2}$ heterodimers retain largely $GABA_{B1}$ 'pharmacology', but one could also expect a mixed or totally different type of pharmacological signature from such a heterodimeric receptor. What about a heterodimer between $sst_5$ and $D_1$ receptors, will it adopt dopaminergic and somatostatinergic features? Heterodimerization of $sst_{2A}$ and $sst_3$ receptors results in a complex that retains $sst_2$ binding properties but lost affinity for $sst_3$-selective ligands (Pfeiffer et al, 2001). By contrast, the $sst_{2A}$ receptor and the related μ-opioid receptor heterodimerize in HEK 293 cells without major changes in ligand binding and coupling properties; the heterodimers show cross-phosphorylation and cross-desensitization upon binding of $sst_2$- or μ-selective ligands suggesting that heterodimer formation contributes to the regulation of SST receptors (Pfeiffer et al, 2002). This may be physiologically relevant as co-expression of the $sst_{2A}$ receptor and the μ-opioid receptor was reported in the periaqueductal gray, involved in pain (see Selmer et al, 2000). Whether there is a synergistic action of somatostatin and opioids on the same neurone remains to be determined. So far SST receptor oligomerization was mostly generated in transfected cells expressing high numbers of certain SST receptor subtypes (Froidevaux and Eberle, 2002). On the other hand, little has been reported with respect to the interaction of SST receptor subtypes and accessory proteins; e.g. it has been established that RAMP1 and 2 binding to the same CGRP-like receptor will profoundly modify the nature of the responses observed, in one case it will be amylin-like, in the other CGRP-like (Tilakaratne *et al.*, 2000; Zumpe *et al.*, 2000).

Thus, the apparent affinity values measured *in vitro* using recombinantly expressed homogeneous receptors may not always reflect the *in vivo* situation. This aspect has a different pharmacological relevance if the drug under study is an agonist or an antagonist, but becomes even more complex when considering the multiplicity of interactions of receptors with their own congeners, and a range of accessory proteins, in addition to the multiple G proteins which may certainly affect function by what is generally referred to as cross-talk and receptor trafficking. Thus, the diversity of responses elicited by SST may not only be explained by the aforementioned multitude of protein-protein interactions involving SST receptors. Additional response modulation may occur at the ligand level, in that SST and certain analogues can interfere with the opioid receptor system. For example, Mundey and colleagues (2000) showed that SST abolishes the inhibitory effects of the $\mu$-opioid receptor antagonist CTOP (DPhe-Cys-Tyr-DTrp-Orn-Thr-Pen-Thr-$NH_2$) on the contraction of isolated guinea-pig ileum.

## Physiology and SST Receptor Subtype Selective Biological Effects

Somatostatins demonstrate a wide spectrum of activities and biological functions. The peptides, and their receptor subtypes, are distributed widely throughout the body. Somatostatinergic neurones are found in high densities in all parts of the cerebral cortex; the anterior pituitary; the neural lobe of the pituitary; the limbic system; brain stem and spinal cord (Reichlin, 1983). There are two types of neurones containing SST; (i) interneurons, present in the olfactory bulb, cerebral cortex, hippocampus, amygdala, hypothalamus, striatum and retina and (ii) long projecting neurones, which are typified by the tuberoinfundibular neurohormonal somatostatinergic system which projects from the anterior periventricular hypothalamic nucleus to the median eminence (Epelbaum *et al.*, 1994). Outside the nervous system, SST is present in the exocrine and endocrine pancreas, adrenals, thyroids, kidneys, immune cells, gastrointestinal tract and salivary glands (Reichlin, 1983; Reisine & Bell, 1995; Patel, 1999).

SST acts via a number of different signaling routes. In the anterior pituitary it acts as a neurohormone, being secreted from neurones into the blood. In the nervous system, it is a neurotransmitter or neuromodulator and in the rest of the body it can act locally as a paracrine or autocrine regulator, or systemically as a true hormone (Reichlin, 1983; Patel, 1997). In the brain, SST acts as a regulator of neocortical, striatal, limbic and hypothalamic neurones by the modulation of the release of various neurotransmitters and neuromodulators (Epelbaum, 1986; Raynor & Reisine, 1992; Epelbaum et al., 1994). It is thought to have effects on motor, sensory, behavioural, cognitive and autonomic systems (Reichlin, 1983; Patel & Srikant, 1997). In peripheral tissues, SST is important in endocrine, gastrointestinal and immune cell functions (Aguila et al., 1991; Patel, 1997). SST regulates endocrine and exocrine secretion in: (i) the anterior pituitary, where it inhibits secretion of GH, thyroid stimulating hormone and prolactin; (ii) the pancreas, where it inhibits the release of amylase from acinar cells, and glucagon and insulin from islet cells; (iii) the gastrointestinal tract, where it inhibits peptide hormone and HCl secretion; (iv) and an inhibitory function in various glands, such as the thyroid and salivary gland (Brazeau et al., 1973; Vale et al., 1974; Mandarino et al., 1981; Reichlin, 1983; Feniuk et al., 1993). SST also inhibits cellular proliferation in the stomach mucosa (McHenry et al., 1991) and inhibits the secretion of cytokines (IL-8 and IL-1$\beta$) from intestinal epithelial cells, thus playing a role in the regulation of mucosal inflammatory responses. These effects are mediated by $sst_2$ and $sst_5$ receptors (Chowers et al., 2000).

The specific physiological role of the SST receptor subtypes is an area of research still under intense investigation, and with the exception of the $sst_2$ receptor, our knowledge on the precise contributions of each receptor subtype is limited. This is mainly due to the lack of selective agonists and antagonists with adequate pharmacokinetic properties. In addition, the expression of several SST receptor subtypes in single cells may lead to confusion over whether a single receptor subtype mediates a response or whether multiple subtypes are involved (Patel & Srikant, 1997).

Particularly in the CNS, the roles of individual SST receptors are uncertain. The presence of $sst_{1-4}$ mRNA in the neocortex, hippocampus and amygdala suggests roles in the regulation of complex integrative functions such as locomotor activity, learning and memory (Meyerhof, 1998). In addition, a preferential involvement of $sst_1$ in the intrahypothalamic regulation of GH secretion by SST has been suggested by Lanneau and colleagues (2000) using antisense approaches in rodents.

In the periphery, the subtype specific functions of SST receptors have been characterized in more detail. However, the subtype selectivity of SST analogues used in most of these studies remains questionable and the results are therefore inconclusive and/or controversial. Specific functions have been suggested for the $sst_1$, $sst_2$, $sst_3$ and $sst_5$ receptor subtypes by studying the effects of selected ligands in in vitro and in vivo models; including SST receptor transfected cell lines and knockout (KO) mice (Zheng et al., 1997; Martinez et al., 1998).

*Role of sst$_1$:* Using sst$_1$ receptor transfected cells, Buscail and colleagues (1994) demonstrated inhibition of cell proliferation by exposure to SST. Furthermore, the sst$_1$ receptor was shown to be upregulated in human vessels and also atherosclerotic arteries express predominantly sst$_1$ receptors (Curtis *et al.*, 2000a; 2000b), indicating the potential for sst$_1$ selective SST analogues to modulate, for example, vessel remodelling post-transplantation. *Role of sst$_2$:* The sst$_2$ subtype is involved in the inhibition of GH release from pituitary (Raynor *et al.*, 1993a), stimulated glucagon release from pancreatic α–cells (Rossowski & Coy, 1994; Coy & Taylor, 1996; Rohrer & Schaeffer, 2000; Strowski *et al.*, 2000), gastric acid secretion (Rossowski *et al.*, 1994; Lloyd *et al.*, 1995), gastrin release from gastric mucosa (Prinz *et al.*, 1994), histamine release in the stomach (Zaki *et al.*, 1996) and ion secretion in the colon (Warhurst *et al.*, 1996). Furthermore, sst$_2$ mediates inhibitory effects on proliferation in normal vascular smooth muscle (Bruns *et al.*, 2000) and transformed cells (Weckbecker *et al.*, 1996; Delesque *et al.*, 1997) with the two isoforms, sst$_{2A}$ and sst$_{2B}$, mediating opposing effects on cell proliferation, inhibition and induction, respectively, as a function of differential cell signalling (Sellers *et al.*, 2000). Moreover, injury after ballooning or vessel transplantation results in upregulation of the sst$_2$ receptor subtype in rats (Curtis *et al.*, 2000a). *Role of sst$_3$:* The sst$_3$ subtype is suggested to mediate the inhibition of gastric smooth muscle relaxation (Gu *et al.*, 1995). Moreover, the sst$_3$ receptor may play a role in the induction of cell apoptosis and inhibition of cell proliferation (Sharma & Srikant, 1998; Weckbecker *et al.*, 1999). Human peripheral B- and T-lymphocytes were shown to exclusively express sst$_3$ receptors, in contrast to human monocytes which express sst$_2$ upon activation with LPS (Oomen *et al.*, 2000); indicating a potential role in cell proliferation, cytokine and immunoglobulin production and natural killer activity (Lichtenauer-Kaligis *et al.*, 2000). *Role of sst$_4$:* The role of the sst$_4$ receptor is currently unclear, although sst$_4$ is expressed in various tissues including placenta, brain, islets and lungs (Schindler *et al.*, 1995; Buscail *et al.*, 1996; Schloos *et al.*, 1997; Caron *et al.*, 1997); and the biochemical properties, such as G-protein coupling, inhibition of adenylate cyclase or activation of PLC are very similar to those of sst$_1$ receptors (Table 1). The limited knowledge on sst$_4$ receptor function relates to the limited availability of appropriate tools, such as selective ligands (see above). *Role of sst$_5$:* The sst$_5$ subtype has been shown to mediate inhibition of amylase release from exocrine pancreas and insulin release from endocrine pancreas (Rossowski *et al.*, 1994; Coy & Taylor, 1996; Rohrer & Schaeffer, 2000; Strowski *et al.*, 2000). Recently, the involvement of sst$_5$ in the inhibition of peptide YY secretion was concluded from studies on the effects of the sst$_5$ selective analogue L-362,855 in fetal rat intestinal cell cultures (Chisholm & Greenberg, 2000). Parmar and colleagues (1999), demonstrated a role of sst$_5$ in the inhibition of GHRH-stimulated GH release from rat pituitary cells using selective SST analogues. Furthermore, prolactin secretion can be inhibited by exposure of cultured prolactinomas to sst$_5$ selective agonists (Shimon *et al.*, 1997b).

The first clinically useful SST analogue was the octapeptide SMS 201-995 (octreotide, Sandostatin®), which was introduced into clinical practice in 1983 for treatment of hormone-producing pituitary, pancreatic, and intestinal tumours, and has remained the mainstay of SST analogue therapy (Bauer *et al.*, 1982; Lamberts *et al.*, 1991; 1996). Octreotide, along with BIM 23014 (lanreotide), RC 160 (Vapreotide), and the hexapeptide MK-678 (seglitide), bind to only three of the five SST receptor subtypes, displaying high affinity for $sst_2$ and $sst_5$ and moderate affinity for $sst_3$ (Reisine & Bell, 1995; Bruns *et al.*, 1996; Shimon *et al.*, 1997a). It should be noted that the affinity of octreotide, BIM 23014, RC 160, and MK-678 for $sst_2$ and $sst_5$ is comparable to that of SST-14, indicating that they are neither selective for these subtypes nor more potent than the endogenous ligand (Patel & Srikant, 1994).

Based on these similarities in profiles, it is clear that a distinction between $sst_2$ and $sst_5$ receptor mediated effects, which can be involved in a number of regulatory processes such as GH release, gastric acid secretion, epileptic seizure, gastrointestinal motility, tumour growth, vascular smooth muscle proliferation or pain induced by colon distension, is not an easy task with the currently available tools. The more recent identification of non-peptide subtype selective agonists by Rohrer and colleagues (1998) has helped to support the concept of $sst_2$ receptors to inhibit glucagon release from mouse pancreatic α-cells and $sst_5$ receptors to mediate insulin release from pancreatic β-cells (see chapter 22). However, it may well be that ultimately, the winner of the race will be these agonists that interact at several SST receptors with similar affinities to provide improved efficacy in the clinical treatment of various conditions where a case can be made for multiple receptor involvement as seen already in certain endocrine conditions or for instance tumors that co-express a variety of SST receptors.

## CONCLUSION

Somatostatins and their receptors are widely distributed throughout the body, whereas cortistatins seem to be limited to the brain, with a few exceptions. Moreover, the $sst_5$ receptor subtype, whilst present in the pituitary, is absent from the CNS. This receptor subtype-specific tissue distribution provides an interesting basis for pharmaco-therapeutic intervention, although there may be evidence for redundancy, since both SST and SST receptor KO animals do not show marked phenotypes. Furthermore, it has been shown that, at least at the receptor level, there is cellular co-localization. There are currently five SST receptor subtypes, which recognize both SST and CST native peptides with very high affinity, and it remains to be seen whether there are more receptors of this family, or whether CST may have additional specific receptors.

SST receptor agonists such as octreotide (and lanreotide) are in clinical use for the treatment of acromegaly, gastroenterological and pancreatic tumours and other gastrointestinal disorders. Also, some analogues are used, or are in development for tumour imaging and radiotherapy. The majority of these effects are mediated by $sst_2$ receptors, which may be important for numerous other potential indications such as pain, epilepsy, neurological disorders, and diabetes (see further chapters). It would seem that only a few functional effects can be unambiguously attributed to the other receptors of this family, such as $sst_1$, $sst_3$ or $sst_5$. However, further confirmation of many SST-mediated physiologic effects are required, and antagonists are still crucial to support the agonist data. It will be interesting to see if the more universal agonists, like SOM230, will have improved efficacy in conditions where several SST receptors may work in concert.

As discussed, selective ligands are in development, but their selectivity is typically determined for cloned rat or human receptors *in vitro*, and species differences are to be expected. Moreover, the assessment of selectivity may be preliminary in view of the possibility that additional receptor subtypes may emerge, be it for SST and/or CST. Recent data using recombinant human receptors, revealed that compounds thought to be selective exhibit distinct pharmacological profiles *in vivo*, with unexpected affinity at certain SST receptors. Thus, although progress has been made in the development of potent and selective SST receptor agonists, there is opportunity for improvement. Finally, the ever increasing complexity of receptor homo- and hetero-dimerisation, interactions with multiple G-proteins and other accessory proteins, may explain why *in vivo* effects are often more complex than those determined *in vitro* with comparatively simple recombinant SST receptor expressing systems.

## ACKNOWLEDGEMENTS
DH and JH were supported by EC Contract QLG3-CT-1999-00908 and Swiss grant BBW 00-0427.

## REFERENCES

1.  Aguila, M.C., Dees, W.L., Haensly, W.E. & McCann, S.M. (1991). Evidence that somatostatin is localized and synthesized in lymphoid organs. *Proc.Natl.Acad.Sci.U.S.A.* **88**, 11485-11489.
2.  Ankersen, M., Crider, M., Liu, S., Ho, B., Andersen, H.S. & Stidsen, C. (1998). Discovery of a Novel Non-Peptide Somatostatin Agonist with SST4 Selectivity. *J.Am.Chem.Soc.* **120**, 1368-1373.
3.  Bass, R.T., Buckwalter, B.L., Patel, B.P., Pausch, M.H., Price, L.A., Strnad, J. & Hadcock, J.R. (1996). Identification and characterization of novel somatostatin antagonists. *Mol.Pharmacol.* **50**, 709-715.
4.  Bauer, W., Briner, U., Doepfner, W., Haller, R., Huguenin, R., Marbach, P., Petcher, T.J. & Pless, J. (1982). SMS 201-995: a very potent and selective octapeptide analog of somatostatin with prolonged action. *Life Sci.* **31**, 1133-1140.
5.  Baumbach, W.R., Carrick, T.A., Pausch, M.H., Bingham, B., Carmignac, D., Robinson, I.C.A.F., Houghten, R., Eppler, C.M., Price, L.A. & Zysk, J.R. (1998). A linear hexapeptide somatostatin antagonist blocks somatostatin activity in vitro and influences growth hormone release in rats. *Mol.Pharmacol.* **54**, 864-873.
6.  Baumeister, H., Kreuzer, O.J., Roosterman, D., Schafer, J. & Meyerhof, W. (1998). Cloning, expression, pharmacology and tissue distribution of the mouse somatostatin receptor subtype 5. *J.Neuroendocrinol.* **10**, 283-290.

7. Bell, G.I. & Reisine, T. (1993). Molecular biology of somatostatin receptors. *Trends Neurosci.* **16**, 34-38.

8. Brazeau, P., Vale, W., Burgus, R., Ling, N., Butcher, M., Rivier, J. & Guillemin, R. (1973). Hypothalamic polypeptide that inhibits the secretion of immunoreactive pituitary growth hormone. *Science* **179**, 77-79.

9. Bruno, J.F., Xu, Y., Song, J. & Berelowitz, M. (1992). Molecular cloning and functional expression of a brain-specific somatostatin receptor. *Proc.Natl.Acad.Sci.U.S.A.* **89**, 11151-11155.

10. Bruns, C., Lewis, I., Briner, U., Meno-Tetang, G. & Weckbecker, G. SOM230: a novel somatostatin peptidomimetic with a broad somatotropin release inhibiting factor (SST) receptor binding and a unique antisecretory profile. Eur. J. Endocrinol. **146**, 707-716 (2002).

11. Bruns, C., Raulf, F., Hoyer, D., Schloos, J., Luebbert, H. & Weckbecker, G. (1996). Binding properties of somatostatin receptor subtypes. *Metab., Clin.Exp.* **44**, 17-20.

12. Bruns, C., Shi, V., Hoyer, D., Schuurman, H. & Weckbecker, G. (2000). Somatostatin receptors and the potential use of Sandostatin to interfere with vascular remodelling. *Eur.J.Endocrinol.* **143**, S3-S7.

13. Buscail, L., Delesque, N., Esteve, J.P., Saint-Laurent, N., Prats, H., Clerc, P., Robberecht, P., Bell, G.I. & Liebow, C. (1994). Stimulation of tyrosine phosphatase and inhibition of cell proliferation by somatostatin analogs: mediation by human somatostatin receptor subtypes SSTR1 and SSTR2. *Proc.Natl.Acad.Sci.U.S.A.* **91**, 2315-2319.

14. Buscail, L., Saint-Laurent, N., Chastre, E., Vaillant, J.C., Gespach, C., Capella, G., Kalthoff, H., Lluix, F., Vaysse, N. & Susini, C. (1996). Loss of sst2 somatostatin receptor gene expression in human pancreatic and colorectal cancer. *Cancer Res.* **56**, 1823-1827.

15. Caron, P., Buscail, L., Beckers, A., Esteve, J.P., Igout, A., Hennen, G. & Susini, C. (1997). Expression of somatostatin receptor SST4 in human placenta and absence of octreotide effect on human placental growth hormone concentration during pregnancy. *J.Clin.Endocrinol.Metab.* **82** 3771-3776.

16. Carruthers, A.M., Warner, A.J., Michel, A.D., Feniuk, W. & Humphrey, P.P.A. (1999). Activation of adenylate cyclase by human recombinant sst5 receptors expressed in CHO-K1 cells and involvement of G.alpha.s proteins. *Br.J.Pharmacol.* **126**, 1221-1229.

17. Chisholm, C. & Greenberg, G.R. (2000). Somatostatin receptor subtype-5 mediates inhibition of peptide YY secretion from rat intestinal cultures. *Am.J.Physiol.* **279**, G983-G989.

18. Chowers, Y., Cahalon, L., Lahav, M., Schor, H., Tal, R., Bar-Meir, S. & Levite, M. (2000). Somatostatin through its specific receptor inhibits spontaneous and TNF-.alpha.- and bacteria-induced IL-8 and IL-1.beta. secretion from intestinal epithelial cells. *J.Immunol.* **165**, 2955-2961.

19. Christopoulos, G., Perry, K.J., Morfis, M., Tilakaratne, N., Gao, Y., Fraser, N.J., Main, M.J., Foord, S.M. & Sexton, P.M. (1999). Multiple amylin receptors arise from receptor activity-modifying protein interaction with the calcitonin receptor gene product. *Mol.Pharmacol.* **56**, 235-242.

20. Corness, J.D., Demchyshyn, L.L., Seeman, P., Van Tol, H.H.M., Srikant, C.B., Kent, G., Patel, Y.C. & Niznik, H.B. (1993). A human somatostatin receptor (SSTR3), located on chromosome 22, displays preferential affinity for somatostatin-14 like peptides. *FEBS Lett.* **321**, 279-284.

21. Coy, D.H. & Taylor, J.E. (1996). Receptor-specific somatostatin analogs: correlations with biological activity. *Metab. Clin.Exp.* **44**, 21-23.

22. Coy, D.H. & Taylor, J.E. (1999). Development of somatostatin agonists with high affinity and specificity for the human and rat type 5 receptor subtype. *Pept.Proc.Am.Pept.Symp.*, **15**, 559-560.

23. Curtis, S.B., Chen, J.C., Winkelaar, G., Turnbull, R.G., Hewitt, J., Buchan, A.M. & Hsiang, Y.N. (2000a). Effect of endothelial and adventitial injury on somatostatin receptor expression. *Surgery* **127**, 577-583.

24. Curtis, S.B., Hewitt, J., Yakubovitz, S., Anzarut, A., Hsiang, Y.N. & Buchan, A.M.J. (2000b). Somatostatin receptor subtype expression and function in human vascular tissue. *Am.J.Physiol.* **278**, H1815-H1822.

25. Dalm, V.A., van Hagen, M., van Koetsveld, P.M., Langerak, A.W., van der Lely, A.-J., Lamberts, S.W. & Hofland, L.J. Cortistatin rather than somatostatin as a potential endogenous ligand for somatostatin receptors in the human immune system. *J. Clin. Endocrinol. Metab.* **88**, 270-276 (2003).

26. Dasgupta, P., Singh, A.T. & Mukherjee, R. (2000). Lipophilization of somatostatin analog RC-160 with long chain fatty acid improves its anti-proliferative activity on human oral carcinoma cells in vitro. *Life Sci.* **66**, 1557-1570.

27. De Lecea, L., Criado, J.R., Prospero-Garcia, O., Gautvik, K.M., Schweitzer, P., Danielson, P.E., Dunlop, C.L., Siggins, G.R., Henriksen, S.J. & Sutcliffe, J.G. (1996). A cortical neuropeptide with neuronal depressant and sleep-modulating properties. *Nature* **381**, 242-245.

28. Delesque, N., Buscail, L., Esteve, J.P., Saint-Laurent, N., Mueller, C., Weckbecker, G., Bruns, C., Vaysse, N. & Susini, C. (1997). SST2 Somatostatin receptor expression reverses tumorigenicity of human pancreatic cancer cells. *Cancer Res.* **57**, 956-962.

29. Demchyshyn, L.L., Srikant, C.B., Sunahara, R.K., Kent, G., Seeman, P., Van Tol, H.H.M., Panetta, R., Patel, Y.C. & Niznik, H.B. (1993). Cloning and expression of a human somatostatin-14-selective receptor variant (somatostatin receptor 4) located on chromosome 20. *Mol.Pharmacol.* **43**, 894-901.

30. Epelbaum, J. (1986). Somatostatin in the central nervous system: physiology and pathological modifications. *Prog.Neurobiol. (Oxford)* **27**, 63-100.

31. Epelbaum, J., Dournaud, P., Fodor, M. & Viollet, C. (1994). The neurobiology of somatostatin. *Crit.Rev.Neurobiol.* **8**, 25-44.

32. Fehlmann, D., Langenegger, D., Schuepbach, E., Siehler, S., Feuerbach, D. & Hoyer, D. (2000). Distribution and characterisation of somatostatin receptor mRNA and binding sites in the brain and periphery. *J.Physiol. (Paris)* **94**, 265-281.

33. Feniuk, W., Dimech, J. & Humphrey, P.P.A. (1993). Characterization of somatostatin receptors in guinea pig isolated ileum, vas deferens and right atrium. *Br.J.Pharmacol.* **110**, 1156-1164.

34. Feniuk, W., Jarvie, E., Luo, J. & Humphrey, P P.A. (2000). Selective somatostatin sst2 receptor blockade with the novel cyclic octapeptide, CYN-154806. *Neuropharmacology* **39**, 1443-1450.

35. Feuerbach, D., Fehlmann, D., Nunn, C., Siehler, S., Langenegger, D., Bouhelal, R., Seuwen, K. & Hoyer, D. (2000). Cloning, expression and pharmacological characterization of the mouse somatostatin sst5 receptor. *Neuropharmacology* **39**, 1451-1462.

36. Foord, S.M. & Marshall, F.H. (1999). RAMPs: accessory proteins for seven transmembrane domain receptors. *Trends Pharmacol.Sci.* **20**, 184-187.

37. Froidevaux, S. & Eberle, A.N. Somatostatin analogs and radiopeptides in cancer therapy. *Biopolymers* **66**, 161-183 (2002).

38. Fukusumi, S., Kitada, C., Takekawa, S., Kizawa, H., Sakamoto, J., Miyamoto, M., Hinuma, S., Kitano, K. & Fujino, M. (1997). Identification and characterization of a novel human cortistatin-like peptide. *Biochem.Biophys.Res.Commun.* **232**, 157-163.

39. Gademann, K., Kimmerlin, T., Hoyer, D. & Seebach, D. Peptide folding induces high and selective affinity of a linear and small ß-peptide to the human somatostatin receptor 4. *J. Med. Chem.* **44**, 2460-2468 (2001).

40. Gu, Z.F., Corleto, V.D., Mantey, S.A., Coy, D.H., Maton, P.N. & Jensen, R.T. (1995). Somatostatin receptor subtype 3 mediates the inhibitory action of somatostatin on gastric smooth muscle cells. *Am.J.Physiol.* **268**, G739-G748.

41. Hocart, S.J., Jain, R., Murphy, W.A., Taylor, J.E. & Coy, D.H. (1999). Highly potent cyclic disulfide antagonists of somatostatin. *J.Med.Chem.* **42**, 1863-1871.

42. Hofland, L.J., van der Hoek, J., Van Koetsveld, P.M., Bruns, C. Weckbecker, Mooij, D., Waaijers, M., van Aken, M., Beckers, A., de Herder, W.W. & Lamberts, S.W.J. The novel somatostatin analog SOM230 inhibits ACTH release by cultured human corticotroph tumors. Endocrine Soc. Meeting, Abstract P2-499 (June 2003).

43. Hoyer, D., Bell, G.I., Berelowitz, M., Epelbaum, J., Feniuk, W., Humphrey, P.P., O'Carroll, A.M., Patel, Y.C., Schonbrunn, A., Taylor, J.E. & Reisine, T. (1995a). Classification and nomenclature of somatostatin receptors. *Trends.Pharmacol.Sci.* **16**, 86-88.

44. Hoyer, D., Dixon, K., Gentsch, C., Vassout, A., Enz, A., Jaton, A., Nunn, C., Schoeffter, P., Neumann, P., Troxler, T. & Pfaeffli, P. NVP-SRA880, a somatostatin sst1 receptor antagonist promotes social interactions, reduces aggressive behaviour and stimulates learning. *The Pharmacologist*, **44**, 2, suppl. 1, A254 (2002).
45. Hoyer, D., Epelbaum, J., Feniuk, W., Humphrey, P.P.A., Meyerhof, W., Patel, Y., Reisine, T., Reubi, J.C., Schonbrunn, A. & Vezzani, A. (2000), "Classification of somatostatin receptors". In: *The IUPHAR compendium of receptor characterisation and classification*, IUPHAR Media.
46. Hoyer, D., Luebbert, H. & Bruns, C. (1994). Molecular pharmacology of somatostatin receptors. *Naunyn-Schmiedeberg's Arch.Pharmacol.* **350**, 441-453.
47. Hoyer, D., Perez, J., Schoeffter, P., Langenegger, D., Schuepbach, E., Kaupmann, K., Luebbert, H., Bruns, C. & Reubi, J.C. (1995b). Pharmacological identity between somatostatin SS-2 binding sites and SSTR-1 receptors. *Eur.J.Pharmacol., Mol.Pharmacol.Sect.* **289**, 151-161.
48. Kaupmann, K., Bruns, C., Raulf, F., Weber, H.P., Mattes, H. & Luebbert, H. (1995). Two amino acids, located in transmembrane domains VI and VII, determine the selectivity of the peptide agonist SMS 201-995 for the SSTR2 somatostatin receptor. *EMBO J.* **14**, 727-735.
49. Kluxen, F.W., Bruns, C. & Luebbert, H. (1992). Expression cloning of a rat brain somatostatin receptor cDNA. *Proc.Natl.Acad.Sci.U.S.A.* **89**, 4618-4622.
50. Koenig, J.A., Edwardson, J.M. & Humphrey, P.P. (1997). Somatostatin receptors in Neuro2A neuroblastoma cells: ligand internalization. *Br.J.Pharmacol.* **120**, 52-59.
51. Krenning, E.P., Valkema, R., Kooij, P.P.M., Breeman, W.A.P., Bakker, W.H., De Herder, W.W., Van Eijck, C.H.J., Kwekkeboom, D.J., De Jong, M., Jamar, F. & Pauwels, S. (2000). The role of radioactive somatostatin and its analogues in the control of tumor growth. *Recent Results Cancer Res.* **153**, 1-13.
52. Lamberts, S.W., Van Der Lely, A.J., De Herder, W.W. & Hofland, L.J. (1996). Octreotide. *N.Eng.J.Med.* **334**, 246-254.
53. Lamberts, S.W.J., Krenning, E.P. & Reubi, J.C. (1991). The role of somatostatin and its analogs in the diagnosis and treatment of tumors. *Endocr.Rev.* **12**, 450-482.
54. Lanneau, C., Bluet-Pajot, M.T., Zizzari, P., Csaba, Z., Dournaud, P., Helboe, L., Hoyer, D., Pellegrini, E., Tannenbaum, G.S., Epelbaum, J. & Gardette, R. (2000). Involvement of the sst1 somatostatin receptor subtype in the intrahypothalamic neuronal network regulating growth hormone secretion: an in vitro and in vivo antisense study. *Endocrinology* **141**, 967-979.
55. Lewis, I., Bauer, W., Albert, R., Chandramouli, N., Pless, J., Weckbecker, G. & Bruns, C. A novel somatostatin mimic with broad somatotropin release inhibitory factor receptor binding and superior therapeutic potential. *J. Med. Chem.* **46**, 2334-2344 (2003).
56. Liapakis, G., Hoeger, C., Rivier, J. & Reisine, T. (1996). Development of a selective agonist at the somatostatin receptor subtype SSTR1. *J.Pharmacol.Exp.Ther.* **276**, 1089-1094.
57. Lichtenauer-Kaligis, E.G.R., Van Hagen, P.M., Lamberts, S.W.J. & Hofland, L.J. (2000). Somatostatin receptor subtypes in human immune cells. *Eur.J.Endocrinol.* **143**, S21-S25.
58. Lloyd, K.C.K., Wang, J., Aurang, K., Gronhed, P., Coy, D.H. & Walsh, J.H. (1995). Activation of somatostatin receptor subtype 2 inhibits acid secretion in rats. *Am.J.Physiol.* **268**, G102-G106.
59. Mandarino, L., Stenner, D., Blanchard, W., Nissen, S., Gerich, J., Ling, N., Brazeau, P., Bohlen, P., Esch, F. & Guillemin, R. (1981). Selective effects of somatostatin-14, -25 and -28 on in vitro insulin and glucagon secretion. *Nature* **291**, 76-77.
60. Marshall, F.H., Jones, K.A., Kaupmann, K. & Bettler, B. (1999). GABAB receptors - the first 7TM heterodimers. *Trends Pharmacol.Sci.* **20**, 396-399.
61. Martin, J.L., Chesselet, M.F., Raynor, K., Gonzales, C. & Reisine, T. (1991). Differential distribution of somatostatin receptor subtypes in rat brain revealed by newly developed somatostatin analogs. *Neuroscience (Oxford)* **41**, 581-593.
62. Martinez, V., Curi, A.P., Torkian, B., Schaeffer, J.M., Wilkinson, H.A., Walsh, J.H. & Tache, Y. (1998). High basal gastric acid secretion in somatostatin receptor subtype 2 knockout mice. *Gastroenterology* **114**, 1125-1132.

63. McHenry, L., Murthy, K.S., Grider, J.R. & Makhlouf, G.M. (1991). Inhibition of muscle cell relaxation by somatostatin: tissue-specific, cAMP-dependent, pertussis toxin-sensitive. *Am.J.Physiol.* **261**, G45-G49.

64. McKeen, E.S., Feniuk, W., Michel, A.D., Kidd, E.J. & Humphrey, P.P.A. (1996). Identification and characterization of heterogeneous somatostatin binding sites in rat distal colonic mucosa. *Naunyn-Schmiedeberg's Arch.Pharmacol.* **354**, 543-549.

65. Meyerhof, W. (1998). The elucidation of somatostatin receptor functions: a current view. *Rev.Physiol., Biochem.Pharmacol.* **133**, 55-105.

66. Meyerhof, W., Wulfsen, I., Schoenrock, C., Fehr, S. & Richter, D. (1992). Molecular cloning of a somatostatin-28 receptor and comparison of its expression pattern with that of a somatostatin-14 receptor in rat brain. *Proc.Natl.Acad.Sci.U.S.A.* **89**, 10267-10271.

67. Moldovan, S., DeMayo, F. & Brunicardi, F.C. (1998). Cloning of the mouse SSTR5 gene. *J.Surg.Res.* **76**, 57-60.

68. Mundey, M.K., Ali, A., Mason, R. & Wilson, V.G. (2000). Pharmacological examination of contractile responses of the guinea-pig isolated ileum produced by μ-opioid receptor antagonists in the presence of, and following exposure to, morphine. *Br.J.Pharmacol.* **131**, 893-902.

69. Nehring, R.B., Richter, D. & Meyerhof, W. (2000). Glycosylation affects agonist binding and signal transduction of the rat somatostatin receptor subtype 3. *J.Physiol. (Paris)* **94**, 185-192.

70. Nunn, C. Rueping, M., Langenegger, D., Schuepbach, E., Kimmerlin, T., Micuh, P., Hurth, K., Seebach, D. & Hoyer, D. $\beta^2/\beta^3$– and $\alpha/\beta^3$–tetrapeptide derivatives as potent agonists at somatostatin sst4 receptors. *Naunyn Schmiedeberg's Arch. Pharmacol.* **367**, 95-103 (2003a).

71. Nunn, C., Langenegger, D., Hurth, K., Fehlmann, D., Schmidt, K. & Hoyer, D. Agonist properties of putative small molecule somatostatin receptor subtype-2 selective antagonists. *Eur. J. Pharmacol.*, **465**, 211-218 (2003c).

72. Nunn, C., Schoeffter, P., Langenegger, D. & Hoyer, D. Functional characterisation of the putative somatostatin sst$_2$ receptor antagonist CYN 154806. *Naunyn Schmiedeberg's Arch Pharmacol.* **367**, 1-9 (2003b).

73. O'Carroll, A.M., Lolait, S.J., Koenig, M. & Mahan, L.C. (1992). Molecular cloning and expression of a pituitary somatostatin receptor with preferential affinity for somatostatin-28. *Mol.Pharmacol.* **42**, 939-946.

74. Oomen, S.P.M.A., Hofland, L.J., Van Hagen, P.M., Lamberts, S.W.J. & Touw, I.P. (2000). Somatostatin receptors in the haematopoietic system. *Eur.J.Endocrinol.* **143**, S9-S14.

75. Panetta, R., Greenwood, M.T., Warszynska, A., Demchyshyn, L.L., Day, R., Niznik, H.B., Srikant, C.B. & Patel, Y.C. (1994). Molecular cloning, functional characterization, and chromosomal localization of a human somatostatin receptor (somatostatin receptor type 5) with preferential affinity for somatostatin-28. *Mol.Pharmacol.* **45**, 417-427.

76. Parmar, R.M., Chan, W.W., Dashkevicz, M., Hayes, E.C., Rohrer, S.P., Smith, R.G., Schaeffer, J.M. & Blake, A.D. (1999). Nonpeptidyl somatostatin agonists demonstrate that sst2 and sst5 inhibit stimulated growth hormone secretion from rat anterior pituitary cells. *Biochem.Biophys.Res.Commun.* **263**, 276-280.

77. Patel, Y.C. & Galanopoulou, A. (1995). Processing and intracellular targeting of prosomatostatin-derived peptides: the role of mammalian endoproteases. *Ciba Found.Symp.* **190**, 26-50.

78. Patel, Y.C. & O'Neil, W. (1988). Peptides derived from cleavage of prosomatostatin at carboxyl- and amino-terminal segments. Characterization of tissue and secreted forms in the rat. *J.Biol.Chem.* **263**, 745-751.

79. Patel, Y.C. & Srikant, C.B. (1994). Subtype selectivity of peptide analogs for all five cloned human somatostatin receptors (hsstr 1-5). *Endocrinology* **135**, 2814-2817.

80. Patel, Y.C. & Srikant, C.B. (1997). Somatostatin receptors. *Trends Endocrinol.Metab.* **8**, 398-405.

81. Patel, Y.C. (1997). Molecular pharmacology of somatostatin receptor subtypes. *J.Endocrinol.Invest.* **20**, 348-367.

82. Patel, Y.C. (1999). Somatostatin and its receptor family. *Front.Neuroendocrinol.* **20**, 157-198.

83. Patel, Y.C., Greenwood, M., Kent, G., Panetta, R. & Srikant, C.B. (1993). Multiple gene transcripts of the somatostatin receptor SSTR2: Tissue selective distribution and cAMP regulation. *Biochem.Biophys.Res.Commun.* **192**, 288-294.

84. Pfeiffer, M., Koch, T., Schroder, H., Klutzny, M., Kirscht, S., Kreienkamp, H.J., Hollt, V., Schulz, S., Homo- and heterodimerization of somatostatin receptor subtypes. Inactivation of sst(3) receptor function by heterodimerization with sst(2A). *J. Biol. Chem.* **276**: 14027-14036 (2001).

85. Pfeiffer, M., Koch, T., Schroder, H., Laugsch, M., Hollt, V. & Schulz S. Heterodimerization of somatostatin and opioid receptors cross-modulates phosphorylation, internalization, and desensitization. *J. Biol. Chem.* **277**, 19762-19772 (2002).

86. Piwko, C., Thoss, V.S. & Hoyer, D. (1995). Localization and pharmacological characterization of somatostatin sst2 sites in the rat cerebellum. *Naunyn-Schmiedeberg's Arch.Pharmacol.* **352**, 607-613.

87. Piwko, C., Thoss, V.S., Probst, A. & Hoyer, D. (1996). Localization and pharmacological characterization of somatostatin recognition sites in human cerebellum. *Neuropharmacology* **35**, 713-723.

88. Piwko, C., Thoss, V.S., Probst, A. & Hoyer, D. (1997). The elusive nature of cerebellar somatostatin receptors: studies in rat, monkey and human cerebellum. *J Recept.Signal Transduction Res.* **17**, 385-405.

89. Poitout, L., Roubert, P., Contour-Galcera, M.-O., Moinet, C., Lannoy, J., Pommier, J., Plas, P., Bigg, D. & Thurieau, C. Identification of potent non-peptide somatostatin antagonists with sst3 selectivity. *J. Med. Chem.* **44**, 2990-3000 (2001).

90. Pradayrol, L., Jornvall, H., Mutt, V. & Ribet, A. (1980). N-terminally extended somatostatin: the primary structure of somatostatin-28. *FEBS Lett.* **109**, 55-58.

91. Prinz, C., Sachs, G., Walsh, J.H., Coy, D.H. & Wu, S.V. (1994). The somatostatin receptor subtype on rat enterochromaffinlike cells. *Gastroenterology* **107**, 1067-1074.

92. Puebla, L., Mouchantaf, R., Sasi, R., Khare, S., Bennett, H.P.J., James, S. & Patel, Y.C. (1999). Processing of rat preprocortistatin in mouse AtT-20 cells. *J.Neurochem.* **73**, 1273-1277.

93. Raynor, K. & Reisine, T. (1992). Somatostatin receptors. *Crit.Rev.Neurobiol.* **6**, 273-289.

94. Raynor, K., Coy, D.C. & Reisine, T. (1992). Analogs of somatostatin bind selectively to brain somatostatin receptor subtypes. *J.Neurochem.* **59**, 1241-1250.

95. Raynor, K., Murphy, W.A., Coy, D.H., Taylor, J.E., Moreau, J.P., Yasuda, K., Bell, G.I. & Reisine, T. (1993a). Cloned somatostatin receptors: Identification of subtype-selective peptides and demonstration of high affinity binding of linear peptides. *Mol.Pharmacol.* **43**, 838-844.

96. Raynor, K., O'Carroll, A.M., Kong, H., Yasuda, K., Mahana, L.C., Bell, G.I. & Reisine, T. (1993b). Characterization of cloned somatostatin receptors SSTR4 and SSTR5. *Mol.Pharmacol.* **44**, 385-392.

97. Reichlin, S. (1983). Somatostatin. *N.Engl.J.Med* **309**, 1556-1563.

98. Reisine, T. & Bell, G.I. (1995). Molecular biology of somatostatin receptors. *Endocr.Rev.* **16**, 427-442.

99. Reubi, J.C. & Maurer, R. (1986). Different ionic requirements for somatostatin receptor subpopulations in the brain. *Regul.Pept.* **14**, 301-311.

100. Reubi, J.C. (1984). Evidence for two somatostatin-14 receptor types in rat brain cortex. *Neurosci.Lett.* **49**, 259-263.

101. Reubi, J.C., Schaer, J.C., Waser, B., Hoeger, C. & Rivier, J. (1998). A selective analog for the somatostatin sst1-receptor subtype expressed by human tumors. *Eur.J.Pharmacol.* **345**, 103-110.

102. Reubi, J.C., Schaer, J.C., Wenger, S., Hoeger, C., Erchegyi, J., Waser, B. & Rivier, J. SST3-selective potent peptidic somatostatin receptor antagonists. *Proc. Natl. Acad. Sci. U.S.A.* **97**, 13973-13978 (2000).

103. Reubi, J:C., Eisenwiener, K.-P., Rink, H., Waser, B., & Mäcke, H.R. A new peptidic somatostatin agonist with high affinity to all five somatostatin receptors. *Eur. J. Pharmacol.* **456**, 45-49 (2002).

104. Rocheville, M., Lange, D.C., Kumar, U., Patel, S.C., Patel, R.C. & Patel, Y.C. (2000a). Receptors for dopamine and somatostatin: Formation of hetero-oligomers with enhanced functional activity. *Science (Washington, D.C.)* **288**, 154-157.

105. Rocheville, M., Lange, D.C., Kumar, U., Sasi, R., Patel, R.C. & Patel, Y.C. (2000b). Subtypes of the somatostatin receptor assemble as functional homo- and heterodimers. *J.Biol.Chem.* **275**, 7862-7869.

106. Rohrer, L., Raulf, F., Bruns, C., Buettner, R., Hofstaedter, F. & Schuele, R. (1993). Cloning and characterization of a fourth human somatostatin receptor. *Proc.Natl.Acad.Sci.U.S.A.* **90**, 4196-4200.

107. Rohrer, S.P. & Berk, S.C. (1999). Development of somatostatin receptor subtype selective agonists through combinatorial chemistry. *Curr.Opin.Drug Discovery Dev.* **2**, 293-303.

108. Rohrer, S.P. & Schaeffer, J.M. (2000). Identification and characterization of subtype selective somatostatin receptor agonists. *J.Physiol. (Paris)* **94**, 211-215.

109. Rohrer, S.P., Birzin, E.T., Mosley, R.T., Berk, S.C., Hutchins, S.M., Shen, D.M., Xiong, Y., Hayes, E.C., Parmar, R.M., Foor, F., Mitra, S.W., Degrado, S.J., Shu, M., Klopp, J.M., Cai, S.J., Blake, A., Chan, W.W.S., Pasternak, A., Yang, L., Patchett, A.A., Smith, R.G., Chapman, K.T. & Schaeffer, J.M. (1998). Rapid identification of subtype-selective agonists of the somatostatin receptor through combinatorial chemistry. *Science (Washington, D.C.)* **282**, 737-740.

110. Rossowski, W.J. & Coy, D.H. (1994). Specific inhibition of rat pancreatic insulin or glucagon release by receptor-selective somatostatin analogs. *Biochem.Biophys.Res.Commun.* **205**, 341-346.

111. Rossowski, W.J., Gu, Z.F., Akarca, U.S., Jensen, R.T. & Coy, D.H. (1994). Characterization of somatostatin receptor subtypes controlling rat gastric acid and pancreatic amylase release. *Peptides (Tarrytown, N.Y.)* **15**, 1421-1424.

112. Schindler, M., Harrington, K.A., Humphrey, P.P.A. & Emson, P.C. (1995). Cellular localization and co-expression of somatostatin receptor messenger RNAs in the human brain. *Mol.Brain Res.* **34**, 321-326.

113. Schloos, J., Raulf, F., Hoyer, D. & Bruns, C. (1997). Identification and pharmacological characterization of somatostatin receptors in rat lung. *Br.J.Pharmacol.* **121**, 963-971.

114. Schoeffter, P., Perez, J., Langenegger, D., Schuepbach, E., Bobirnac, I., Luebbert, H., Bruns, C. & Hoyer, D. (1995). Characterization and distribution of somatostatin SS-1 and SRIF-1 binding sites in rat brain: identity with SSTR-2 receptors. *Eur.J.Pharmacol., Mol.Pharmacol.Sect.* **289**, 163-173.

115. Schwabe, W., Brennan, M.B. & Hochgeschwender, U. (1996). Isolation and characterization of the mouse (Mus musculus) somatostatin receptor type-4-encoding gene (mSSTR4). *Gene* **168**, 233-235.

116. Seebach, D., Schaeffer, L., Brenner, M. & Hoyer, D. Design and synthesis of □-dipeptide derivatives with submicromolar affinities for human somatostatin receptors. *Angew. Chemie, Int. Ed.* **42**, 776-778 (2003).

117. Sellers, L.A., Alderton, F., Carruthers, A.M., Schindler, M. & Humphrey, P.P.A. (2000). Receptor isoforms mediate opposing proliferative effects through Gβγ-activated p38 or Akt pathways. *Mol.Cell.Biol.* **20**, 5974-5985.

118. Selmer, I-S., Schindler, M., Allen, J. P., Humphrey, P. P. A. & Emson, P. C. Advances in understanding neuronal somatostatin receptors. *Reg. Pep.* **90**, 1-18 (2000).

119. Sharma, K. & Srikant, C.B. (1998). Induction of wild-type p53, Bax, and acidic endonuclease during somatostatin-signaled apoptosis in MCF-7 human breast cancer cells. *Int.J.Cancer* **76**, 259-266.

120. Shimon, I., Taylor, J.E., Dong, J.Z., Bitonte, R.A., Kim, S., Morgan, B., Coy, D.H., Culler, M.D. & Melmed, S. (1997a). Somatostatin receptor subtype specificity in human fetal pituitary cultures. Differential role of SSTR2 and SSTR5 for growth hormone, thyroid-stimulating hormone, and prolactin regulation. *J.Clin.Invest.* **99**, 789-798.

121. Shimon, I., Yan, X., Taylor, J.E., Weiss, M.H., Culler, M.D. & Melmed, S. (1997b). Somatostatin receptor (SSTR) subtype-selective analogs differentially suppress in vitro growth

hormone and prolactin in human pituitary adenomas: novel potential therapy for functional pituitary tumors. *J.Clin.Invest.* **100**, 2386-2392.

122. Siehler, S. & Hoyer, D. (1999a). Characterization of human recombinant somatostatin receptors. 2. Modulation of GTP.gamma.S binding. *Naunyn-Schmiedeberg's Arch.Pharmacol.* **360**, 500-509.

123. Siehler, S. & Hoyer, D. (1999b). Characterization of human recombinant somatostatin receptors. 3. Modulation of adenylate cyclase activity. *Naunyn-Schmiedeberg's Arch.Pharmacol.* **360**, 510-521.

124. Siehler, S. & Hoyer, D. (1999c). Characterization of human recombinant somatostatin receptors. 4. Modulation of phospholipase C activity. *Naunyn-Schmiedeberg's Arch.Pharmacol.* **360**, 522-532.

125. Siehler, S., Seuwen, K. & Hoyer, D. (1998a). [125I][Tyr3]octreotide labels human somatostatin sst2 and sst5 receptors. *Eur.J.Pharmacol.* **348**, 311-320.

126. Siehler, S., Seuwen, K. & Hoyer, D. (1998b). [125I]Tyr10-cortistatin14 labels all five somatostatin receptors. *Naunyn-Schmiedeberg's Arch.Pharmacol.* **357**, 483-489.

127. Siehler, S., Seuwen, K. & Hoyer, D. (1999a). Characterization of human recombinant somatostatin receptors. 1. Radioligand binding studies. *Naunyn-Schmiedeberg's Arch.Pharmacol.* **360**, 488-499.

128. Siehler, S., Zupanc, G.K.H., Seuwen, K. & Hoyer, D. (1999b). Characterization of the fish sst3 receptor, a member of the SRIF1 receptor family: atypical pharmacological features. *Neuropharmacology* **38**, 449-462.

129. Spier, A.D. & De Lecea, L. (2000). Cortistatin: a member of the somatostatin neuropeptide family with distinct physiological functions. *Brain Res.Rev.* **33**, 228-241.

130. Stolz, B., Weckbecker, G., Smith-Jones, P.M., Albert, R., Raulf, F. & Bruns, C. (1998). The somatostatin receptor-targeted radiotherapeutic [90Y-DOTA-dPhe1,Tyr3]octreotide (90Y-SMT 487) eradicates experimental rat pancreatic CA 20948 tumors. *Eur.J.Nucl.Med.* **25**, 668-674.

131. Strowski, M.Z., Parmar, R.M., Blake, A.D. & Schaeffer, J.M. (2000). Somatostatin inhibits insulin and glucagon secretion via two receptor subtypes: an in vitro study of pancreatic islets from somatostatin receptor 2 knockout mice. *Endocrinology* **141**, 111-117.

132. Thoss, V.S., Perez, J., Duc, D. & Hoyer, D. (1995). Embryonic and postnatal mRNA distribution of five somatostatin receptor subtypes in the rat brain. *Neuropharmacology* **34**, 1673-1688.

133. Thoss, V.S., Piwko, C. & Hoyer, D. (1996). Somatostatin receptors in the rhesus monkey brain: localization and pharmacological characterization. *Naunyn-Schmiedeberg's Arch.Pharmacol.* **353**, 648-660.

134. Thoss, V.S., Piwko, C., Probst, A. & Hoyer, D. (1997). Autoradiographic analysis of somatostatin SRIF1 and SRIF2 receptors in the human brain and pituitary. *Naunyn-Schmiedeberg's Arch.Pharmacol.* **355**, 168-176.

135. Tilakaratne, N., Christopoulos, G., Zumpe, E.T., Foord, S.M. & Sexton, P.M. (2000). Amylin receptor phenotypes derived from human calcitonin receptor/RAMP coexpression exhibit pharmacological differences dependent on receptor isoform and host cell environment. *J.Pharmacol.Exp.Ther.* **294**, 61-72.

136. Tran, V.T., Beal, M.F. & Martin, J.B. (1985). Two types of somatostatin receptors differentiated by cyclic somatostatin analogs. *Science (Washington, D.C., 1883-)* **228**, 492-495.

137. Vale, W., Rivier, C., Brazeau, P. & Guillemin, R. (1974). Effects of somatostatin on the secretion of thyrotropin and prolactin. *Endocrinology* **95**, 968-977.

138. Vanetti, M., Kouba, M., Wang, X., Vogt, G. & Hoellt, V. (1992). Cloning and expression of a novel mouse somatostatin receptor (SSTR2B). *FEBS Lett.* **311**, 290-294.

139. Warhurst, G., Higgs, N.B., Fakhoury, H., Warhurst, A.C., Garde, J. & Coy, D.H. (1996). Somatostatin receptor subtype 2 mediates somatostatin inhibition of ion secretion in rat distal colon. *Gastroenterology* **111**, 325-333.

140. Weckbecker, G., Briner, U., Lewis, I. & Bruns, C. SOM230: a new somatostatin peptidomimetic with potent inhibitory effects on the growth hormone/insulin-like growth factor-I axis in rats, primates, and dogs. *Endocrinology* **143**, 4123-4130 (2002).

141. Weckbecker, G., Raulf, F., Tolesvai, L. & Bruns, C. (1996). Potentiation of the anti-proliferative effects of anti-cancer drugs by octreotide in vitro and in vivo. *Digestion* **57**, 22-28.

142. Weckbecker, G., Stolz, B., Susini, C. & Bruns, C. (1999), "Antiproliferative somatostatin analogues with potential in oncology". In: *Octreotide, the next decade*. St. Lamberts, ed. BioScientifica Bristol, pp. 339-352.

143. White, J.H., Wise, A., Main, M.J., Green, A., Fraser, N.J., Disney, G.H., Barnes, A.A., Emson, P., Foord, S.M. & Marshall, F.H. (1998). Heterodimerization is required for the formation of a functional GABAB receptor. *Nature* **396**, 679-682.

144. Wilkinson, G.F., Feniuk, W. & Humphrey, P.P. (1997). Characterization of human recombinant somatostatin sst5 receptors mediating activation of phosphoinositide metabolism. *Br.J.Pharmacol.* **121**, 91-96.

145. Xu, Y., Song, J., Bruno, J.F. & Berelowitz, M. (1993). Molecular cloning and sequencing of a human somatostatin receptor, hSSTR4. *Biochem.Biophys.Res.Commun.* **193**, 648-652.

146. Yamada, Y., Kagimoto, S., Kubota, A., Yasuda, K., Masuda, K., Someya, Y., Ihara, Y., Li, Q. & Imura, H. (1993a). Cloning, functional expression and pharmacological characterization of a fourth (hSSTR4) and a fifth (hSSTR5) human somatostatin receptor subtype. *Biochem.Biophys.Res.Commun.* **195**, 844-852.

147. Yamada, Y., Post, S.R., Wang, K., Tager, H.S., Bell, G.I. & Seino, S. (1992a). Cloning and functional characterization of a family of human and mouse somatostatin receptors expressed in brain, gastrointestinal tract, and kidney. *Proc.Natl.Acad.Sci.U.S.A.* **89**, 251-255.

148. Yamada, Y., Reisine, T., Law, S.F., Ihara, Y., Kubota, A., Kagimoto, S., Seino, M., Seino, Y., Bell, G.I. & Seino, S. (1992b). Somatostatin receptors, an expanding gene family: Cloning and functional characterization of human SSTR3, a protein coupled to adenylyl cyclase. *Mol.Endocrinol.* **6**, 2136-2142.

149. Yamada, Y., Stoffel, M., Espinosa, R E. III, Xiang, K.S., Seino, M., Seino, S., Le Beau, M.M. & Bell, G.I. (1993b). Human somatostatin receptor genes: Localization to human chromosomes 14, 17, and 22 and identification of simple tandem repeat polymorphisms. *Genomics* **15**, 449-452.

150. Yang, L., Berk, S.C., Rohrer, S.P. Mosley, R.T., Guo, L., Underwood, D.J., Arison, B.H., Birzin, E.T., Hayes, E.C., Mitra, S.W., Parmar, R.M., Cheng, K., Wu, T.J., Butler, B.S., Foor, F., Pasternak, A., Pan, Y., Silva, M., Freidinger, R.M., Smith, R.G., Chapman, K., Schaeffer, J.M. & Patchett, A.A. (1998). Synthesis and biological activities of potent peptidomimetics selective for somatostatin receptor subtype 2. *Proc.Natl.Acad.Sci.U.S.A.* **95**, 10836-10841.

151. Yasuda, K., Rens-Domiano, S., Breder, C.D., Law, S.F., Saper, C.B., Reisine, T. & Bell, G.I. (1992). Cloning of a novel somatostatin receptor, SSTR3, coupled to adenylyl cyclase. *J.Biol.Chem.* **267**, 20422-20428.

152. Zaki, M., Harrington, L., McCuen, R., Coy, D.H., Arimura, A. & Schubert, M.L. (1996). Somatostatin receptor subtype 2 mediates inhibition of gastrin and histamine secretion from human, dog, and rat antrum. *Gastroenterology* **111**, 919-924.

153. Zheng, H., Bailey, A., Jiang, M.H., Honda, K., Chen, H.Y., Trumbauer, M.E., Van der Ploeg, L.H., Schaeffer, J.M., Leng, G. & Smith, R.G. (1997). Somatostatin receptor subtype 2 knockout mice are refractory to growth hormone-negative feedback on arcuate neurons. *Mol.Endocrinol.* **11**, 1709-1717.

154. Zumpe, E.T., Tilakaratne, N., Fraser, N.J., Christopoulos, G., Foord, S.M. & Sexton, P.M. (2000). Multiple Ramp Domains Are Required for Generation of Amylin Receptor Phenotype from the Calcitonin Receptor Gene Product. *Biochem.Biophys.Res.Commun.* **267**, 368-372.

# 7
# EXPRESSION OF SOMATOSTATIN RECEPTORS IN HUMAN TISSUES IN HEALTH AND DISEASE

Jean Claude Reubi, Beatrice Waser, Jean-Claude Schaer
*Division of Cell Biology and Experimental Cancer Research, Institute of Pathology, University of Berne, Switzerland*

## INTRODUCTION

In the last one and a half decade, we have learned that somatostatin receptors (SSTR) can be overexpressed in selected human tumors. This information has been a trigger to evaluate the potential use of somatostatin (SST) analogs in cancer a) as long-term drugs, b) as diagnostic tools, c) as radiotherapeutic tools. A prerequisite for the success of these clinical applications (see chapter by Krenning et al.) is a strong expression of somatostatin receptors in these tumors. In the present chapter, we would like to summarize basic in vitro information about somatostatin receptor expression in normal human tissues, in non-neoplastic diseases as well as in cancer. As there is much more information available about expression of somatostatin receptors in cancer than in normal tissues, we will discuss cancer first.

## SOMATOSTATIN RECEPTORS IN CANCER

### General
The prevalence of in vitro investigations on human neoplasms rather than on normal tissues is due to the fact that resected fresh tumor material is much more easily available than samples of non-neoplastic diseases or of normal tissue. Moreover, the density of receptors in several cancers turned out to be much higher than in normal tissues, making cancers particularly adequate for detailed investigations. The first evidence for the expression of SSTRs in human primary tumors was presented as early as 1984 in GH-producing pituitary adenomas (1). Since then, receptor autoradiographic studies and binding studies have shown that many other tumor types express SSTRs. First evidence for the existence of distinct SSTR subtypes in tumors was given with autoradiography in 1987 for neuroendocrine tumors (2). After the discovery of the five SSTR subtypes in 1992 (for review, see (3)), mRNA investigations (RT-PCR, Northern blots, in situ hybridization) and immunohistochemical studies were performed in order to identify more precisely the involved SSTR subtypes.

*Receptor binding*

Early studies using in vitro receptor autoradiography have permitted to identify human tumors expressing SSTRs, but without giving at that time detailed information about subtypes. A very high incidence, and often also high density of SSTRs have been found in neuroendocrine tumors, in particular in GH-secreting pituitary adenomas and gastroenteropancreatic tumors. But also other tumors including brain tumors, breast carcinomas, lymphomas, renal cell cancers, mesenchymal tumors, prostatic, ovarian, gastric, hepatocellular and nasopharyngeal carcinomas express SSTRs (4). The main tumors shown to express SSTRs are listed in Table 1, with literature references for further information. Various radioligands were used in binding studies, either based on natural somatostatin, e.g. $^{125}$I-analogs of SST-14 or SST-28 28 as universal radioligands, or based on synthetic, small-sized analogs, such as $^{125}$I-[Tyr$^3$]-octreotide, $^{125}$I-MK-678 or $^{125}$I-RC-160, which label only selected SSTR subtypes. In general, these studies revealed that SSTR expression, was highly variable from one individual to another and from one tumor type to another. While there are tumors which frequently have a high density of receptors, such as meningiomas or medulloblastomas, some others have a much lower density, such as lymphomas; two such examples of SSTR-positive cancers are shown in Figure 1. Moreover, some tumors have a rather homogeneous SSTR distribution, like most neuroendocrine tumors, in particular gastroentero-pancreatic tumors; others, such as breast carcinomas, are characterized by a highly heterogeneous SSTR distribution, with regions of high density next to regions lacking the receptor (4). Receptor homogeneity is of considerable importance with regard to potential targeting of these SSTRs for diagnosis or therapy.

An important problem of in vitro analysis of SSTRs in tumor samples has been that of false positive results due to contamination of the tumor samples with non-neoplastic tissues, themselves expressing frequently SSTRs (see below). For instance, most of the colorectal cancers have only minimal expression of octreotide binding sites while the vessels located around the tumors (peritumoral vessels) have a high density of those sites. Such an example is shown in Figure 1. Therefore, analysis of SSTRs in tumors without a morphological control of their precise location (tissue homogenates) may be misleading. In glial tumors, even morphological methods, such as receptor autoradiography, may not be able to sufficiently distinguish between SSTRs in tumor cells and in remaining nerve fibers (5, 6). These glial tumors are known, indeed, to heavily infiltrate the normal brain, which itself strongly expresses SSTRs. The high proportion of glial tumor samples "contaminated" by SSTR-positive nervous tissues makes it difficult to assess precisely the degree of SSTR expression by these tumors, although both Dutour et al. (7) and Held-Feindt et al. (8) could identify glial tumor cells expressing sst$_2$ with high resolution techniques

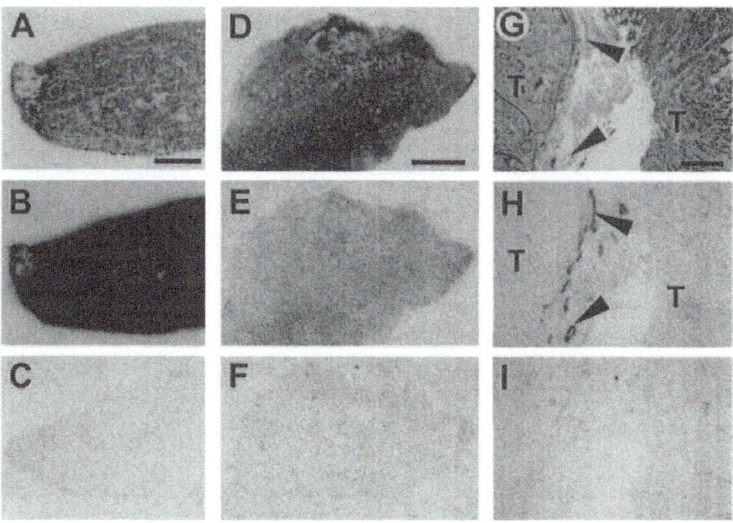

*Figure 1.* Receptor autoradiography showing (left) a medulloblastoma with very high density of receptors, (middle) a non-Hodgkin lymphoma with low receptor density and (right) a colorectal cancer without SSTRs in the tumor (T) but with SSTRs in peritumoral vessels (arrowhead).
A, D, G: Hematoxylin-eosin stained sections; Bars = 1 mm. B, E, H: Autoradiograms showing total binding of $^{125}$I-[Tyr$^3$]-octreotide; C, F, I: Autoradiograms showing nonspecific binding (in presence of $10^{-6}$ M SST-28).

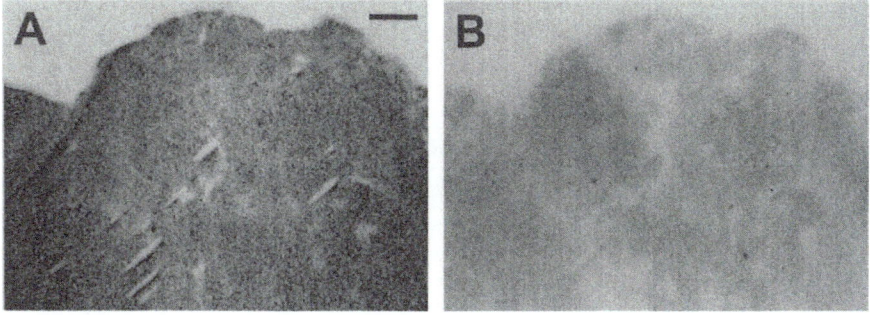

*Figure 2.* sst$_1$ mRNA in a leiomyoblastoma.
A: Hematoxylin-eosin stained section; Bar = 1 mm. B: Autoradiogram showing sst$_1$ mRNA in the tumor.

## Receptor mRNA

Receptor binding data were confirmed and extended by studies measuring SSTR mRNA. Human tumors often express multiple SSTR subtype mRNAs, as reported first in pituitary adenomas (9-13) and gastroenteropancreatic tumors (13-15). In the past few years, a profusion of papers have appeared identifying mRNA for the various SSTR subtypes in a large variety of other human cancers: In Figure 2, an example of a $sst_1$-expressing mesenchymal tumor with abundant $sst_1$ mRNA is shown. In Table 1, a non-exhaustive list of references of studies measuring sst mRNA is given for various cancer types. In many of the studies using RT-PCR methodologies, the incidence of the various receptor mRNAs in tumors appears to be higher than predicted by the receptor binding studies; moreover, mRNAs for most SSTR subtypes appear to be frequently expressed concomitantly in individual tumors. It is presently not completely clear whether 1) these RT-PCR data reflect an overestimation of the real contribution of the various sst mRNAs due to the outstanding sensitivity of the method, 2) whether these mRNAs originate in part from non-tumoral adjacent

*Table 1.* Human tumors expressing SSTRs: Overview of the literature on receptor binding, receptor mRNA and receptor immunohistochemistry. Numbers in parenthesis correspond to the cited literature references. Ca, carcinoma.

| Tumor type | Receptor binding | Receptor mRNA | Receptor immunohistochemistry |
|---|---|---|---|
| Pituitary adenomas | (1, 32, 71, 72) | (9-13, 73) | - |
| Gastroenteropancreatic tumors | (2, 33, 74) | (13-15, 75, 76) | (22-26) |
| Medullary thyroid Ca | (77) | (75, 78, 79) | (25) |
| Pheochromocytomas/ Paragangliomas | (32, 50, 80) | (50) | (19, 24, 25) |
| Neuroblastomas | (69, 81) | (75, 82, 83) | (19, 83) |
| Medulloblastomas | (6, 84) | (6, 84) | (6, 19) |
| Meningiomas | (85) | (7, 75) | (7, 20) |
| Gliomas | (5, 6) | (6, 7, 86) | (6-8) |
| Pancreatic Ca | (16, 18) | (16, 17) | (87) |
| Small cell lung Ca | (88) | (75) | (19, 25) |
| Breast Ca | (32, 89, 90) | (13, 76, 91) | (21, 92) |
| Prostate Ca | (32, 93-95) | (94-96) | - |
| Ovarian Ca | (97, 98) | (75, 98) | - |
| Renal cell Ca | (99) | (76) | - |
| Lymphomas | (100) | (75, 101) | - |
| Sarcomas | (102) | (102) | - |
| Gastric Ca | (103, 104) | - | - |
| Hepatocellular Ca | (105, 106) | - | - |
| Colorectal Ca | (34, 107, 108) | (17, 51, 109) | - |
| Nasopharyngeal Ca | (110) | - | - |

tissues (see below) or 3) whether these mRNAs may not always be translated into significant amounts of the respective receptor subtype proteins. Thus, one may caution against an overestimation of mRNA data using ultra-sensitive methods such as RT-PCR. But there are examples that the sst mRNAs detected in tumors indeed do not always reflect the amount of receptor protein measurable with binding techniques. This is the case for instance for pancreatic carcinomas, which can express $sst_2$ mRNA without a significant amount of $sst_2$ protein (16-18). Obviously, rather than the mRNA, it is the receptor protein located on the cell membrane that is the main target for the current clinical applications of somatostatin ligands.

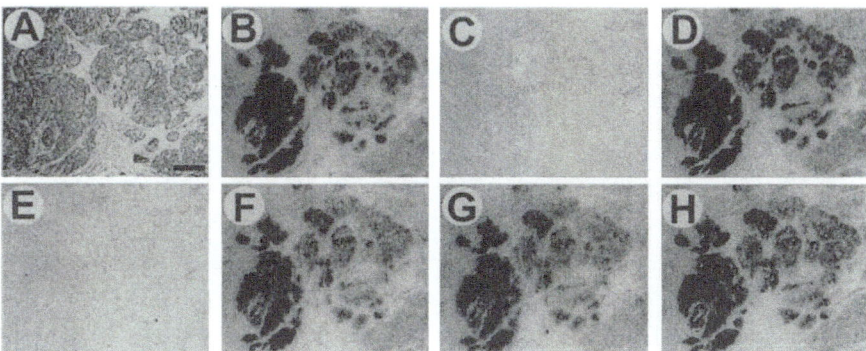

*Figure 3*. Representative example of a breast carcinoma expressing predominantly $sst_2$. Receptor autoradiography with $^{125}I$-[LTT]-SST28 as radioligand and displacement with five somatostatin analogs selective for each of the five ssts. A: Hematoxylin-eosin stained section; Bar = 1 mm. B: Autoradiogram showing total binding of $^{125}I$-[LTT]-SST28. Note the strong but partly non-homogeneous labeling of tumor tissue. C: Nonspecific binding (in presence of 100 nM SST28). D-H: Binding of $^{125}I$-[LTT]-SST28 in presence of an $sst_1$(D)-, $sst_2$(E)-, $sst_3$(F)-, $sst_4$(G)- and $sst_5$(H)-selective analog. The $sst_2$ analog, but not the others, displaces completely the binding in the tumor tissue

*Receptor immunohistochemistry*

An emerging new technique to detect SSTR subtypes is the immunohistochemical detection, which has the advantage of a high cellular resolution. The results are, however, highly dependent on the quality and specificity of the developed antibodies. Several antibodies are currently available. Strong selectivity and specificity criteria are, however, necessary to assess the immunohistochemical staining of receptors in human tissues. Up to now, carefully controlled immunohistochemical studies have been performed in cancer primarily with $sst_2$ antibodies. A high density of $sst_2$ was found in neuroblastomas, medulloblastomas, paragangliomas and small cell lung cancers using R2-88 (19) as well as in meningiomas (20) and breast cancers (21) using Schulz's antibodies. The majority of gastroenteropancreatic tumors were shown to have $sst_2$ immunohistochemically (22-26). In most of these tumors, $sst_2$ was shown to have a preferential membrane-bound location. An example of an $sst_{2A}$-expressing carcinoid is shown in Figure 4A. There have been few reports investigating other sst subtypes with immunohistochemistry,

112

such as $sst_1$ in a few gastroenteropancreatic tumors (24) and $sst_3$ in breast cancers (21). The observation of an intracellular location of some of these receptors (24) is intriguing and still not fully understood. An intracellular location may explain why they are only rarely detected with receptor autoradiography. Table 1 summarizes the main immunohistochemical studies of SSTR subtypes performed in cancer tissues.

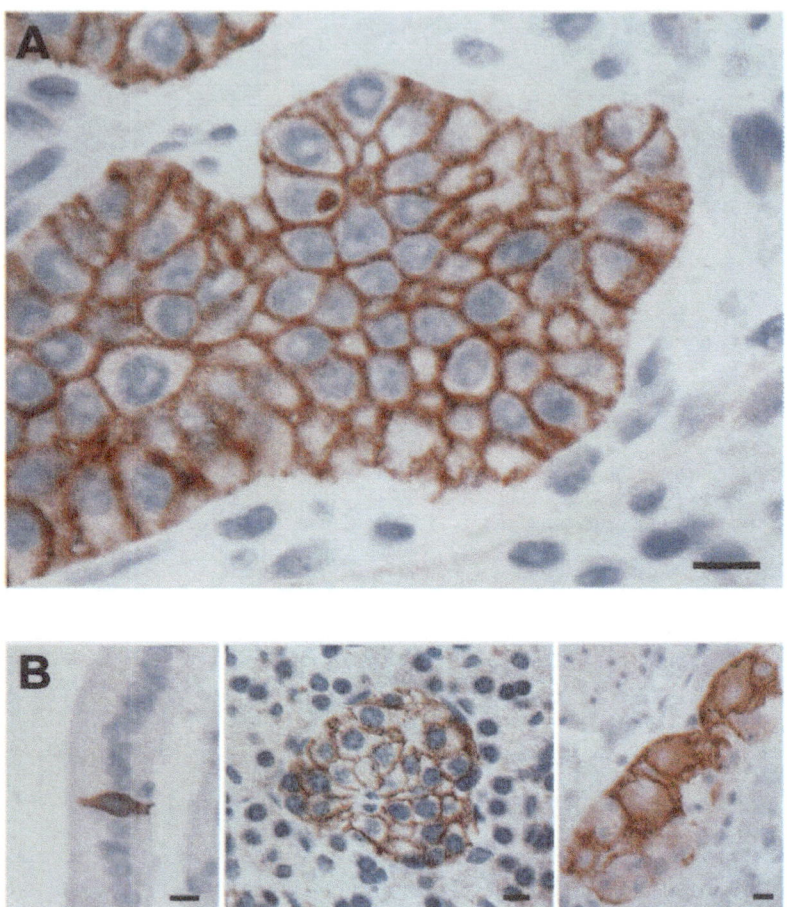

*Figure 4.* A: Immunohistochemical detection of $sst_{2A}$ in a human gut carcinoid. Note the strong membrane-bound staining in all tumor cells with R2-88 antibody. Bar = 0.01 mm.
B: Immunohistochemical detection of $sst_{2A}$ in normal human tissues: left: endocrine cell in the gastrointestinal mucosa; middle: pancreatic islet; right: nerve cells of the plexus myentericus. Bars = 0.01 mm.

Analysis of sst by immunohistochemistry may become an additional, useful parameter for the clinician, to evaluate a tumor's biology and choose for therapeutic options, in particular since standard formalin-fixed material is sufficient for this type of investigation (22).

*Binding studies with subtype-selective analogs*
The recent development of SSTR subtype-selective analogs, both as peptides and non-peptides (27-29), represents an important advance permitting hopefully to 1) evaluate the distribution of various receptor subtype proteins in tissue, 2) determine the specific biological effects mediated by the various subtypes, 3) design new drugs for specific therapeutic strategies. Some of these analogs were already used to refine SSTR binding studies for the detection of receptor subtypes in tissues (30, 31). Moreover, in a study using receptor autoradiography with five different subtype-selective analogs, we evaluated SSTR subtypes expressed in cancers (32); these data suggest that in many SSTR-positive tumors there is a predominance of the proteins of one or two ssts. A preponderance of $sst_2$ binding sites is seen in the majority of neuroblastomas, medulloblastomas, breast cancers, meningiomas, paragangliomas, renal cell carcinomas, lymphomas, hepatocellular carcinomas and small cell lung cancers. An example of a $sst_2$-expressing breast cancer identified with this method is shown in Figure 3. Conversely, $sst_1$ is frequent in sarcomas and prostate cancers, while $sst_3$ is frequent in inactive pituitary adenomas (29, 32). A large subtype variability is seen, among others, in growth hormone- producing pituitary adenomas (specially $sst_2$ + $sst_5$), pheochromocytomas, hormone-producing gastroenteropancreatic tumors and gastric cancers. A more recent study (33) documents the wide and differential expression of $sst_1$-$sst_5$ receptors in gut and lung neuroendocrine pancreatic tumors. Interestingly, $sst_4$ is not expressed predominantly in the human cancers tested.

# SOMATOSTATIN RECEPTORS IN OTHER DISEASES

SSTRs have been shown to be expressed in vessels, either in peritumoral vessels (34, 35) or in vessels located in chronically inflamed diseases such as inflammatory bowel disease or arthritis (36, 37). In receptor autoradiography experiments, these vessels can be preferentially labeled by $^{125}$I-[Tyr$^3$]-octreotide, suggesting the presence of $sst_2$ and/or $sst_5$. Pharmacological evidence has clearly indicated that human vascular receptors can express the $sst_2$ subtype (38). Immunohistochemical data reporting vascular location of sst (39, 40) have to be interpreted with caution since some sst antibodies may crossreact with myosin in the smooth muscles of the vessel wall (41). The potential functional and therapeutical roles of SSTRs in human vessels are discussed in detail in the chapter by Hajri et al.. It should further be mentioned that granulomas of various origins express SSTRs, primarily located in the epitheloid cells (42). Finally, activated lymphocytes, for instance within reactive lymph nodes, also express SSTRs.

# SOMATOSTATIN RECEPTORS IN NORMAL TISSUES

*General*

Available information on the distribution of SSTRs and their subtypes in normal human tissues is still incomplete, mainly due to the limited access to normal human material. Unfortunately, it is not possible to simply extrapolate animal data to humans, as there are numerous indications of species variability in SSTR expression (43, 44). Human SSTRs were detected as early as 1986 in the human brain using receptor autoradiography (45) or homogenate binding assays (46). Shortly after, pharmacological evidence for two distinct SSTR subtypes in the human brain was given by comparing $^{125}$I-[Tyr$^3$]-octreotide and $^{125}$I-[LTT]-SST28 (47). High density of SSTRs was found, among others, in the human cortex, hippocampus, amygdala, striatum (48) (see chapter by Epelbaum for details). While the majority of SSTR distribution concentrated on the central nervous system, only limited binding studies were performed in another main somatostatin target tissue, the human gastrointestinal tract. Nevertheless, in the course of receptor autoradiographic studies of human gastrointestinal tumors, the presence of SSTRs was also reported in the human gastrointestinal mucosa, in the myenteric plexus and in submucosal vessels (34). Other SSTR-expressing human tissues include the human kidney (49) and the adrenal (50).

*Subtype distribution*

In both the human CNS and peripheral tissues, a complex pattern of sst subtype expression has been observed, with often overlapping coexpression of multiple subtypes in a tissue-specific pattern. Moreover, a great individual variability in subtype expression and in receptor density has been observed in normal human organs (51). However, by far not all human tissues have been evaluated for subtype expression. Moreover, the great majority of studies have measured receptor mRNAs rather than receptor proteins. As already mentioned above, mRNAs may not necessarily reflect the presence of the mature receptor proteins, as mRNAs may not always be processed to proteins (16). Moreover, most mRNA measurements are so sensitive (RT-PCR; RNase protection assay) that they may falsely identify a receptor-negative organ or tissue (homogenate) as receptor-positive, as they may detect mRNA originating from SSTR-positive systems present ubiquitously throughout the body (immune, vascular, nervous systems), rather than originating from the epithelial cell elements composing the main part of the organ under investigation.

The mRNAs for all ssts have been identified in human brain (48, 52), most parts of the gastrointestinal tract (14, 51, 53, 54) as well as the endocrine system, in particular pituitary and adrenal (12, 50, 52). Four ssts were detected in the adult pituitary, whereas all subtypes were found in the fetal gland (12). In other organs, selected subtypes have been identified: the human placenta

(55) as well as the fetal and adult lung (56, 57) have a predominance of $sst_4$, the human heart has $sst_5$ (58), the kidney has predominantly $sst_2$ mRNA (59).

SSTR subtype-specific antibodies, for immunohistochemical detection of the receptor proteins, have recently been developed and allow to start clarifying the sst distribution (41, 60) in normal tissues. Immunohistochemical studies show $sst_1$-$sst_5$ receptors in the normal human pancreatic islets (61). While $sst_{2A}$ is found abundantly in the islets, the human acini are lacking $sst_{2A}$ (44), in contrast to the rat where the acini have abundant $sst_{2A}$ (62). $sst_{2A}$ and $sst_4$ have recently been detected by immunohistochemistry in specific regions of the human brain (60, 63). $sst_{2A}$ is also present in the peripheral nervous system (plexus myentericus and submucosus ) (41). It is found in the immune system, i.e. in the germinal centers of lymphoid follicles and in human peripheral blood lymphocytes (41, 64). While all five subtypes are found in mononuclear leucocytes (65), sst3 and sst5 were identified in T-lymphocytes (29, 66). Figure 4B is an illustration of $sst_{2A}$ immunostaining in three different types of human tissues, namely an endocrine cell in the gastrointestinal tract, pancreatic islet cells and nerve cells of the plexus myentericus.

## CONCLUSIONS

The knowledge of SSTR expression in tumors and normal tissues is a prerequisite for clinical investigations with somatostatin analogs. Indeed, the high expression of SSTRs in tumors represents the molecular basis for novel diagnostical and radiotherapeutical applications of somatostatin analogs (67, 68). Moreover, the presence of SSTRs may, in some instances, such as neuroblastomas, represent a good prognostic factor for patient survival (69, 70), while the absence of receptors is related to early death. Furthermore, the absence of SSTRs, such as sst2 in pancreatic cancers, may represent a significant growth advantage to these tumors (17). One of the main challenges for the future will be to unravel the various physiological and pathophysiological roles of somatostatin mediated by SSTR subtypes, and identify potential therapeutical indications for the newly developed subtype-selective somatostatin analogs (27-29).

## REFERENCES

1.      Reubi JC, Landolt AM. High density of somatostatin receptors in pituitary tumors from acromegalic patients. J. Clin. Endocrinol. Metab. 1984;59:1148-1151.
2.      Reubi JC, Häcki WH, Lamberts SWJ. Hormone-producing gastrointestinal tumors contain a high density of somatostatin receptors. J Clin Endocrinol Metab 1987;65:1127-1134.
3.      Patel Y. Somatostatin and its receptor family. Front. Neuroendocrinol. 1999;20:157-198.
4.      Reubi JC, Krenning E, Lamberts SWJ, Kvols L. In vitro detection of somatostatin receptors in human tumors. Metabolism 1992;41 Suppl.2:104-110.

5.     Reubi JC, Lang W, Maurer R, Koper JW, Lamberts SWJ. Distribution and biochemical characterization of somatostatin receptors in tumors of the human central nervous system. Cancer Res. 1987;47:5758-5764.

6.     Cervera P, Videau C, Viollet C, Pettrucci C, Lacombe J, Winsky-Sommeren R, et al. Comparison of somatostatin receptor expression in human gliomas and medulloblastomas. J Neuroendocrinol. 2002;14(6):458-471.

7.     Dutour A, Kumar U, Panetta R, Ouafik LH, Fina F, Sasi R. Expression of somatostatin receptor subtypes in human brain tumors. Int. J. Cancer 1998;76:620-627.

8.     Held-Feindt J, Krisch B, Mentlein R. Molecular analysis of the somatostatin receptor subtype 2 in human glioma cells. Brain Res Mol Brain Res 1999;64(1):101-7.

9.     Miller GM, Alexander JM, Bikkal HA, Katznelson L, Zervas NT, Klibanski A. Somatostatin receptor subtype gene expression in pituitary adenomas. J. Clin. Endocrinol. Metab. 1995;80:1386-1392.

10.    Greenman Y, Melmed S. Expression of three somatostatin receptor subtypes in pituitary adenomas: Evidence for preferential SSTR5 expression in the mammosomatotroph lineage. J. Clin. Endocrinol. Metab. 1994;79:724-729.

11.    Greenman Y, Melmed S. Heterogeneous expression of two somatostatin receptor subtypes in pituitary tumors. J. Clin. Endocrinol. Metab. 1994;78:398-403.

12.    Panetta R, Patel YC. Expression of mRNA for all five human somatostatin receptors (hSSTR1-5) in pituitary tumors. Life Sci. 1995;56:333-342.

13.    Schaer JC, Waser B, Mengod G, Reubi JC. Somatostatin receptor subtypes sst1, sst2, sst3, and sst5 expression in human pituitary, gastroenteropancreatic and mammary tumors: Comparison of mRNA analysis with receptor autoradiography. Int. J. Cancer 1997;70:530-537.

14.    Jaïs P, Terris B, Ruszniewski P, LeRomancer M, Reyl-Desmars F, Vissuzaine C, et al. Somatostatin receptor subtype gene expression in human endocrine gastroentero-pancreatic tumours. Eur. J. Clin. Invest. 1997;27:639-644.

15.    Wulbrand U, Wied M, Zöfel P, Göke B, Arnold R, Fehmann H-C. Growth factor receptor expression in human gastroenteropancreatic neuroendocrine tumours. Eur. J. clin. Invest. 1998;28:1038-1049.

16.    Fisher WE, Doran TA, Muscarella II P, Boros LG, Ellison EC, Schirmer WJ. Expression of somatostatin receptor subtype 1-5 genes in human pancreatic cancer. J. Natl. Cancer Inst. 1998;90:322-324.

17.    Buscail L, Saint-Laurent N, Chastre E, Vaillant JC, Gespach C, Capella G, et al. Loss of sst2 somatostatin receptor gene expression in human pancreatic and colorectal cancer. Cancer Res. 1996;56:1823-1827.

18.    Reubi JC, Horisberger U, Essed CE, Jeekel J, Klijn JGH, Lamberts SWJ. Absence of somatostatin receptors in human exocrine pancreatic adenocarcinomas. Gastroenterology 1988;95:760-763.

19.    Reubi JC, Waser B, Liu Q, Laissue JA, Schonbrunn A. Subcellular distribution of somatostatin sst2A receptors in human tumors of the nervous and neuroendocrine systems: membranous versus intracellular location. J. Clin. Endocrinol. Metab. 2000;85:3882-3891.

20.    Schulz S, Pauli SU, Schulz S, Handel M, Dietzmann K, Firsching R, et al. Immunohistochemical determination of five somatostatin receptors in meningioma reveals frequent overexpression of somatostatin receptor subtype sst2A. Clin Cancer Res 2000;6(5):1865-74.

21.    Schulz S, Schulz S, Schmitt J, Wiborny D, Schmidt H, Olbricht S, et al. Immunocytochemical detection of somatostatin receptors sst1, sst2A, sst2B and sst3 in paraffin-embedded breast cancer tissue using subtype-specific antibodies. Clin. Cancer Res. 1998;4:2047-2052.

22.    Reubi JC, Kappeler A, Waser B, Laissue JA, Hipkin RW, Schonbrunn A. Immunohistochemical localization of somatostatin receptors sst2A in human tumors. Am. J. Pathol. 1998;153:233-245.

23.    Janson ET, Stridsberg M, Gobl A, Weslin J-E, Oeberg K. Determination of somatostatin receptor subtype 2 in carcinoid tumors by immunohistochemical

investigation with somatostatin receptor subtype 2 antibodies. Cancer Res. 1998;58:2375-2378.

24.  Hofland LJ, Liu Q, Van Koetsveld PM, Zuijderwijk J, Van der Ham F, De Krijger RR, et al. Immunohistochemical detection of somatostatin receptor subtypes sst1 and sst2A in human somatostatin receptor positive tumors. J. Clin. Endocrinol. Metab. 1999;84:775-780.

25.  Kimura N, Pilichowska M, Date F, Kimura I, Schindler M. Immunohistochemical expression of somatostatin type 2A receptor in neuroendocrine tumors. Clin. Cancer Res. 1999;5:3483-3487.

26.  Kulaksiz H, Eissele R, Rossler D, Schulz S, Hollt V, Cetin Y, et al. Identification of somatostatin receptor subtypes 1, 2A, 3, and 5 in neuroendocrine tumours with subtype specific antibodies. Gut 2002;50(1):52-60.

27.  Rohrer SP, Birzin ET, Mosley RT, Berk SC, Hutchins SM, Shen D, et al. Rapid identification of subtype-selective agonists of the somatostatin receptor through combinatorial chemistry. Science 1998;282:737-740.

28.  Liapakis G, Hoeger C, Rivier J, Reisine T. Development of a selective agonist at the somatostatin receptor subtype SSTR1. J. Pharmacol. Exp. Ther. 1996;276:1089-1094.

29.  Reubi JC, Schaer JC, Wenger S, Hoeger C, Erchegyi J, Waser B, et al. sst3-selective potent peptidic somatostatin receptor antagonists. Proc. Natl. Acad. Sci. USA 2000;97(25):13973-13978.

30.  Leroux P, Bucharles C, Bologna E, Vaudry H. des-AA-1,2,5[D-Trp8, IAmp9]somatostatin-14 allows the identification of native rat somatostatin sst1 receptor subtype. Eur J Pharmacol 1997;337(2-3):333-6.

31.  Reubi JC, Schaer JC, Waser B, Hoeger C, Rivier J. A selective analog for the somatostatin receptor subtype sst1 expressed by human tumors. Eur. J. Pharmacol. 1998;345:103-110.

32.  Reubi JC, Waser B, Schaer JC, Laissue JA. Somatostatin receptor sst1-sst5 expression in normal and neoplastic human tissues using receptor autoradiography with subtype-selective ligands. Eur. J. Nucl. Med. 2001;28:836-846.

33.  Reubi JC, Waser B. Concomitant expression of several peptide receptors in neuroendocrine tumors as molecular basis for in vivo multireceptor tumor targeting. Eur J Nucl Med 2003;30(5):781-793.

34.  Reubi JC, Mazzucchelli L, Hennig I, Laissue J. Local upregulation of neuropeptide receptors in host blood vessels around human colorectal cancers. Gastroenterology 1996;110:1719-1726.

35.  Denzler B, Reubi JC. Expression of somatostatin receptors in peritumoral veins of human tumors. Cancer 1999;85:188-198.

36.  Reubi JC, Mazzucchelli L, Laissue J. Intestinal vessels express a high density of somatostatin receptors in human inflammatory bowel disease. Gastroenterology 1994;106:951-959.

37.  Reubi JC, Waser B, Krenning EP, Markusse HM, Vanhagen M, Laissue JA. Vascular somatostatin receptors in synovium from patients with rheumatoid arthritis. Eur. J. Pharmacol. 1994;271:371-378.

38.  Dimech J, Feniuk W, Latimer RD, Humphrey PPA. Somatostatin-induced contraction of human isolated saphenous vein involves sst2 receptor-mediated activation of L-type calcium channels. J. Cardiovasc. Pharmacol. 1995;26:721-728.

39.  ten Bokum AM, Melief MJ, Schonbrunn A, van der Ham F, Lindeman J, Hofland LJ, et al. Immunohistochemical localization of somatostatin receptor sst2A in human rheumatoid synovium. J Rheumatol 1999;26(3):532-5.

40.  Watson JC, Balster DA, Gebhardt BM, O'Dorisio TM, O'Dorisio MS, Espenan GD, et al. Growing vascular endothelial cells express somatostatin subtype 2 receptors. Br J Cancer 2001;85(2):266-72.

41.  Reubi JC, Laissue JA, Waser B, Steffen DL, Hipkin RW, Schonbrunn A. Immunohistochemical detection of somatostatin sst2a receptors in the lymphatic, smooth muscular, and peripheral nervous systems of the human gastrointestinal tract: Facts and artifacts. J. Clin. Endocrinol. Metab. 1999;84:2942-2950.

118

42.     van Hagen PM, Krenning EP, Reubi JC, Kwekkeboom D, Oei HY, Mulder AH, et al. Somatostatin analogue scintigraphy in granulomatous diseases. Eur. J. Nucl. Med. 1994;21:497-502.

43.     Maurer R, Reubi JC. Somatostatin receptors in the adrenal. Mol. Cell. Endocrinol. 1986;45:81-90.

44.     Reubi JC, Kappeler A, Waser B, Schonbrunn A, Laissue JA. Immunohistochemical localization of somatostatin receptor sst2A in human pancreatic islets. J. Clin. Endocrinol. Metab. 1998;83:3746-3749.

45.     Reubi JC, Cortès R, Maurer R, Probst A, Palacios JM. Distribution of somatostatin receptors in the human brain: an autoradiographic study. Neuroscience 1986;18:329-346.

46.     Beal MF, Tran VT, Mazurek MF, Chattha G, Martin JB. Somatostatin binding sites in human and monkey brain: localization and characterization. J. Neurochem. 1986;46:359-365.

47.     Reubi JC, Probst A, Cortès R, Palacios JM. Distinct topographical localisation of two somatostatin receptor subpopulations in the human cortex. Brain Res. 1987;406:391-396.

48.     Csaba Z, Dournaud P. Cellular biology of somatostatin receptors. Neuropeptides 2001;35(1):1-23.

49.     Reubi JC, Horisberger U, Studer UE, Waser B, Laissue JA. Human kidney as target for somatostatin: High affinity receptors in tubules and vasa recta. J. Clin. Endocrinol. Metab. 1993;77:1323-1328.

50.     Epelbaum J, Bertherat J, Prévost G, Kordon C, Meyerhof W, Wulfsen I, et al. Molecular and pharmacological characterization of somatostatin receptor subtypes in adrenal, extraadrenal, and malignant pheochromocytomas. J. Clin. Endocrinol. Metab. 1995;80:1837-1844.

51.     Laws S, Gough AC, Evans AA, Bains MA, Primerose JN. Somatostatin receptors subtype mRNA expression in human colorectal cancer and normal colic mucosa. Br. J. Cancer 1997;75:360-366.

52.     Thoss VS, Pérez J, Probst A, Hoyer D. Expression of five somatostatin receptor mRNAs in the human brain and pituitary. Naunyn-Schmiedeberg's Arch. Pharmacol. 1996;354:411-419.

53.     Le Romancer M, Cherifi Y, levasseur S, Laigneau J-P, Peranzi G, Jaïs P, et al. Messenger RNA expression of somatostatin receptor subtypes in human and rat gastric mucosae. Life Sci. 1996;58:1091-1098.

54.     Vuaroqueaux V, Dutour A, Briard N, Monges G, Grino M, Oliver C, et al. No loss of sst receptors gene expression in advanced stages of colorectal cancer. Eur. J. Endocrinol. 1999;140:362-366.

55.     Caron P, Buscail L, Beckers A, Esteve J-P, Igout A, Susini C. Expression of somatostatin receptor sst4 in human placenta and absence of octreotide effect on human placental growth hormone concentration during pregnancy. J. Clin. Endocrinol. Metab. 1997;82:3771-3776.

56.     Bruns C, Raulf F, Hoyer D, Schloos J, Lübbert H, Weckbecker G. Binding properties of somatostatin receptor subtypes. Metabolism 1996;45, Suppl. 1:17-20.

57.     Rohrer L, Raulf F, Bruns C, Buettner R, Hofstaedter F, Schüle R. Cloning and characterization of a fourth human somatostatin receptor. Proc. Natl. Acad. Sci. USA 1993;90:4196-4200.

58.     OCarroll A, Raynor K, Lolait SJ, Reisine T. Characterization of cloned human somatostatin receptor SSTR5. Molec. Pharmacol. 1994;46:291-298.

59.     Yamada Y, Post SR, Wang K, Tager H, Bell G, Seino S. Cloning and functional characterization of a family of human and mouse somatostatin receptors expressed in brain, gastro-intestinal tract and kidney. Proc. Natl. Acad. Sci. USA 1992;89:251-255.

60.     Schindler M, Holloway S, Humphrey PPA, Waldvogel H, Faull RLM, Berger W, et al. Localization of the somatostatin sst2(a) receptor in human cerebral cortex, hippocampus and cerebellum. Neuroreport 1998;9:521-525.

61.     Kumar U, Sasi R, Suresh S, Patel A, Thangaraju M, Metrakos P, et al. Subtype-selective expression of the five somatostatin receptors (hSSTR1-5) in human

pancreatic islet cells: A quantitative double-label immunohistochemical analysis. Diabetes 1999;48:77-85.

62.    Hunyady B, Hipkin RW, Schonbrunn A, Mezey E. Immunohistochemical localization of somatostatin receptor sst2A in the rat pancreas. Endocrinology 1997;138:2632-2635.

63.    Selmer I, Schindler M, Humphrey PP, Waldvogel HJ, Faull RL, Emson PC. First localisation of somatostatin sst(4) receptor protein in selected human brain areas: an immunohistochemical study. Brain Res Mol Brain Res 2000;82(1-2):114-25.

64.    Tsutsumi A, Takano H, Ischikawa K, Kobayashi S, Koike T. Expression of somatostatin receptor subtype 2 mRNA in human lymphoid cells. Cell. Immunol. 1997;181:44-49.

65.    Ishihara S, Hassan S, Kinoshita Y, Moriyama N, Fukuda R, Maekawa T, et al. Growth inhibitory effects of somatostatin on human leukemia cell lines mediated by somatostatin receptor subtype1. Peptides 1999;20:313-318.

66.    El Ghamrawy C, Rabourdin-Combe C, Krantic S. sst5 somatostatin receptor mRNA induction by mitogenic activation of human T-lymphocytes. Peptides 1999;20:305-311.

67.    Kwekkeboom DJ, Krenning EP, de Jong M. Peptide receptor imaging and therapy. J. Nucl. Med. 2000;41:1704-1713.

68.    Heppeler A, Froidevaux S, Eberle AN, Maecke HR. Receptor targeting for tumor localisation and therapy with radiopeptides. Curr. Med. Chem. 2000;7:971-994.

69.    Moertel CL, Reubi JC, Scheithauer BS, Schaid DJ, Kvols LK. Expression of somatostatin receptors in childhood neuroblastoma. Amer. J. Clin. Path. 1994;102:752-756.

70.    Schilling FH, Bihl H, Jacobsson H, Ambros PF, Martinsson T, Borgstrom P, et al. Combined [111]In-pentetreotide scintigraphy and [123]I-MIBG scintigraphy in neuroblastoma provides prognostic information. Med Pediatr Oncol 2000;35(6):688-91.

71.    Moyse E, Le Dafniet M, Epelbaum J, Pagesy P, Peillon F, Kordon C, et al. Somatostatin receptors in human growth hormone and prolactin-secreting pituitary adenomas. J. Clin. Endocrinol. Metab. 1985;61:98-103.

72.    Reubi JC, Heitz PU, Landolt AM. Visualization of somatostatin receptors and correlation with immunoreactive growth hormone and prolactin in human pituitary adenomas: Evidence for different tumor subclasses. J. Clin. Endocrinol. Metab. 1987;65:65-73.

73.    Jaquet P, Saveanu A, Gunz G, Fina F, Zamora AJ, Grino M, et al. Human somatostatin receptor subtypes in acromegaly: distinct patterns of messenger ribonucleic acid expression and hormone suppression identify different tumoral phenotypes. J. Clin. Endocrinol. Metab. 2000;85:781-792.

74.    Reubi JC, Kvols LK, Waser B, Nagorney D, Heitz PU, Charboneau JW, et al. Detection of somatostatin receptors in surgical and percutaneous needle biopsy samples of carcinoids and islet cell carcinomas. Cancer Res. 1990;50:5969-5977.

75.    Reubi JC, Schaer JC, Waser B, Mengod G. Expression and localization of somatostatin receptor SSTR1, SSTR2 and SSTR3 mRNAs in primary human tumors using in situ hybridization. Cancer Res. 1994;54:3455-3459.

76.    Vikic-Topic S, Raisch KP, Kvols LK, Vuk-Pavlovic S. Expression of somatostatin receptor subtypes in breast carcinoma, carcinoid tumor, and renal cell carcinoma. J. Clin. Endocrinol. Metab. 1995;80:2974-2979.

77.    Reubi JC, Chayvialle JA, Franc B, Cohen R, Calmettes C, Modigliani E. Somatostatin receptors and somatostatin content in medullary thyroid carcinomas. Lab. Invest. 1991;64:567-573.

78.    Mato E, Matias-Guiu X, Chico A, Webb SM, Cabezas R, Berna L, et al. Somatostatin and somatostatin receptor subtype gene expression in medullary thyroid carcinoma. J. Clin. Endocrinol. Metab. 1998;83:2417-2420.

79.    Forssell-Aronsson EB, Nilsson O, Bejegard SA, Kolby L, Bernhardt P, Molne J, et al. 111In-DTPA-D-Phe1-octreotide binding and somatostatin receptor subtypes in thyroid tumors. J Nucl Med 2000;41(4):636-42.

80. Reubi JC, Waser B, Khosla S, Kvols L, Goellner JR, Krenning E, et al. In vitro and in vivo detection of somatostatin receptors in pheochromocytomas and paragangliomas. J. Clin. Endocrinol. Metab. 1992;74:1082-1089.

81. Prevost G, Veber N, Viollet C, Roubert V, Roubert P, Bénard J, et al. Somatostatin-14 mainly binds the somatostatin receptor subtype 2 in human neuroblastoma tumors. Neuroendocrinology 1996;63:188-197.

82. Sestini R, Orlando C, Peri A, Tricarico C, Pazzagli M, Serio M, et al. Quantitation of somatostatin receptor type 2 gene expression in neuroblastoma cell lines and primary tumors using competitive reverse transcription-polymerase chain reaction. Clin. Cancer Res. 1996;2:1757-1765.

83. Albers AR, ODorisio MS, Balster DA, Caprara M, Gosh P, Chen F, et al. Somatostatin receptor gene expression in neuroblastoma. Regul. Peptides 2000;88:61-73.

84. Frühwald MC, O'Dorisio MS, Pietsch T, Reubi JC. High expression of somatostatin receptor subtype 2 (sst2) in medulloblastoma: Implications for diagnosis and therapy. Pediatr. Res. 1999;45:697-708.

85. Reubi JC, Maurer R, Klijn JGM, Stefanko SZ, Foekens JA, Blaauw G, et al. High incidence of somatostatin receptors in human meningiomas: Biochemical characterization. J. Clin. Endocrinol. Metab. 1986;63:433-438.

86. Feindt J, Becker I, Blomer U, Hugo HH, Mehdorn HM, Krisch B, et al. Expression of somatostatin receptor subtypes in cultured astrocytes and gliomas. J Neurochem 1995;65(5):1997-2005.

87. Pilichowska M, Kimura N, Schindler M, Kobari M. Somatostatin type 2A receptor immunoreactivity in human pancreatic adenocarcinomas. Endocr. Pathol. 2001;12(2):144-155.

88. Sagman U, Mullins J, Ginsberg R, Kovacs K, Reubi JC. Identification of somatostatin receptors in human small cell lung carcinomas. Cancer 1990;66:2129-2133.

89. Srkalovic G, Cai RZ, Schally AV. Evaluation of receptors for somatostatin in various tumors using different analogs. J. Clin. Endocrinol. Metab. 1990;70:661-669.

90. Reubi JC, Krenning E, Lamberts SWJ, Kvols L. Somatostatin receptors in malignant tissues. J. Steroid Biochem. Molec. Biol. 1990;37:1073-1077.

91. Evans AA, Crook T, Laws SAM, Gough AC, Royle GT, Primrose JN. Analysis of somatostatin receptor subtype mRNA expression in human breast cancer. Br. J. Cancer 1997;75:798-803.

92. Pilichowska M, Kimura N, Schindler M, Suzuki A, Yoshida R, Nagura H. Expression of somatostatin type 2A receptor correlates with estrogen receptor in human breast carcinoma. Endocr. Pathol. 2000;11(1):57-67.

93. Fekete M, Redding TW, Comaru-Schally AM, Pontes JE, Connelly RW, Srkalovic G, et al. Receptors for luteinizing hormone-releasing hormone, somatostatin, prolactin and epidermal growth factor in rat and human prostate cancer and in benign prostate hyperplasia. Prostate 1989;14:191-208.

94. Reubi JC, Waser B, Schaer JC, Markwalder R. Somatostatin receptors in human prostate and prostate cancer. J. Clin. Endocrinol. Metab. 1995;80:2806-2814.

95. Halmos G, Schally AV, Sun B, Davis R, Bostwick DG, Plonowski A. High expression of somatostatin receptors and messenger ribonucleic acid for its receptor subtypes in organ-confined and locally advanced human prostate cancers. J Clin Endocrinol Metab 2000;85(7):2564-71.

96. Sinisi AA, Bellastella A, Prezioso D, Nicchio MR, Lotti T, Salvatore M, et al. Different expression patterns of somatostatin receptor subtypes in cultured epithelial cells from human normal prostate and prostate cancer. J. Clin. Endocrinol. Metab. 1997;82:2566-2569.

97. Reubi JC, Horisberger U, Klijn JGM, Foekens JA. Somatostatin receptors in differentiated ovarian tumors. Amer J Pathol 1991;138:1267-1272.

98. Halmos G, Sun B, Schally AV, Hebert F, Nagy A. Human ovarian cancers express somatostatin receptors. J Clin Endocrinol Metab 2000;85(10):3509-12.

99. Reubi JC, Kvols L. Somatostatin receptors in human renal cell carcinomas. Cancer Res. 1992;52:6074-6078.

100. Reubi JC, Waser B, van Hagen M, Lamberts SWJ, Krenning EP, Gebbers J, et al. In vitro and in vivo detection of somatostatin receptors in human malignant lymphomas. Int. J. Cancer 1992;50:895-900.

101. Raderer M, Valencak J, Pfeffel F, Drach J, Pangerl T, Kurtaran A, et al. Somatostatin receptor expression in primary gastric versus nongastric extranodal B-cell lymphoma of mucosa-associated lymphoid tissue type. J Natl Cancer Inst 1999;91(8):716-8.

102. Reubi JC, Waser B, Laissue JA, Gebbers J-O. Somatostatin and vasoactive intestinal peptide receptors in human mesenchymal tumors: In vitro identification. Cancer Res. 1996;56:1922-1931.

103. Miller GV, Farmery SM, Woodhouse LF, Primrose JN. Somatostatin binding in normal and malignant human gastrointestinal mucosa. Br. J. Cancer 1992;66:391-395.

104. Reubi JC, Waser B, Schmassmann A, Laissue JA. Receptor autoradiographic evaluation of cholecystokinin, neurotensin, somatostatin, and vasoactive intestinal peptide receptors in gastro-intestinal adenocarcinoma samples: where are they really located? Int. J. Cancer 1999;81:376-386.

105. Kouroumalis E, Skordilis P, Thermos K, Vasilaki A, Moschandrea J, Manousos ON. Treatment of hepatocellular carcinoma with octreotide: a randomised controlled study. Gut 1998;42:442-447.

106. Reubi JC, Zimmermann A, Jonas S, Waser B, Läderach U, Wiedenmann B. Regulatory peptide receptors in human hepatocellular carcinomas. Gut 1999;45:766-774.

107. Iftikhar SY, Thomas WM, Rooney PS, Morris DL. Somatostatin receptors in human colorectal cancer. Eur. J. Surg. Oncol. 1992;18:27-30.

108. Radulovic SS, Milovanovic SR, Cai R, Schally AV. The binding of bombesin and somatostatin and their analogs to human colon cancers. Proc. Soc. Exp. Biol. Med. 1992;200:394-401.

109. Casini Raggi C, Calabro A, Renzi D, Briganti V, Cianchi F, Messerini L, et al. Quantitative evaluation of somatostatin receptor subtype 2 expression in sporadic colorectal tumor and in the corresponding normal mucosa. Clin Cancer Res 2002;8(2):419-27.

110. Loh KS, Waser B, Tan LK, Ruan RS, Stauffer E, Reubi JC. Somatostatin receptors in nasopharyngeal carcinoma. Virchows Arch 2002;441(5):444-8.

# 8
# COEXPRESSION OF MULTIPLE SOMATO-STATIN RECEPTORS IN INDIVIDUAL CELLS

Robert Gardette, Florence Petit, Stéphane Peineau, Christophe Lanneau and Jacques Epelbaum
*INSERM U549, 75014 Paris, France*

## INTRODUCTION

Since the cloning of five distinct somatostatin receptor (sst1-5) genes (Bruno et al., 1992, Meyerhof et al., 1992, O'Carroll et al., 1992, Yamada et al., 1992a,b), extensive studies and several reviews (Krantic et al., 1992, Epelbaum et al., 1994, Schindler et al., 1996, Dournaud et al., 1999, Patel, 1999, Schultz et al., 2000, Selmer et al., 2000a, Csaba and Dournaud, 2001) have focused on tissue localization of ssts. However, much less reports provide precise information on the coexpression of the different receptors at the single cell level. Several studies on the anatomical localization of ssts, using either in situ hybridization or immunocytochemistry, provided only indirect evidence in favor of such coexpressions. The present lack of information on the coexpression of ssts at the single cell level is surprising given 1) the physiological relevance of such coexpressions leading to putative activation of different second messenger systems and hence to putative differential cellular responses to the endogenous agonists somatostatin and cortistatin, and 2) the increasing bulk of data demonstrating the importance of homo- or heterodimerization of G protein-coupled receptors (GPCRs) opening a wide new field for receptology (Overton and Blumer, 2000).

In this review, we first present indirect, then direct, evidence for the colocalization of two or more ssts at the regional and cellular levels for brain, pituitary and peripheral tissues. We then discuss the physiological consequences of such coexpressions, and, in particular, the effects of sst1 and sst2 activation in single mouse hypothalamic neurons.

The cellular expression of ssts in various tissues or cell lines has been characterized by different techniques evidencing either sst mRNAs using northern blot, RT-PCR or in situ hybridization, or sst proteins using subtype selective sst antibodies and immunocytochemistry. Taken together, these studies unequivocally demonstrated that several sst subtypes are found in a given central nervous system structure or peripheral tissue, thus suggesting putative coexpression in the same cells. However, mRNA distributions do not necessarily reflect that of the proteins visualized with specific antibodies.

## PUTATIVE COEXPRESSION OF MULTIPLE SOMATOSTATIN RECEPTOR SUBTYPES

A review of the literature focused on immunocytochemical detection of sst expression in the rat brain is presented in Table 1. It is organized from rostral to caudal structures and shows, in decreasing order, the number of detected receptors in each structure.

*Table 1* : *Regional distribution of ssts expression detected by immunocytochemistry in rat neural structures. Data are organized from rostral to caudal regions and listed in alphabetical order within each region. First column on the left indicates the number of detected receptors in each structure. The two columns on the right of the table subdivide sst2 data between the two isoforms whenever available. Each letter in columns corresponds to the references as indicated at the bottom of the table.*

| | n | sst1 | sst2 | sst3 | sst4 | sst5 | | sst2A | sst2B |
|---|---|---|---|---|---|---|---|---|---|
| **RETINA** | 2 | c | c/f/s | | | | | f/s | s |
| **OLFACTORY BULB** | | | | | | | | | |
| Internal granular layer | 3 | | j | b | m | | | j | |
| Glomerular layer | 2 | | | b | m | | | | |
| External plexiform layer | 1 | | | | m | | | | |
| Internal plexiform layer | 1 | | | b | | | | | |
| Mitral cell layer | 1 | | | | m | | | | |
| **TELENCEPHALON (1)** | | | | | | | | | |
| Cingulate cortex | 4 | | l | b | q | r | | | l |
| Hippocampal dentate gyrus | 4 | e | j | b | m/o/q | | | j | |
| Piriform cortex | 4 | e | a/j/l | b | q | | | a/j | l |
| Ammon's horn, stratum oriens | 3 | | a/j | | q | r | | a/j | |
| Ammon's horn, stratum pyr. | 3 | | a/j | b | | r | | a/j | |
| Amygdaloid basolateral nucleus | 3 | | a/j | b | m | | | a/j | |
| Amygdaloid central nucleus | 3 | | a/j | b | m | | | a/j | |
| Amygdaloid cortical nucleus | 3 | | a/j | b | m | | | a/j | |
| Caudate Putamen | 3 | | a/j | | m/o | r | | a/j | |
| Diagonal band of Broca | 3 | | a/j | b | | r | | a/j | |
| Entorhinal cortex | 3 | | j/l | b | q | | | j | l |
| Frontal cortex | 3 | | l | | q | r | | | l |
| Globus pallidus | 3 | o | | | m | r | | | |

*Table 1 (continued)*

| | n | sst1 | sst2 | sst3 | sst4 | sst5 | sst2A | sst2B |
|---|---|---|---|---|---|---|---|---|
| **TELENCEPHALON (2)** | | | | | | | | |
| Islands of Calleja | 3 | | | b | m | r | | |
| Lateral septal nucleus | 3 | o | a/j/l | b | | | a/j | l |
| Medial septal nucleus | 3 | | j/l | b | | r | j | l |
| Nucleus accumbens | 3 | o | a | | m | | a | |
| Olfactory tubercle | 3 | | a/j | | m | r | a/j | |
| Orbital cortex | 3 | | l | | q | r | | l |
| Parietal cortex | 3 | | l | | q | r | | l |
| Subiculum | 3 | | j | b | q | | j | |
| Ventral pallidum | 3 | o | l | b | | | | l |
| Amygdaloid medial nucleus | 2 | | a/j | b | | | a/j | |
| Amygdalohippocampal area | 2 | | | b | | r | | |
| Anterior olfactory nucleus | 2 | | | b | m | | | |
| Bed nucleus of the stria termi. | 2 | | a | b | | | a | |
| Fundus striati | 2 | | a | | | r | a | |
| Indusium griseum | 2 | | l | b | | | | l |
| Occipital cortex | 2 | | l | | q | | | l |
| Retrosplenial cortex | 2 | | | b | q | | | |
| Somatosensory cortex | 2 | e | j | | | | j | |
| Temporal cortex | 2 | | l | | q | | | l |
| Agranular insular cortex | 1 | | | | | r | | |
| Amygdaloid anterior area | 1 | | | b | | | | |
| Amygdaloid basomedial nucl. | 1 | | | b | | | | |
| Amygdaloid intercalated nucl. | 1 | | | b | | | | |
| Amygdaloid lateral nucleus | 1 | | | b | | | | |
| Amygdaloid posterolateral nucleus | 1 | | | | | r | | |
| Amygdaloid posteromedial nucleus | 1 | | | | | r | | |
| Auditory cortex | 1 | | j | | | | j | |
| Claustrum | 1 | | | b | | | | |
| Dorsal peduncular cortex | 1 | | | | | r | | |
| Endopeduncular nucleus | 1 | | | b | | | | |
| Endopiriform nucleus | 1 | | | b | | | | |
| Granular insular cortex | 1 | | | | | r | | |
| Insular cortex | 1 | | | b | | | | |
| Magnocellular preoptic nucl. | 1 | | | | | r | | |
| Motor cortex | 1 | e | | | | | | |
| Nucleus of the dorsal hippocampal commissure | 1 | | | | | r | | |
| Prelimbic and infralimbic cortices | 1 | | | b | | | | |
| Septofimbrial nucleus | 1 | | | b | | | | |
| Septohippocampal nucleus | 1 | | | b | | | | |
| Substantia innominata | 1 | | | | | r | | |
| Tenia tecta | 1 | | | b | | | | |
| Triangular septal nucleus | 1 | | | b | | | | |

*Table 1 (continued)*

| | n | sst1 | sst2 | sst3 | sst4 | sst5 | | sst2A | sst2B |
|---|---|---|---|---|---|---|---|---|---|
| **DIENCEPHALON (1)** | | | | | | | | | |
| Hypothalamic arcuate nucleus | 5 | e/h | a/h/j | h | h | h | | a/j | |
| Habenula | 4 | o | a/j/l | b | m/q | | | a/j | l |
| Hypothalamic ventromedial nucleus | 4 | e/h | a/h/j | b/h/o | h | | | a | |
| Hypothalamic paraventricular nucleus | 3 | h/e | h/j | h | | | | j | |
| Hypothalamic periventricular nucleus | 3 | d/h | a/h/j | | h | | | a/j | |
| Thalamic lateral geniculate nucleus | 3 | e | j | b | | | | j | |
| *Table 1 (continued)* | | | | | | | | | |
| Thalamic reticular nucleus | 3 | | | b | q | r | | | |
| Hypothalamic lateral area | 2 | | | b | m | | | | |
| Hypothalamic medial preoptic area | 2 | | a/j | | | r | | a/j | |
| Medial geniculate body | 2 | | l | | q | | | | l |
| Thalamic laterodorsal nucleus | 2 | | | b | q | | | | |
| Thalamic mediodorsal nucl. | 2 | | | b | q | | | | |
| Zona incerta | 2 | o | | b | | | | | |
| Hypothalamic anterior area | 1 | | | b | | | | | |
| Hypothalamic dorsal area | 1 | | | b | | | | | |
| Hypothalamic lateral preoptic area | 1 | | | | | r | | | |
| Hypothalamic lateroanterior nucleus | 1 | | | | | r | | | |
| Hypothalamic magnocellular preoptic nucleus | 1 | | j/l | | | | | j | l |
| Hypothalamic posterior area | 1 | | | b | | | | | |
| Hypothalamic posterior nucl. | 1 | e | | | | | | | |
| Hypothalamic suprachiasmatic nucleus | 1 | | | | | r | | | |
| Hypothalamic supraoptic nucl | 1 | | | b | | | | | |
| Hypothalamic tuberomammillary nucleus | 1 | | a | | | | | a | |
| Subthalamus | 1 | | | b | | | | | |
| Thalamic anterodorsal nucl. | 1 | | | b | | | | | |
| Thalamic anteroventral nucl. | 1 | | | b | | | | | |
| Thalamic lateroposterior nucl. | 1 | | | b | | | | | |
| Thalamic paratenial nucleus | 1 | | | b | | | | | |
| Thalamic paraventricular nucleus | 1 | | | b | | | | | |
| Thalamic periventricular nucl. | 1 | e | | | | | | | |
| Thalamic rhomboid nucleus | 1 | | | b | | | | | |
| Thalamic ventrolateral nucl. | 1 | | | b | | | | | |
| Thalamic ventromedial nucl. | 1 | | | b | | | | | |

*Table 1(continued)*

| | n | sst1 | sst2 | sst3 | sst4 | sst5 | | sst2A | sst2B |
|---|---|---|---|---|---|---|---|---|---|
| **MEDIAN EMINENCE** | 5 | e/h/o | h | h | h | h | | | |
| **PITUITARY, ANTERIOR LOBE** | 5 | g | g/i | g | g | g/i/o | | i | |
| **MESENCEPHALON** | | | | | | | | | |
| Substantia nigra | 4 | e/o | j/l | b | q | | | j | l |
| Superior colliculus | 4 | e | a/j/l | b | q | | | a/j | l |
| Inferior colliculus | 3 | e | l | | q | | | | l |
| Interpeduncular nucleus | 3 | o | j | | q | | | j | |
| Mesencephalic trigeminal nucleus | 3 | | l | b | q | | | | l |
| Periacquaeductal gray | 3 | o | a/j/k | b | | | | a/j/k | |
| Oculomotor nucleus | 2 | | l | b | | | | | l |
| Oral pontine reticular nucleus | 2 | | l | | q | | | | l |
| Red nucleus | 2 | | l | | q | | | | l |
| Trochlear nucleus | 2 | | l | b | | | | | l |
| Ventral tegmental area | 2 | | j | b | | | | j | |
| Anterior pretectal nucleus | 1 | | | b | | | | | |
| Darkschewitsch nucleus | 1 | | | b | | | | | |
| Deep mesencephalic nucleus | 1 | | l | | | | | | l |
| *Table 1 (continued)* | | | | | | | | | |
| Intercollicular nucleus | 1 | | | b | | | | | |
| Nucleus of the optic tract | 1 | | | b | | | | | |
| Nucleus raphe dorsalis | 1 | | | b | | | | | |
| Parabigeminal nucleus | 1 | | | b | | | | | |
| Peripeduncular nucleus | 1 | | | b | | | | | |
| Posterior pretectal nucleus | 1 | | | b | | | | | |
| Pretectal olivary nucleus | 1 | | | b | | | | | |
| Retrorubral field | 1 | | | b | | | | | |
| Zona incerta | 1 | o | | | | | | | |
| **METENCEPHALON (1)** | | | | | | | | | |
| Deep cerebellar nuclei | 3 | e | | b | q | | | | |
| Motor trigeminal nucleus | 3 | | j/l | b | q | | | j | l |
| Nucleus raphe median | 3 | | l | b | q | | | | l |
| Vestibular nuclear complex | 3 | | l | b | q | | | | l |
| Caudal pontine reticular nucl. | 2 | | l | | q | | | | l |
| Cerebellar Purkinje cells | 2 | | l | | m | | | | l |
| Lateral reticular nucleus | 2 | | | b | q | | | | |
| Laterodorsal tegmental nucl. | 2 | | a | b | | | | a | |
| Superior olivary complex | 2 | | l | | q | | | | l |
| Cerebellar granular layer | 1 | e/o | | | | | | | |
| Cerebellar molecular layer | 1 | | | | q | | | | |
| Cochlear nuclear complex | 1 | | l | | | | | | l |
| Dorsal nucleus of the lateral lemniscus | 1 | | | b | | | | | |
| Dorsal tegmental nucleus | 1 | | | b | | | | | |
| Nucleus raphe pontis | 1 | | | b | | | | | |
| Locus coeruleus | 1 | | a/j | | | | | a/j | |
| Pedunculopontine tegmental nucleus | 1 | | | b | | | | | |

*Table 1 (continued)*

| | n | sst1 | sst2 | sst3 | sst4 | sst5 | sst2A | sst2B |
|---|---|---|---|---|---|---|---|---|
| **METENCEPHALON (2)** | | | | | | | | |
| Pontine nuclei | 1 | | | b | | | | |
| Pontine reticular nucleus | 1 | | | b | | | | |
| Parabrachial nucleus | 1 | | a | | | | a | |
| Sensory trigeminal nucleus | 1 | | | b | | | | |
| Tegmental reticular nucleus | 1 | | | | q | | | |
| Trapezoid body | 1 | | | | q | | | |
| **MYELENCEPHALON** | | | | | | | | |
| Facial nucleus | 4 | o | l | b | q | | | l |
| Cuneate nucleus | 3 | | l | b | q | | | l |
| Ambiguus nucleus | 2 | | l | b | | | | l |
| Dorsal motor nucleus of the vagal nerve | 2 | | l | b | | | | l |
| Nucleus of the solitary tract | 2 | o | | b | | | | |
| Abducens nucleus | 1 | | | b | | | | |
| Dorsal paragigantocellular nucleus | 1 | | l | | | | | l |
| Gracile nucleus | 1 | | | b | | | | |
| Hypoglossal nucleus | 1 | | l | | | | | l |
| Linear nucleus | 1 | | | b | | | | |
| Medullary reticular nucleus | 1 | | | | q | | | |
| Paragigantocellular reticular nucleus | 1 | | | b | | | | |
| Parvocellular reticular nucleus | 1 | | | b | | | | |
| *Table 1 (continued)* | | | | | | | | |
| Spinal trigeminal tract | 1 | o | | | | | | |
| Spinal trigeminal nucleus | 1 | | l | | | | | l |
| **SPINAL CORD** | | | | | | | | |
| Dorsal horn | 3 | p | j/k/n o/p | b/p | | | j/k/n o | |
| Intermediolateral cell column | 2 | | o | b | | | o | |
| Superficial layers | 2 | o | o | | | | o | |
| Ventral horn | 2 | | l/n/o | b/p | | | | l/n/o |
| Dorsal areas | 1 | o | | | | | | |
| Dorsolateral funiculus | 1 | | k | | | | k | |
| Spinal gray matter | 1 | | n/o | | | | | n/o |
| **DORSAL ROOT GANGLIA** | 1 | | n | | | | n | |

*a : Dournaud et al., 1996; b : Händel et al., 1999; c : Helboe and Moller, 1999; d : Helboe et al., 1998; e : Hervieu and Emson, 1998; f : Johnson et al., 1999; g : Kumar et al., 1997; h : Kumar et al., 1999a; i : Mezey et al., 1998; j : Schindler et al., 1997; k : Schindler et al., 1998; l : Schlinder et al., 1999; m : Schreff et al., 2000; n : Schulz et al., 1998; o : Schulz et al., 2000; p : Segond von Banchet et al., 1999; q : Selmer et al., 2000b; r : Stroh et al., 1999; s : Vasilaki et al., 2001.*

Out of the 165 neural structures listed in Table 1, detection of the five subtypes has only been reported in three structures: i) the hypothalamic arcuate nucleus (sst1 : Hervieu and Emson, 1998; Kumar et al., 1999a, sst2 : Dournaud et al., 1996; Schindler et al., 1997; Kumar et al., 1999a; sst3,4,5 : Kumar et al., 1999a), ii) the median eminence (sst1 : Hervieu and Emson, 1998; Kumar et al., 1999a; Schulz et al., 2000; sst2,3,4,5 : Kumar et al., 1999a) iii) the anterior lobe of the pituitary (sst1: Kumar et al., 1997; sst2: Kumar et al., 1997; Mezey et al., 1998; sst3,4: Kumar et al., 1997; sst5: Kumar et al., 1997; Mezey et al., 1998; Schulz et al., 2000). Interestingly, these three structures belong to the hypothalamo-hypophyseal tract and are primarily involved in neuroendocrine regulations. At the periphery, the five subtype mRNAs are all expressed in spleen (Bruno et al., 1993), jejunum, stomach and pancreas (Krempels et al, 1997). Nevertheless, taken as a whole, the putative coexpression of the five subtypes simultaneously in a same cell is likely to occur with a very low probability. The occurrence of the putative coexpression of four different ssts is also very low, being found only in 8 out of 165 structures. The number of regions expressing three or two different subtypes increases markedly (34 in each case), whereas demonstration of the presence of one sst subtype is reported in 86 brain areas. At the periphery, kidney expresses sst1-4 (Bruno et al., 1993) and colon sst1-3 and 5 mRNAs (Krempels et al, 1997).

A rostro-caudal gradient of possible coexpression can also be infered from these data about the probability of detection of three or more subtypes in a given structure, reaching 40 % in telencephalic structures (3 and 4 subtypes detected in 22 out of 55 telencephalic areas), and decreasing to 21 % (7/33) in the diencephalon, 26 % in the mesencephalon and 17 % in the metencephalon (6/23 and 4/23, respectively for these two areas) and only 13 % in myelencephalic zones (2/15). However, it must be noted that, for several of these brain areas, immunocytochemical detection of one or another specific sst may have not been undertaken as yet. This might greatly increase the number of sst subtypes coexpression in the forthcoming years .

## COEXPRESSION OF MULTIPLE SOMATOSTATIN RECEPTOR SUBTYPES IN INDIVIDUAL CELLS

Studies dedicated to the detection of multiple ssts in individual cells can be classified in different categories : i) immunocytochemical analyses of the expression of one or another specific sst, but coming from different groups or laboratories, ii) non-simultaneous detection of several ssts in clearly defined cells using in situ hybridization or immunocytochemistry within the same study, and iii) simultaneous detection of several ssts in well categorized cells by double immunocytochemistry, double in situ hybridization combining isotopic and non-isotopic detection or single cell RT-PCR.

### Putative sst Coexpression In Individual Cells As Evidenced By Immunocytochemistry

A review of the literature on the immunocytochemical detection of ssts in individual cells is presented in Table 2. Several studies indicated the presence of two or three different sst subtypes in defined neuronal cells. Neocortical (in layers III and V) and hippocampal (in various subfields of Ammon's horn) pyramidal cells express sst1, sst2 and sst4 receptors, although this expression may not be simultaneous in a given individual cell (sst1: Hervieu and Emson, 1998; sst2 : Dournaud et al., 1996; Schindler et al., 1997, 1999; sst4: Schreff et al., 2000). Both sst2 (Schindler et al., 1999), sst3 (Händel et al., 1999) and sst4 (Schreff et al., 2000) have also been detected in cerebellar Purkinje cells.

*Table 2*. *Putative sst coexpressions in individual cells as evidenced by immunocytochemistry. Data are organized according to the extent of putative colocalizations. All data are from rat but for amacrine cells in rabbit retina, as indicated in the table. The two columns on the right of the table subdivide sst2 data between the two isoforms whenever available. Each letter in columns corresponds to the references as indicated at the bottom of the table.*

| | sst1 | sst2 | sst3 | sst4 | sst5 | | sst2A | sst2B |
|---|---|---|---|---|---|---|---|---|
| **FIVE SUBTYPES** | | | | | | | | |
| **PITUITARY** | | | | | | | | |
| Somatotrophs | g | g | g | g | g | | | |
| **THREE SUBTYPES** | | | | | | | | |
| **NEOCORTEX** | | | | | | | | |
| Pyramidal cells, layer III | f | b/j | | k | | | b | j |
| Pyramidal cells, layer V | f | i/j | | k | | | i | j |
| **HIPPOCAMPUS** | | | | | | | | |
| Ammon's horn, Pyramidal cells | f | b/i/j | | k/n | | | b/i | j |
| **CEREBELLUM** | | | | | | | | |
| Purkinje cells | | j | D | k | | | | j |
| **PANCREAS** | | | | | | | | |
| Beta cells | h | h | | | h | | | |
| **TWO SUBTYPES** | | | | | | | | |
| **RETINA** | | | | | | | | |
| Amacrine cells (rat) | e | e | | | | | | |
| Amacrine cells (rabbit) | a | c | | | | | c | |
| **OLFACTORY BULB** | | | | | | | | |
| Mitral cells | | j | | k | | | | j |
| **NEOCORTEX** | | | | | | | | |
| Pyramidal cells, layer VI | f | i | | | | | i | |
| **HIPPOCAMPUS** | | | | | | | | |
| Ammon's horn, interneurons | f | | | k/n | | | | |
| Dentate gyrus, granular cells | f | i | | | | | i | |
| **SPINAL CORD** | | | | | | | | |
| Ventral horn, motoneurons | | l/j | M | | | | | l/j |
| **PANCREAS** | | | | | | | | |
| Alpha cells | h | h | | | | | | |
| Alpha cells | | h | | | h | | | |

a : *Christiani et al., 2000;* b : *Dournaud et al., 1996;* c : *Fontanesi et al., 2000;* d : *Händel et al., 1999;* e : *Helboe and Moller, 1999;* f : *Hervieu and Emson., 1998;* g : *Kumar et al., 1997;* h : *Kumar et al., 1999b;* i : *Schindler et al., 1997;* j : *Schindler et al., 1999;* k : *Schreff et al., 2000;* l : *Schulz et al., 1998;* m : *Segond von Banchet et al.., 1999;* n : *Selmer et al., 2000b*

Several studies indicated the presence of two or three different sst subtypes in defined neuronal cells. Neocortical (in layers III and V) and hippocampal (in various subfields of Ammon's horn) pyramidal cells express sst1, sst2 and sst4 receptors, although this expression may not be simultaneous in a given individual cell (sst1: Hervieu and Emson, 1998; sst2 : Dournaud et al., 1996; Schindler et al., 1997, 1999; sst4: Schreff et al., 2000). Both sst2 (Schindler et al., 1999), sst3 (Händel et al., 1999) and sst4 (Schreff et al., 2000) have also been detected in cerebellar Purkinje cells. sst1 and sst2 expression were reported in retina amacrine cells, displaced pyramidal cells in layer VI of the cortex and granular cells from the dentate gyrus of the hippocampus. Finally, in the olfactory bulb sst2 and sst4 expression has been observed in mitral cells and sst2 and sst3 are present in motoneurons of the ventral horn of the spinal cord.

## Non-Simultaneous Detection of Multiple ssts in Individual Cells

Clonal cell lines represent homogenous cell populations. It was therefore of interest to detect the expression of sst receptors in such models using different techniques such as nuclear protection assay, RT-PCR or northern blots. GH3 cells (Bruno et al., 1994) and PC12 cells (Traina et al., 1998) coexpress the five subtypes, AtT-20 cells four subtypes (sst1,2,4,5: Patel et al., 1994), GC (Traina et al., 1998) and GH4C1 (Xu et al., 1995) cells three subtypes (sst1,2,5 for GC cells, and sst1,2,3 for GH4C1), whereas the human neuroblastoma cell line LA-N-2 expresses only the subtypes 2 and 5 (Nilsson and Folkesson, 1997). It is noteworthy that the various cell lines derived from rat pituitary somatomammotroph cells, i.e. GH3, GH4C1 and GC cells, present with a different repertoire of sst expression. Comparing these three cell lines for the presence of selective transcription factors might provide information on the regulation of sst gene transcription.

Out of the rare structures expressing all five sst subtypes, described in the first part of this chapter, only one study confirmed this fact for pituitary somatotroph cells (Kumar et al, 1997). The same authors also reported the distribution of the sst subtypes in human pancreatic islet cells (Kumar et al, 1999b). sst1 and sst5 were strongly colocalized in 87% of insulin synthesizing beta cells while sst1/sst2 labeled beta cells represent only 46% of the population. Colocalization between sst2 and sst5, and sst2 and sst1, were less abundant (about a third) in glucagon producing alpha cells. Only a few colocalizations between sst5 and sst1-3 were observed in somatostatin containing delta cells. In rat testis sst1-3 displayed a similar distribution in the Sertoli and germ cells, which depended on the stage of the cycle of the seminiferous epithelium (Zhu et al, 1998). In rat brain, an indirect demonstration of the co expression of two different ssts has only been brought by labeling either GHRH and sst1 or sst2 mRNAs by double in situ hybridization, in arcuate GHRH neurons (Tannenbaum et al, 1998). Finally, it is interesting to compare the results presented in three different publications and devoted to the analysis of sst expression in pituitary secretory cells, using either immunocytochemistry (Kumar et al., 1997) or in situ hybridization (Day et al., 1995; O'Carroll and Krempels, 1995) (Figure 1A). The five subtypes are indeed detected in the three studies although in a very different proportion of somatotroph cells for the sst2 and sst5 subtypes. Similarly, important discrepancies can be noted in the expression of sst2 and sst5 receptors detected in other pituitary cell types by two groups (Day et al., 1995; O'Carroll and Krempels, 1995) using a similar *in situ* hybridization technique (Figure 1B).

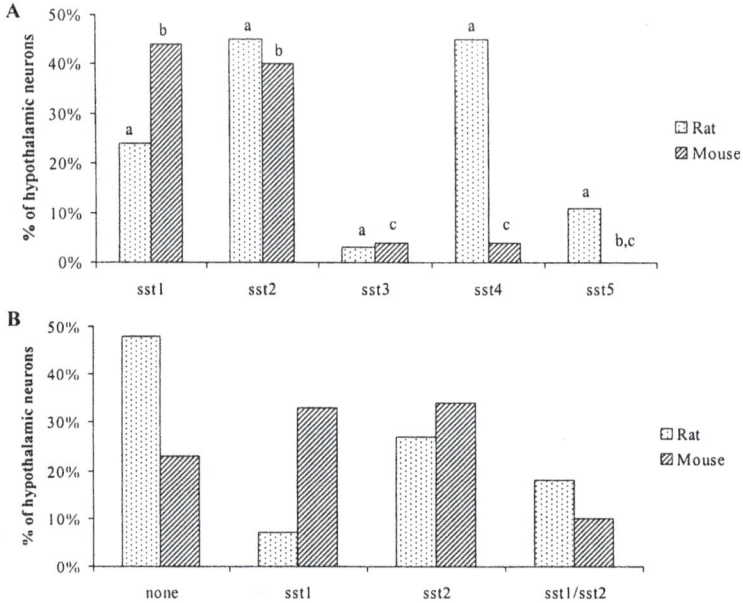

**Figure 1.** *Distribution of the expression of sst receptors in rat anterior pituitary cells as determined by immunocytochemistry (c) or in situ hybridization (a,b). A : Distribution of sst1-5 receptors in rat somatotrophs. Note the agreement between ICC and one of the ISH studies and the discrepancy for sst2 and sst5 receptors for the second ISH study (nd : not determined in the study). B, C, D: Distribution of sst2 and sst5 expression in rat thyrotrophs (B), lactotrophs (C) and corticotrophs (D) from two different ISH studies. Note important discrepancies between the two studies (a : Day et al., 1995 ; b : O'Carroll and Krempels, 1995 ; c : Kumar et al., 1997).*

## Simultaneous Detection of Multiple ssts In Individual Cells

Very few direct demonstrations of the presence of two or more sst subtype mRNAs or proteins are available as yet. Using double immunocytochemistry techniques, the coexpression of sst2B and sst4 has been evidenced in rat cortical pyramidal cells of the layer V (Schreff et al., 2000), and that of sst2A and sst5 in rat anterior pituitary cells (Mezey et al., 1998).

In rat brain, Perez and Hoyer (1995) conducted a large study on the coexpression of sst3 and sst4 mRNA subtypes using combined isotopic and non-isotopic in situ hybridization, leading to the demonstration of such a coexpression in numerous neuronal categories such as cortical neurons, hippocampal pyramidal cells, subicular neurons, amygdalohippocampal neurons, hypothalamic paraventricular and medial preoptic nuclei neurons, anterior olfactory nucleus neurons and thalamic laterodorsal nucleus neurons. They also found the simultaneous presence of the two subtypes in periglomerular and granular cells in the olfactory bulb. Using a similar approach, Schindler et al. (1995) showed the simultaneous expression of sst3 and sst4 mRNAs in human cortical and hippocampal CA4 neurons, and that of sst1 and sst3 mRNAs in this last population.

Finally, data on the coexpression of ssts in individual cells arise from the work of Lanneau et al. (1998, 2000a,b) and Petit et al. (unpublished results) through the use of single cell multiplex RT-PCR technique in cultured mouse and rat hypothalamic neurons, evidencing a striking difference between the two species. As shown in Figure 2A, mouse hypothalamic neurons are almost devoid of sst3-5 expression whereas a high percentage of rat hypothalamic neurons expresses the sst4 subtype. Due to the absence of sst3-5 in mouse, only coexpression of sst1 and sst2 mRNA subtypes have been studied, and compared to those observed in rat hypothalamic neurons.

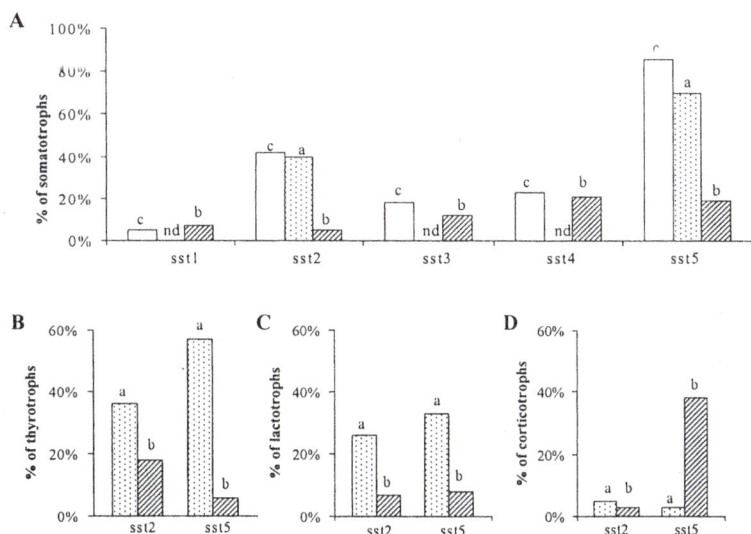

*Figure 2.* Expression of sst1-5 receptors in rat and mouse hypothalamic cultures as determined by single cell RT-PCR. A: Distribution of the expression of sst receptors. Note the lack of sst3-5 expression in the mouse and the high occurrence of sst4 in the rat [Data from : Petit et al., unpublished data (a); Lanneau et al., 1998 (b); Lanneau et al., 2000a (c)]. B : Comparison of the expression and the coexpression of sst1 and sst2 in rat and mouse hypothalamic cultures. Note the almost equivalent sst1/sst2 coexpression in the two species despite a higher sst1 expression in the mouse [Data from : Petit et al., unpublished data (rat); Lanneau et al., 2000b (mouse)].

As shown in Figure 2B, the occurrence of sst1/sst2 coexpression is almost equivalent in rat and in mouse (18% vs 10% of neurons) despite a higher sst1 expression in the mouse. The use of single cell multiplex RT-PCR technique also allowed to analyze sst expression in phenotypically characterized hypothalamic neurons and to compare the expression and the coexpression of sst1 and sst2 subtypes in neurohormone synthesizing hypothalamic neurons. As illustrated in Figure 3, the percentage of mouse SRIF- and GHRH-expressing neurons in which a coexpresssion of sst1/sst2 mRNAs had been detected is distinctly higher than that for the total mouse neuronal population (27% and 42% respectively vs 10%), a phenomenon which is not evident in rat SRIF- and GHRH-expressing neurons (22% and 9% respectively vs 18%). Such differences between these two murine species might be of some relevance for the mechanisms controlling GH secretion.

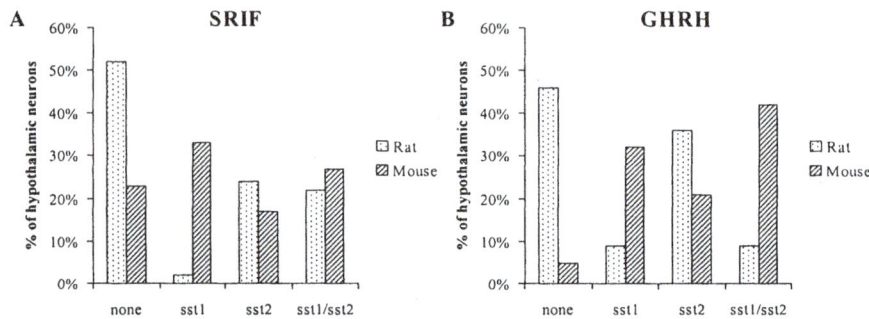

*Figure 3.* Comparison, by single cell multiplex RT-PCR, of the expression and the coexpression of sst1 and sst2 receptors in phenotypically characterized rat and mouse SRIF- or GHRH-expressing hypothalamic neurons. Note the high level of sst1 expression in SRIF-expressing mouse neurons as compared to rat neurons (A), and the high level of coexpression of sst1 and sst2 receptors in GHRH-expressing rat neurons as compared to mouse neurons (B) [Data from : Petit et al., unpublished data (rat); Lanneau et al., 2000a (mouse)].

# PHYSIOLOGICAL IMPLICATIONS OF THE COEXPRESSION OF MULTIPLE SOMATOSTATIN RECEPTOR SUBTYPES IN INDIVIDUAL CELLS

As early as 1983, using intracellular recordings on rat cortical neurons in culture, it was reported that somatostatin effects were predominantly excitatory but application of the peptide also resulted occasionally in inhibitory actions (Delfs and Dichter, 1983). Moreover, unusual response characteristics such as an inverted U-shaped dose response curve for membrane depolarization were also observed. More recently, on salamander photoreceptors, it was shown that somatostatin can inhibit L-type $Ca^{2+}$ currents in rods but enhances it in cones

(Akopian et al, 2000). Previously, it had been shown that somatostatin could either enhance or decrease the glutamate sensitivity of cultured mouse hypothalamic neurons (Gardette et al., 1995). These data suggested a considerable complexity of the actions of somatostatin at the cellular level which might now be interpreted in terms of receptor coexpression and their consequences.

To our knowledge, the only reports on the direct effects of the activation of two different sst subtypes on a single cell come from the work of Lanneau et al. (1998, 2000a) on mouse hypothalamic neurons in culture. By coupling electrophysiological recordings of glutamate-induced responses in the presence of octreotide (a preferentially sst2 agonist) or of CH-275 (a sst1 agonist) to the detection of sst mRNAs present in the recorded neuron, it was demonstrated that the activation of sst1 or sst2 led respectively to increase or decrease of the glutamate-induced current in the same cell (Figure 4) (Lanneau et al., 1998).

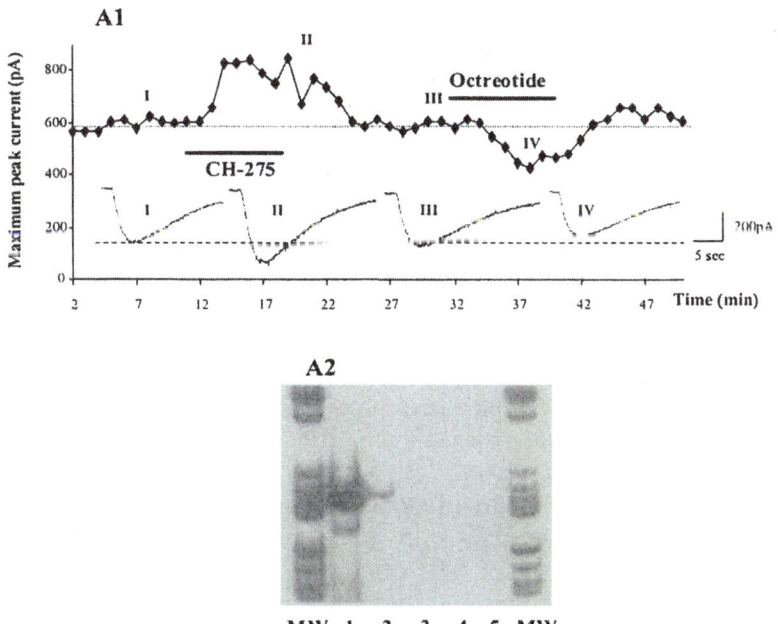

*Figure 4.*: *Example of the consequence of the simultaneous expression of sst1 and sst2 receptor subtypes in a single mouse hypothalamic neuron on the modulation of neuronal glutamate sensitivity. In the same neuron, the glutamate response can be either increased by the sst1 agonist, CH-275 (A1, II), or decreased by the sst2 agonist, octreotide (A1, IV). Typical electrophysiological recordings of the glutamate-induced currents at times indicated by roman numbers are shown under the time plot (I: control, II: CH-275, III: recovery, IV: octreotide). After recording, the cell cytoplasm was harvested and processed for single cell multiplex RT-PCR. As shown in A2, only sst1 and sst2 mRNAs were detected in this neuron, in correlation with the physiological modulations [lanes 1-5 : sst1-sst5, lanes MW : mol wt markers (Roche V), adapted from Lanneau et al., 2000a].*

These results were further confirmed by antisense experiments in which blockade of sst1 or sst2 proteins inhibited octreotide- or CH-275-induced glutamate modulation (Lanneau et al., 2000a). Such observations indicated that the activation of two different sst subtypes, expressed in a same neuron, can result in an opposite modulation of a same effector.

The first hypothesis that could explain this differential modulation relies on an activation of different second messenger pathways, such as coupling to different G proteins. Indeed, the sst2-mediated inhibitory effect is abolished by pertussis toxin contrary to the sst1-excitatory one (Lanneau et al., 1998). In another study, both sst1 and sst2 agonists were able to inhibit adenylyl cyclase activity from cell membranes (Viollet et al., 1997). Such results can be compared to those obtained on stably transfected mouse fibroblast Ltk-cells or transiently transfected human embryonic kidney cells HEK 293 (Hou et al, 1994). In these models as in mouse hypothalamic neuronal cultures, both sst1 and sst2 receptor subtypes inhibited inhibition of cAMP accumulation by a pertussis dependent mechanism. However, the sst1 receptor was able to activate the $Na^+/H^+$ ion exchanger and this action was insensitive to pertussis toxin. Activation of tyrosine phosphatase activities through sst1 (Buscail et al, 1994; Florio et al, 1999) and sst2 (Buscail et al, 1994; Reardon et al, 1997) receptors were also found to be pertusiss toxin-sensitive or insensitive (see also chapter 10), suggesting that in this pathway also, different subtypes of G proteins might couple the receptors in individual cells. Therefore, the simultaneous setting of different second messenger pathways leading to differential effects on a same cellular effector through the activation of different receptor subtypes born in a same cell may turn out to be a more general rule than expected.

A second hypothesis is based on recent data, obtained by bioluminescence-, fluorescence- or photobleaching fluorescence-energy transfer, which provide evidences that several seven transmembrane GPCRs (Devi, 2000) including sst receptors (Rocheville et al, 2000a,b) can function as homo- or heterodimers. In the absence of agonist, a mixture of monomers and dimers was evident in sst5 transfected cells expressing high levels of membrane bound receptors while monomers were predominant in cells expressing at a lower level. In presence of somatostatin, receptor dimerization dose-dependently increased. sst5 dimers display different properties from the wild type one, such as binding affinity, ligand-induced internalization or up-regulation (Rocheville et al., 2000b). Indirect evidences on transfected cells also suggest that sst5 can heterodimerize with sst1, but not sst4, as well as with the longer form of the dopaminergic D2 receptor (Rocheville et al, 2000a,b). In that last case, heterodimerization was strongly dependent on the presence of either one or the other agonist. sst2A and sst3 receptors can also form heterodimers when coexpressed in HEK 293 cells and this results in inactivation of sst3 receptor function (Pfeiffer et al, 2001). Finally, dimerization of sst2A and mu-opioid receptor (MOR1) induced phosphorylation, internalization, and desensitization of both sst2A and MOR1 after exposure of the sst2A-MOR1 heterodimer to a sst2A-selective ligand. Reciprocally, exposure of sst2A-MOR1 heterodimers to a mu-selective ligand induced phosphorylation and desensitization of both MOR1 and sst2A without internalization of sst2A. (Pfeiffer et al, 2002). sst dimerization could also have further functional consequences, already described in other receptor systems, such as a different agonist selectivity (George et al., 1998), down- (Abdalla et al., 1999) or up- (Hebert et al., 1996; Overton and Blumer, 2000) regulation of coupled second messengers, or changes in the functional response of one of the dimerized receptors (Charles et al, 2003). In this last study, however, transfection of mu- and delta opioid receptors in GH3 cells changed the functional response to mu agonists from inhibitory to excitatory but did not interfere with the coupling of endogenously expressed somatostatin receptors to calcium signalling in those cells.

Finally, several recent studies have demonstrated that sst receptors do not only interact with multiple G proteins or several GPCRs but that they can also physically bind, through their C-terminal domain, to a new class of proteins displaying anchoring and scaffolding functions (Zitzer et al, 1999a, b; Schwarzler et al, 2000). Cortactin binding protein 1 (Cort-BP1, also named SHANK2), a brain specific binding protein involved in the restructuring of the cytoskeleton through its interaction with cortactin, binds sst2 receptors through the receptor PSD-95/discs large/ZO-1 (PDZ) recognizing domain located at its C-terminal. While SHANK2 expressed alone in HEK cells displayed a cytoplasmic staining, coexpression with sst2A induced a somatostatin-dependent redistribution of both proteins at the cell membrane (Zitzer et al, 1999a). Interestingly, there is a striking overlap in the expression pattern of sst2 and SHANK2 by in situ hybridization. Another protein, exhibiting strong homologies with SHANK2 and named SSTRIP (sst receptor interacting protein, or SHANK1), which binds to NMDA and metabotropic glutamate receptors (Lim et al, 1999; Naisbitt et al, 1999; Tu et al, 1999; Yao et al, 1999), also binds to sst2A in co-transfected cells (Zitzer et al, 1999b). The fact that SHANK1 can bind both sst2A and glutamate receptors may be of importance in the physiological interactions described above. However, the subcellular localizations of sst2 and SHANK1 appear different, the former being mainly distributed at extrasynaptic sites (Dournaud et al, 1998; Csaba and Dournaud, 2001) and the latter in post-synaptic densities (Lim et al, 1999; Naisbitt et al, 1999; Tu et al, 1999). More recently, sst1 receptors were found to interact with Skb1Hs (Schwarzler et al, 2000), the human homolog of the yeast Skb1 protein, for which the function is not yet clearly established. Interestingly, HEK cells expressing both sst1 and Skb1Hs displayed a higher number of somatostatin binding sites as compared to cells expressing only sst1, suggesting that sKb1Hs is involved in the cell surface targetting of sst1 receptors.

In summary, it appears that coexpression of somatostatin receptors could be the rule rather than the exception. Such coexpressions of two or more sst subtypes may lead, when activated by endogenous agonists, to crosstalk between either the receptors themselves, directly or indirectly to SSTRIP proteins, or second messenger systems that will interfere with effector responses. The growing importance of homo- or heterodimerizations and receptor-protein interactions, if confirmed in physiologically ssts-expressing cells, should bring even more attention on physiological implications of receptor coexpressions. Finally, the addressing of different receptor subtypes to different subcellular compartments (see for instance Dournaud et al., 1998 ; Schreff et al., 2000 ; Schulz et al., 2000) may assign different functions to these molecules such as mediating pre- or post-synaptic effects of somatostatin within one given neuron.

## ACKNOWLEDGMENTS

This study was supported by the "Institut National de la Santé et de la Recherche Médicale," the "Centre National de la Recherche Scientifique," and European Community contract QLG-1999-0098. CL was supported by a grant from Novartis. The authors are grateful to Dr. P. Dournaud for helpful discussion, Drs J. Rivier and C. Hoeger for the gift of CH-275, and to Novartis for octreotide.

## REFERENCES

1. Abdalla, S., Zaki, E., Lother, H., Quitterer, U. Involvement of the amino terminus of the B(2) receptor in agonist-induced receptor dimerization. J Biol Chem 1999, 274:26079-26084.
2. Akopian, A., Johnson, J., Gabriel, R., Brecha, N., Witkovsky, P. Somatostatin modulates voltage-gated K(+) and Ca(2+) currents in rod and cone photoreceptors of the salamander retina. J Neurosci 2000, 20:929-936.
3. Bruno, J.F., Xu, Y., Berelowitz, M. Somatostatin regulates somatostatin receptor subtype mRNA expression in GH3 cells. Biochem Biophys Res Commun 1994, 202:1738-1743.
4. Bruno, J.F., Xu, Y., Song, J., Berelowitz, M. Molecular cloning and functional expression of a brain-specific somatostatin receptor. Proc Natl Acad Sci USA 1992, 89:11151-11155.
5. Bruno, J.F., Xu, Y., Song, J., Berelowitz, M. Tissue distribution of somatostatin receptor subtype messenger ribonucleic acid in the rat. Endocrinology 1993, 133:2561-2567.
6. Buscail, L., Delesque, N., Esteve, J.P., Saint-Laurent, N., Prats, H., Clerc, P., Robberecht, P., Bell, G.I., Liebow, C., Schally, A.V., et al. Stimulation of tyrosine phosphatase and inhibition of cell proliferation by somatostatin analogues: mediation by human somatostatin receptor subtypes SSTR1 and SSTR2. Proc Natl Acad Sci USA 1994, 91:2315-2319.
7. Charles, A.C., Mostovskaya, N., Asas, K., Evans, C.J., Dankovich, M.L., Hales, T.G. Coexpression of delta-opioid receptors with micro receptors in GH3 cells changes the functional response to micro agonists from inhibitory to excitatory. Mol Pharmacol. 2003, 63:89-95.
8. Cristiani, R., Fontanesi, G., Casini, G., Petrucci, C., Viollet, C., Bagnoli, P. Expression of somatostatin subtype 1 receptor in the rabbit retina. Invest Ophthalmol Vis Sci 2000, 41:3191-3199.

9. Csaba Z., Dournaud, P. Cellular biology of somatostatin receptors. Neuropeptides 2001, 35 :1-23.
10. Day, R., Dong, W., Panetta, R., Kraicer, J., Greenwood, M.T., Patel, Y.C. Expression of mRNA for somatostatin receptor (sstr) types 2 and 5 in individual rat pituitary cells. A double labeling in situ hybridization analysis. Endocrinology 1995, 136:5232-5235.
11. Delfs, J.R., Dichter, M.A. Effects of somatostatin on mammalian cortical neurons in culture: physiological actions and unusual dose response characteristics. J Neurosci 1983, 3:1176-1188.
12. Devi, L.A. G-protein-coupled receptor dimers in the lime light. Trends Pharmacol Sci 2000, 21:324-326.
13. Dournaud, P., Boudin, H., Schonbrunn, A., Tannenbaum, G.S., Beaudet, A. Interrelationships between somatostatin sst2A receptors and somatostatin-containing axons in rat brain: evidence for regulation of cell surface receptors by endogenous somatostatin. J Neurosci 1998, 18:1056-1071.
14. Dournaud, P., Gu, Y.Z., Schonbrunn, A., Mazella, J., Tannenbaum, G.S., Beaudet, A. Localization of the somatostatin receptor SST2A in rat brain using a specific anti-peptide antibody. J Neurosci 1996, 16:4468-4478.
15. Dournaud, P., Slama, A., Beaudet, A., Epelbaum, J. "Somatostatin receptors." In Handbook of Chemical Neuroanatomy, Vol. 16:Peptide receptors Part I, R. Quirion, A. Björklund, T. Hökfelt, eds. Amsterdam: Elsevier, 1999.
16. Epelbaum, J., Dournaud, P., Fodor, M., Viollet, C. The neurobiology of somatostatin. Crit Rev Neurobiol 1994, 8:25-44.
17. Florio, T., Yao, H., Carey, K.D., Dillon, T.J., Stork, P.J. Somatostatin activation of mitogen-activated protein kinase via somatostatin receptor 1 (SSTR1). Mol Endocrinol 1999, 13:24-37.
18. Fontanesi, G., Gargini, C., Bagnoli, P. Postnatal development of somatostatin 2A (sst2A) receptors expression in the rabbit retina. Dev Brain Res 2000, 123:67-80.
19. Gardette, R., Faivre-Bauman, A., Loudes, C., Kordon, C., Epelbaum, J. Modulation by somatostatin of glutamate sensitivity during development of mouse hypothalamic neurons in vitro. Dev Brain Res 1995, 86:123-133.
20. George, S.R., Lee, S.P., Varghese, G., Zeman, P.R., Seeman, P., Ng, G.Y., O'Down, B.F. A transmembrane domain-derived peptide inhibits D1 dopamine receptor function without affecting receptor oligomerization. J Biol Chem 1998, 273:30244-30248.
21. Händel, M., Schulz, S., Stanarius, A., Schreff, M., Erdtmann-Vourliotis, M., Schmidt, H., Wolf, G., Höllt, V. Selective targeting of somatostatin receptor 3 to neuronal cilia. Neuroscience 1999, 89:909-926.
22. Hebert, T.E., Moffett, S., Morello, J.P., Loisel, T.P., Bichet, D.G., Barret, C., Bouvier, M. A peptide derived from a beta2-adrenergic receptor transmembrane domain inhibits both receptor dimerization and activation. J Biol Chem 1996, 271:16384-16392.
23. Helboe, L., Moller, M. Immunohistochemical localization of somatostatin receptor subtypes sst1 and sst2 in the rat retina. Invest Ophthalmol Vis Sci 1999, 40:2376-2382.
24. Helboe, L., Stidsen, C.E., Moller, M. Immunohistochemical and cytochemical localization of the somatostatin receptor subtype sst1 in the somatostatinergic parvocellular neuronal system of the rat hypothalamus. J Neurosci 1998, 18:4938-4945.
25. Hervieu, G., Emson, P.C. The localization of somatostatin receptor 1 (sst1) immunoreactivity in the rat brain using an N-terminal specific antibody. Neuroscience 1998, 85:1263-1284.
26. Hou, C., Gilbert, R.L., Barber, D.L. Subtype-specific signaling mechanisms of somatostatin receptors SSTR1 and SSTR2. J Biol Chem 1994, 269:10357-10362.
27. Johnson, J., Wu, V., Wong, H., Walsh, J.H., Brecha, N.C. Somatostatin receptor subtype 2A expression in the rat retina. Neuroscience 1999, 94:675-683.
28. Krantic, S., Quirion, R., Uhl, G. "Somatostatin receptors." In Handbook of Chemical Neuroanatomy, Vol. 11:Neuropeptide receptors in the CNS, A. Björklund, T. Hökfelt, M.J. Kuhar, eds. Amsterdam: Elsevier, 1992.

29. Krempels, K., Hunyady, B., O'Carroll, A.M., Mezey, E. Distribution of somatostatin receptor messenger RNAs in the rat gastrointestinal tract. Gastroenterology 1997, 112:1948-1960.
30. Kumar, U., Laird, D., Srikant, C.B., Escher, E., Patel, Y.C. Expression of the five somatostatin receptor (SSTR1-5) subtypes in rat pituitary somatotrophs: quantitative analysis by double-layer immunofluorescence confocal microscopy. Endocrinology 1997, 138:4473-4476.
31. Kumar, U., Ong, W., Patel, S., Patel, Y.C. Cellular expression of the five somatostatin receptor subtypes (sstr1-5) in rat hypothalamus : a comparative immunohistochemical analysis. Proceedings of The Endocrine Society's 81st Annual Meeting, 1999 June 12-15; San Diego, 1999a.
32. Kumar, U., Sasi, R., Suresh, S., Patel, A., Thangaraju, M., Metrakos, P., Patel, S.C., Patel, Y.C. Subtype-selective expression of the five somatostatin receptors (hSSTR1-5) in human pancreatic islet cells : a quantitative double-label immunohistochemical analysis. Diabetes 1999b, 48:77-85.
33. Lanneau, C., Bluet-Pajot, M. T., Zizzari, P., Csaba, Z., Dournaud, P., Helboe, L., Hoyer, D., Pellegrini, E., Tannenbaum, G. S., Epelbaum, J., Gardette, R. Involvement of the Sst1 somatostatin receptor subtype in the intrahypothalamic neuronal network regulating growth hormone secretion: an in vitro and in vivo antisense study. Endocrinology 2000a, 141:967-979.
34. Lanneau, C., Peineau, S., Petit, F., Epelbaum, J., Gardette, R. Somatostatin modulation of excitatory synaptic transmission between periventricular and arcuate hypothalamic nuclei in vitro. J Neurophysiol 2000b, 84:1464-1474.
35. Lanneau, C., Viollet, C., Faivre-Bauman, A., Loudes, C., Kordon, C., Epelbaum, J., Gardette, R. Somatostatin receptor subtypes sst1 and sst2 elicit opposite effects on the response to glutamate of mouse hypothalamic neurones: an electrophysiological and single cell RT-PCR study. Eur J Neurosci 1998, 10:204-212.
36. Lim, S., Naisbitt, S., Yoon, J., Hwang, J.I., Suh, P.G., Sheng, M., Kim, E. Characterization of the Shank family of synaptic proteins. Multiple genes, alternative splicing, and differential expression in brain and development. J Biol Chem 1999, 274:29510-29518.
37. Meyerhof, W., Wulfsen, I., Schonrock, C., Fehr, S., Richter, D. Molecular cloning of a somatostatin-28 receptor and comparison of its expression pattern with that of a Mezey, E., Hunyady, B., Mitra, S., Hayes, E., Liu, Q., Schaeffer, J., Schonbrunn, A. Cell specific expression of the sst2A and sst5 somatostatin receptors in the rat anterior pituitary. Endocrinology 1998, 139:414-419.
38. Naisbitt, S., Kim, E., Tu, J.C., Xiao, B., Sala, C., Valtschanoff, J., Weinberg, R.J., Worley, P.F., Sheng, M. Shank, a novel family of postsynaptic density proteins that binds to the NMDA receptor/PSD-95/GKAP complex and cortactin. Neuron 1999, 23:569-582.
39. Nilsson, L., Folkesson, R. Coexistence of somatostatin receptor subtypes in the human neuroblastoma cell line LA-N-2. FEBS Lett 1997, 401:83-88.
40. O'Carroll, A.M., Krempels, K. Widespread distribution of somatostatin receptor messenger ribonucleic acids in rat pituitary. Endocrinology 1995, 136:5224-5227.
41. O'Carroll, A.M., Lolait, S.J., Konig, M., Mahan, L.C. Molecular cloning and expression of a pituitary somatostatin receptor with preferential affinity for somatostatin-28. Mol Pharmacol 1992, 42, 939-946.
42. Overton, M.C., Blumer, K.J. G-protein-coupled receptors function as oligomers in vivo. Curr Biol 2000, 10:341-344.
43. Patel, Y.C. Somatostatin and its receptor family. Front Neuroendocrinol 1999, 20:157-198.
44. Patel, Y.C., Panetta, R., Escher, E., Greenwood, M., Srikant, C.B. Expression of multiple somatostatin receptor genes in AtT-20 cells. Evidence for a novel somatostatin-28 selective receptor subtype. J Biol Chem 1994, 269:1506-1509.
45. Perez, J., Hoyer, D. Co-expression of somatostatin SSTR-3 and SSTR-4 receptor messenger RNAs in the rat brain. Neuroscience 1995, 64:241-253.

46. Pfeiffer, M., Koch, T., Schroder, H., Klutzny, M., Kirscht, S., Kreienkamp, H.J., Höllt, V., Schulz, S. Homo- and heterodimerization of somatostatin receptor subtypes. Inactivation of sst3 receptor function by heterodimerization with sst2A. J Biol Chem 2001, 276:14027-14036.

47. Pfeiffer, M., Koch, T., Schroder, H., Laugsch, M., Hollt, V., Schulz, S. Heterodimerization of somatostatin and opioid receptors cross-modulates phosphorylation, internalization, and desensitization. J Biol Chem 2002, 277:19762-19772.

48. Reardon, D.B., Dent, P., Wood, S.L., Kong, T., Sturgill, T.W. Activation in vitro of somatostatin receptor subtypes 2, 3, or 4 stimulates protein tyrosine phosphatase activity in membranes from transfected Ras-transformed NIH 3T3 cells: coexpression with catalytically inactive SHP-2 blocks responsiveness. Mol Endocrinol 1997, 11:1062-1069.

49. Rocheville, M., Lange, D.C., Kumar, U., Patel, S.C., Patel, R.C., Patel, Y.C. Receptors for dopamine and somatostatin: formation of hetero-oligomers with enhanced functional activity. Science 2000a, 288:154-157.

50. Rocheville, M., Lange, D.C., Kumar, U., Sasi, R., Patel, R.C., Patel, Y.C. Subtypes of the somatostatin receptor assemble as functional homo- and heterodimers. J Biol Chem 2000b, 275:7862-7869.

51. Schindler, M., Harrington, K.A., Humphrey, P.P., Emson, P.C. Cellular localization and co-expression of somatostatin receptor messenger RNAs in the human brain. Mol Brain Res 1995, 34:321-326.

52. Schindler, M., Holloway, S., Hathway, G., Woolf, C.J., Humphrey, P.P., Emson, P.C. Identification of somatostatin sst2(a) receptor expressing neurones in central regions involved in nociception. Brain Res 1998, 798:25-35.

53. Schindler, M., Humphrey, P. P., Emson, P. C. Somatostatin receptors in the central nervous system. Prog Neurobiol 1996, 50:9-47.

54. Schindler, M., Humphrey, P.P., Lohrke, S., Friauf, E. Immunohistochemical localization of the somatostatin sst2(b) receptor splice variant in the rat central nervous system. Neuroscience 1999, 90:859-874.

55. Schindler, M., Sellers, I. A., Humphrey, P.P., Emson, P.C. Immunohistochemical localization of the somatostatin SST2(A) receptor in the rat brain and spinal cord. Neuroscience 1997, 76:225-240.

56. Schreff, M., Schulz, S., Händel, M., Keilhoff, G., Braun, H., Pereira, G., Klutzny, M., Schmidt, H., Wolf, G., Höllt, V. Distribution, targeting, and internalization of the sst4 somatostatin receptor in rat brain. J Neurosci 2000, 20:3785-3797.

57. Schulz, S., Händel, M., Schreff, M., Schmidt, H., Höllt, V. Localization of five somatostatin receptors in the rat central nervous system using subtype-specific antibodies. J Physiol Paris 2000, 94:259-264.

58. Schulz, S., Schmidt, H., Händel, M., Schreff, M., Höllt, V. Differential distribution of alternatively spliced somatostatin receptor 2 isoforms (sst2A and sst2B) in rat spinal cord. Neurosci Lett 1998, 257:37-40.

59. Schwarzler, A., Kreienkamp, H.J., Richter, D. Interaction of the somatostatin receptor subtype 1 with the human homolog of the Shk1 kinase-binding protein from yeast. J Biol Chem 2000, 275:9557-9562.

60. Segond von Banchet, G., Schindler, M., Hervieu, G.J., Beckmann, B., Emson, P.C., Heppelmann, B. Distribution of somatostatin receptor subtypes in rat lumbar spinal cord examined with gold-labelled somatostatin and anti-receptor antibodies. Brain Res 1999, 816:254-257.

61. Selmer, I., Schindler, M., Allen, J.P., Humphrey, P.P., Emson, P.C. Advances in understanding neuronal somatostatin receptors. Regul Peptides 2000a, 90:1-18.

62. Selmer, I.S., Schindler, M., Humphrey, P.P., Emson, P.C. Immunohistochemical localization of the somatostatin sst(4) receptor in rat brain. Neuroscience 2000b, 98:523-533.

63. Stroh, T., Kreienkamp, H.J., Beaudet, A. Immunohistochemical distribution of the somatostatin receptor subtype 5 in the adult rat brain: predominant expression in the basal forebrain. J Comp Neurol 1999, 412:69-82.

64. Tannenbaum, G.S., Zhang, W.H., Lapointe, M., Zeitler, P., Beaudet, A. Growth hormone-releasing hormone neurons in the arcuate nucleus express both Sst1 and Sst2 somatostatin receptor genes. Endocrinology 1998, 139:1450-1453.

65. Traina, G., Lanneau, C., Arnoux, A., Porokhov, B., Bagnoli, P., Epelbaum, J. Expression and coupling of somatostatin receptors in rat adrenal (PC12) and pituitary (GC) cell lines. Neurosci Lett 1998, 252:131-134.

66. Tu, J.C., Xiao, B., Naisbitt, S., Yuan, J.P., Petralia, R.S., Brakeman, P., Doan, A., Aakalu, V.K., Lanahan, A.A., Sheng, M., Worley, P.F. Coupling of mGluR/Homer and PSD-95 complexes by the Shank family of postsynaptic density proteins. Neuron 1999, 23:583-592.

67. Vasilaki, A., Gardette, R., Epelbaum, J., Thermos, K. NADPH-diaphorase colocalizes with somatostatin receptor subtypes sst2A and sst2B in the retina. Invest Ophthalmol Vis Sci, in press.

68. Viollet, C., Lanneau, C., Faivre-Bauman, A., Zhang, J., Djordjijevic, D., Loudes, C., Gardette, R., Kordon, C., Epelbaum, J. Distinct patterns of expression and physiological effects of sst1 and sst2 receptor subtypes in mouse hypothalamic neurons and astrocytes in culture. J Neurochem 1997, 68:2273-2280.

69. Xu, Y., Berelowitz, M., Bruno, J.F. Dexamethasone regulates somatostatin receptor subtype messenger ribonucleic acid expression in rat pituitary GH4C1 cells. Endocrinology 1995, 136:5070-5075.

70. Yao, I., Hata, Y., Hirao, K., Deguchi, M., Ide, N., Takeuchi, M., Takai, Y. Synamon, a novel neuronal protein interacting with synapse-associated protein 90/postsynaptic density-95-associated protein. J Biol Chem 1999, 274:27463-27466.

71. Yamada, Y., Post, S.R., Wang, K., Tager, H.S., Bell, G.I., Seino, S. Cloning and functional characterization of a family of human and mouse somatostatin receptors expressed in brain, gastrointestinal tract, and kidney. Proc Natl Acad Sci USA 1992a, 89:251-255.

72. Yamada, Y., Reisine, T., Law, S.F., Ihara, Y., Kubota, A., Kagimoto, S., Seino, M., Seino, Y., Bell, G.I., Seino, S. Somatostatin receptors, an expanding gene family: cloning and functional characterization of human SSTR3, a protein coupled to adenylyl cyclase. Mol Endocrinol 1992b, 6:2136-2142.

73. Zhu, L.J., Krempels, K., Bardin, C.W., O'Carroll, A.M., Mezey, E. The localization of messenger ribonucleic acids for somatostatin receptors 1, 2, and 3 in rat testis. Endocrinology 1998, 139:350-357

74. Zitzer, H., Honck, H.H., Bachner, D., Richter, D., Kreienkamp, H.J. Somatostatin receptor interacting protein defines a novel family of multidomain proteins present in human and rodent brain. J Biol Chem 1999a, 274:32997-33001.

75. Zitzer, H., Richter, D., Kreienkamp, H.J. Agonist-dependent interaction of the rat somatostatin receptor subtype 2 with cortactin-binding protein 1. J Biol Chem 1999b, 274:18153-18156.

# 9

# SOMATOSTATIN RECEPTOR SUBTYPE SELECTIVITY FOR CYTOTOXIC AND CYTOSTATIC SIGNALING

Coimbatore B. Srikant
*Fraser Laboratories, McGill University Health Centre and Royal Victoria Hospital, Montreal, Quebec, Canada , H3A 1A1*

Somatostatin (SST), originally identified as a hypothalamic peptide inhibitor of growth hormone secretion, was subsequently shown to be widely distributed, to exist in two biologically active forms (SST-14 and SST-28) and to function as an inhibitor of secretion of almost every hormone and growth factor. Initially all the cellular effects of SST were viewed as the consequence of its ability to inhibit hormonal secretion. This notion became untenable with the realization that SST could directly inhibit tumor cell proliferation as initially demonstrated by Liebow et al., [1] and confirmed by numerous subsequent studies that have been reviewed earlier [2-7]. Today SST is best regarded as a unique, pleuripotent, hormone which modulates the functions of central and peripheral nervous systems, exocrine and endocrine organs as well as immune and vascular systems to influence such diverse functions as secretion, motility, cell growth and proliferation. A growing body of evidence has now clearly established that the direct antiproliferative action of SST in tumor cells can induce cytotoxic signals to promote apoptotic cell death or activate signals promoting cytostatic growth arrest. Functional characterization of the five cloned SST receptors (SSTR) have revealed that SST-induced apoptosis and growth arrest occur in a SSTR subtype-selective manner that utilizes tyrosine phosphatase-dependent activation of cytotoxic or cytostatic signaling. This review presents a historical perspective of somatostatin and its receptors, our current understanding of their antiproliferative actions and their clinical relevance.

## Regulation of Cellular Signaling by SST

SST acts by binding to its receptors (SSTR), which are members of the heptahelical G protein-coupled receptors. Pertussis toxin-sensitive Gα subfamily of heterotrimeric guanine nucleotide binding (G) proteins couple ligand activated SSTR to several effector pathways including adenylylcyclase-cAMP, protein tyrosine phosphatases, phospholipase A2 and phospholipase C and to Na$^+$/H$^+$ exchanger and, K$^+$ and Ca$^{2+}$ channels to regulate cellular functions (Table 1). SSTRs are coupled to several subsets of K$^+$ channels (delayed rectifier, inward rectifier, ATP-sensitive K$^+$ channels and large conductance Ca$^{2+}$ -activated BK channels). SSTR-mediated activation of K$^+$ channels causes reversible hyper polarization of the membrane, cessation of spontaneous action potential activity, inhibition of normal depolarization induced Ca$^{2+}$ influx via voltage-sensitive Ca$^{2+}$ channels thereby contributing to a reduction in intracellular Ca$^{2+}$. In addition, SSTRs act directly to inhibit high voltage-dependent Ca$^{2+}$ channels and block Ca$^{2+}$ currents by inducing cGMP-dependent protein kinase. Its antisecretory effects arise principally from the reduction in intracellular cAMP and Ca$^{2+}$ while its antiproliferative effects are mediated through regulation of kinases and phosphatases. SSTRs activate a

**TABLE 1**
*Regulation of principal intracellular signaling modulators by SST\**

| **Induction** | **Suppression** |
|---|---|
| cGMP | cAMP |
| Na$^+$/H$^+$ exchange | Na$^+$/H$^+$ exchange |
| K$^+$ | Ca$^{2+}$ |
| Proteases | PKA |
| Endonucleases | PKC |
| Tyrosine phosphatases | Tyrosine kinases |
| Serine/Threonine phosphatases | Serine/Threonine kinases |
| MAPK | MAPK |
| Calcineurin | Phospholipase A2 |
| Phospholipase C | Phospholipase C |

*\*Compiled from data reported in references 1,8-23 and numerous other publications*

number of phosphatases including serine threonine phosphatases [8], calcineurin [9] and protein tyrosine phosphatases [1,10-15]. We showed that membrane associated PTP is not directly activated in breast cancer cells (e.g., MCF-7, T47D and ZR75-1) contradicting earlier claims to that effect [11,16,17] and identified the SH2 domain bearing phosphotyrosine phosphatase SHP-1 (PTP1C/SHPTP1/HCP) as the intracellular PTP recruited to the membrane in cells preincubated with the SST analog octreotide (OCT) in a time- and concentration-dependent manner [17]. The association of SHP-1 with SSTR has been confirmed by other laboratories [18-23].

The antiproliferative actions of SST analogs have been demonstrated both *in vivo* and *in vitro*. For example, SST analogs such as OCT, lanreotide (BIM23014) and vapreotide (RC-160) inhibit the growth of DMBA-induced or transplanted rat mammary carcinomas and cultured cells derived from endocrine, neuroendocrine and exocrine tumors (brain, breast, colon, lung, pancreas, pituitary, prostate and thyroid [24-26]. These analogs retain the biologically active core amino acid residues (Phe[7], Trp[8], Lys[9] and Thr[10]) of SST and are rendered metabolically more stable than SST by virtue of the substitution L-Trp by D-Trp moiety. Their antiproliferative actions *in vivo* are exerted in part through inhibition of the secretion of hormones and growth factors that promote tumor growth, angiogenesis, vasoconstriction, and modulate immune cell function. In addition to their antisecretory effects, they do exert direct cytostatic or cytotoxic actions in tumor cells to induce growth arrest and apoptosis respectively [20,26-39]. For instance, SST treatment causes apoptosis in some human breast cancer cells (MCF-7, T47D, ZR75-1), mouse pituitary tumor (AtT-20) cells and in transplantable murine colon38 cancer cell line. By contrast, it triggers $G_1$ or $G_2$ cell cycle arrest in tumor cells derived from the thyroid (FRTL), rat pituitary ($GH_3$) and mouse pancreas (MIN6) [29,31,32,40-42]. We identified the redistribution of cytosolic SHP-1 to the membrane, induction of wild type (wt) tumor suppressor protein p53, the proapoptotic protein Bax, caspases (cysteine-aspartate specific proteases), disruption of pH homeostasis and activation of a cation-insensitive acidic endonuclease as key events in SST-induced apoptosis in MCF-7 breast cancer cells [17,33,43]. SST-induced SHP-1-dependent disruption of pH homeostasis occurs distal to caspase-8, p53 and Bax [20,35,36]. We have further characterized the role of SHP-1 in SSTR mediated acidification-dependent apoptosis and more recently established that SHP-1 abrogates the anti-apoptotic activity of Bcl-2 [17,20,35,36,44-46]. SST-induced acidification and apoptosis was abrogated by Bcl-2 at the level of the endoplasmic reticulum, but not mitochondria [45]. SST-induced acidification could be mediated by the inhibition of $Na^+/H^+$ exchanger and vacuolar $H^+$ATPase, since chemical inhibitors of these proton transport mechanisms decrease the cell pH to the same extent as SST and trigger acidification-dependent apoptosis [36]. By comparing the effects of ectopically expressed wild type and catalytically inactive mutant of SHP-1 we documented that SHP-1 is necessary not only for SST-induced, caspase-8-dependent reduction in cytoplasmic pH, but also for acidification-dependent activation of effector caspases and apoptosis to occur [36]. The antiproliferative signaling of SST is not dependent on its inhibition of cAMP, although when elevated it prevents SST-induced acidification by both PKA-dependent and -independent mechanisms [37]. These findings demonstrate that the antiproliferative and antisecretory actions of SST are distinct and are not interdependent. Interestingly, SHP-1 has also been implicated in SST-induced cytostatic cell growth inhibition [18,21,22,47]. Such cytostatic action has been shown to inhibit cell cycle progression at $G_1/G_0$ and or $G_2/M$ phases by inducing p21 and/or p27 in tumor cells derived from pancreas, pituitary, thyroid, colon, lung and prostate [28-32,48-50].

## Somatostatin Receptor Family

Five distinct SSTR subtypes encoded by nonallelic genes present on different chromosomes have been identified, characterized and shown to be variably expressed in both normal and tumor cells [7,51,52]. Four of these genes are intronless while the fifth gene has a single intron and generates alternately spliced forms of SSTR2A and SSTR2B that differ only in the length of the cytoplasmic C-tail [5]. SSTRs have now been classified into two distinct subgroups on the basis of high (SSTRs 2,3 and 5) or extremely poor (SSTRs1 and 4) binding affinities for octa- and hexa-peptide SST analogs. Such a classification holds good on a structural basis as well since SSTRs 2,3 and 5 share >50% sequence homology whereas SSTRs 1 and 4 are 60% identical and 78% similar [53-57]. SSTRs 2,3,4 and 5 functionally couple to $G\alpha_{i1}$, $G\alpha_{i2}$, $G\alpha_{i3}$ and $G\alpha_{o2}$ [58,59] whereas SSTR3 associates with $G\alpha_{i1}$, $G\alpha_{i3}$ and $G\alpha_o$ [60,61]. One or more of these pertussis toxin sensitive G proteins couple all five SSTRs negatively to adenylylcyclase and positively to phosphotyrosine phosphatase [13,27,43,44,62]. Several SSTR subtypes have been shown to modulate MAP kinases [13,28,41,63,64]. While SSTR3 was shown to inhibit Raf-1-dependent activation of MAPK in NIH3T3 cells [13], it was found to exert a dual effect in MIN-6 cells characterized by a transient increase followed by subsequent decrease [41]. Cordelier et al., reported that SSTR5 is negatively coupled to MAPK through cGMP and not tyrosine phosphatase [64], a claim that remains unsubstantiated by others.

## Expression of SSTR Subtypes in Tumors

SSTRs are expressed in many tumor cell lines including those derived from the pituitary (AtT-20, GH4C1) islet (Rinm5f, HIT), pancreatic (AR42J, MIN), breast (MCF-7, T47D, ZR75-1) and brain (LAN-2 neuroblastoma) [5,52,65]. SSTR expression and/or binding has been reported in a vast number of benign as well as malignant human tumors including the functioning and nonfunctioning pituitary tumors, carcinoid tumors, insulinomas, glucagonomas, pheochromocytoma, breast, renal and prostate carcinomas, meningioma and glioma. Given the diverse actions of SST and the heterogeneity of its receptors, its ability to differentially regulate the growth and survival of different tumor cells may arise from the variable expression of SSTR subtypes and/or subtype-selective activation of signaling events that elicit cytostatic or cytotoxic response. As discussed in a previous review [6] SSTR2 mRNA is the most abundantly expressed, followed by those of SSTRs 1,3 and 4. Interestingly SSTR5 mRNA expression appears to be somewhat specific in that it is abundant in tumors of the breast but is absent in islet cell tumors. While the fact that these tumors express multiple SSTR subtypes to varying degrees at the level of mRNA has been established by RT-PCR analysis, it remains to be confirmed if the pattern of expression of SSTR subtypes at the mRNA level reflects the pattern of expression of functional receptors at the protein level. To date immunoblot analysis has been described only for SSTR2 in a small number of tumors. Such analysis is semi-quantitative at best due to limitations of antibody specificity and sensitivity of detection. Quantitation of their relative abundance remains a daunting task and requires detailed binding analysis using agonists possessing exquisite SSTR subtype specificity. The peptidic and non-peptidic agonists currently available exhibit SSTR subtype selectivity, but lack the absolute specificity required for such a task.

## SSTR Subtype Selective Antiproliferative Signaling

Insights into agonist dependent, SSTR subtype-selective, antiproliferative signaling have been obtained from the heterologous expression of individual subtypes in SSTR negative cells like CHO-K1, COS-7 and HEK293 [44,64,66,67]. In order to determine the SSTR subtype selectivity for SST-induced cytostatic and cytotoxic signaling we compared the effect of agonist treatment on cell cycle parameters, intracellular pH, p53, Rb, p21, caspases, endonuclease activation and the integrity of genomic DNA in CHO-K1 cells stably expressing each of the five human (h) SSTR subtypes. We first compared the changes in cell cycle parameters following treatment with OCT (SSTRs 2,3 and 5) or D-Trp[8] SST-14 (SSTRs1 and 4). The findings presented in fig. 1 shows that hSSTR3-mediated changes in cell cycle parameters were distinct from that induced through other subtypes: a marked decrease in $G_1$ phase and a concomitant appearance of cells with hypodiploid DNA content typing the presence of cells undergoing apoptosis. By contrast, the growth regulatory effects of SST agonists exerted through other four SSTRs induced an increase in the cell population in $G_1$. The rank order potency of these subtypes for cytostatic signaling was hSSTR5~hSSTR2 >hSSTR4>hSSTR1. In parallel experiments we determined the proliferative index of these cells from the ratio of cells in $G_1$ and S phases in the absence and presence of 100 nM D-Trp[8] SST-14 for 24 h. A decrease in the $G_1$/S ratio observed in cells expressing hSSTRs 1,2,4 and 5 signified the inhibition of proliferation (fig. 2). By contrast, there was pronounced reduction in $G_1$/S ratio in hSSTR3 expressing cells indicating that the cytotoxic signaling of SST occurred in the absence of cell cycle block. hSSTR3-mediated cytotoxic signaling and the cytostatic effects elicited through other hSSTR subtypes were pertussis toxin-sensitive G protein mediated and tyrosine phosphatase-dependent [43,44,66]. The hSSTR3-mediated cytotoxic signaling and hSSTR5-mediated cytostatic signaling were abrogated by the truncation of their respective C-terminal domains as well as by point mutations at Ser residues within the C-tails [44,67]. Our finding that hSSTRs 1 and 4 display rather poor coupling to the cytostatic signaling pathway, but not stimulate cell proliferation contradicts previous claims that SSTR4-mediated effect of SST is mitogenic and not growth inhibitory [68,69]. These findings have been confirmed by others [21,22,27,70-72]. However, Guillermet et al., [19] reported that apoptosis could be induced in BxPC-3 pancreatic cancer cells by enforced expression of SSTR2 in its native state (i.e., the absence of exogenous SST) [19]. These authors also claimed that SSTR2

148

transfection in BxPC-3 cells up-regulates TNFR expression thereby enhancing the apoptotic responsiveness to death receptor activation [19]. The claim that this results from an autocrine feedback effect through "*SSTR2-induced up regulation of SST expression*" as previously hypothesized [73] remains unsubstantiated. Moreover, the degree of apoptosis reported in this study was <2%, a level well below the sensitivity of detection in contrast to >30% apoptosis induced through SSTR3.

*hSSTR3-signaled apoptosis is p53 mediated*

We demonstrated that hSSTR3-mediated cytotoxic signaling is associated with the induction of wild type p53. Dual label analysis of cells labeled with epitope-specific, wild type selective, anti-p53 antibody (pAb 1802) and PI revealed OCT-induced increase in p53 and, in particular, correlated with the hypodiploid DNA content in apoptotic cells (fig.3). Likewise, an increase in the proapoptotic protein Bax was also observed [66]. hSSTR3-mediated apoptosis occurred in the absence of any change in p21 and c-Myc [66]. Other biochemical changes that typify CHO-K1/hSSTR3 cells undergoing SST-induced apoptosis included cytoplasmic acidification and activation of a cation-insensitive acidic endonuclease [43]. hSSTR-3-mediated cytostatic effects also induced distinct morphological changes indicative of nuclear shrinkage [44,66].

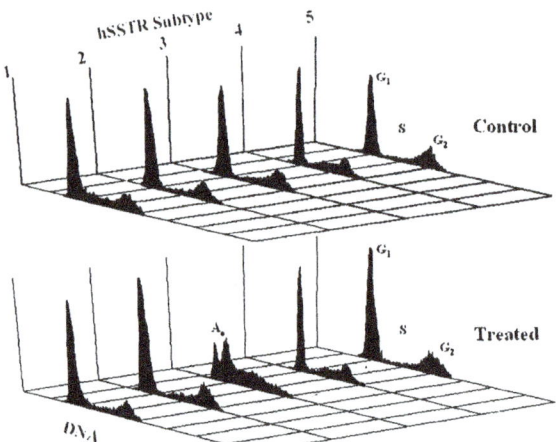

FIGURE 1. *Subtype-selective hSSTR-mediated regulation of cell cycle progression. CHO-K1 cells expressing individual hSSTRs 1-5 were incubated for 24 h in the absence (top panel, Control) and presence of (bottom panel, treated) of 100 nM D-Trp$^8$ SST-14. Cell cycle profiles were determined by propidium iodide stained nuclear DNA in a Coulter EPICS 750 series flow cytometer. A 5 watt argon laser lamp that emits light at 351-363 nm excited the propidium iodide fluorescence that whose emission was detected through a 610 nm long pass filter. Data was analyzed using the WinMDI (http://facs.scripps.edu/software.html) software.*

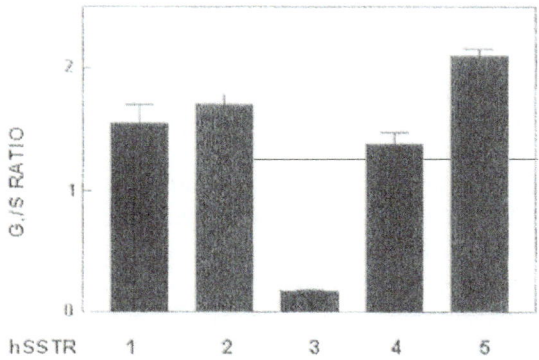

FIGURE 2. hSSTR subtype selective regulation of cell cycle progression in CHO-K1 cells. Proliferative index was calculated from ratio of cells in $G_1$ and S phase in following treatment with 100 nM D-Trp[8] SST-14 for 24 h. The change in $G_1/S$ ratio in the treated cells is expressed relative to the $G_1/S$ ratios of the corresponding untreated cells taken as 1. The increase in $G_1/S$ ratios in cells expressing hSSTRs 2,5, 4 and 1 indicates that peptide treatment elicits cytostatic response through these subtypes. By contrast, the absence of cytostatic effect was indicated by the striking decrease in $G_1/S$ ratio in hSSTR3 expressing cells.

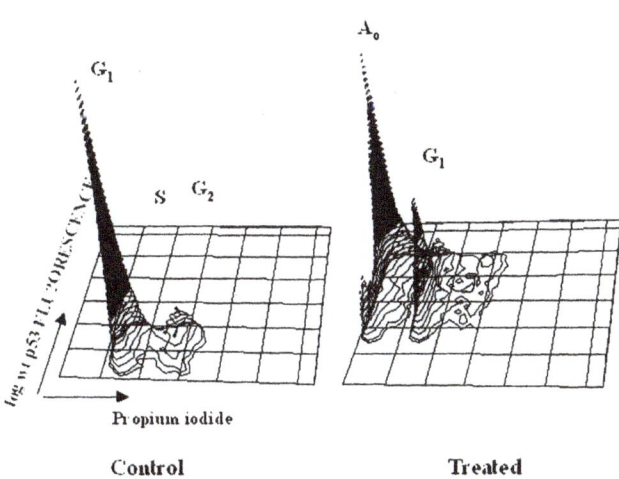

FIGURE 3. Bivariate analysis of hSSTR3 expressing CHO-K1 cells dual labeled with pAb 1801 (wt p53) and propidium iodide (DNA). OCT (100 nM) induced increase in wt p53 occurs in all phases of the cell cycle. Additionally, the correlation between induction of wt p53 and apoptosis is seen from the increase in this tumor suppressor protein in cells with hypodiploid DNA content in the $A_0$ region.

*Implications of SSTR subtype selective differential antiproliferative actions in the rationalization of SST therapy in cancer*

The use of SST therapy has been established in several types of tumors including pituitary adenomas (acromegaly, TSHoma), neuroendocrine and islet

150

cell metastatic tumors and, in chemotherapy-induced diarrhea and acute esophageal vericea bleeding. Agonist-induced internalization of SSTR occurs maximally via SSTR3 followed by SSTR5, 4 and 2 [74]. The high rate of mediated internalization of SSTR3-mediated internalization coupled with the induction of p53 and Bax provides an opportunity for targeted delivery of tagged radionuclides or chemotoxins to maximize the cytotoxic responsiveness of SSTR3 positive tumor cells. Derivatives of octreotide, tyr3-octreotide, tyr3-octreotate, vapreotide and lanreotide containing the chelator moieties of DTPA (diethylene triamine penta-acetic acid) or DOTA (1,4,7,10-tetraazacyclododecane-1,4,7,10-tetra-acetic acid) and tagged with □- and □-emitting radionuclides such as $^{177}$Lu, $^{111}$In, $^{99}$Tc, $^{90}$Y, $^{67/68}$Ga, $^{64}$Cu and $^{18}$F have been used in the diagnosis and treatment of SSTR positive tumors [4,75-78].

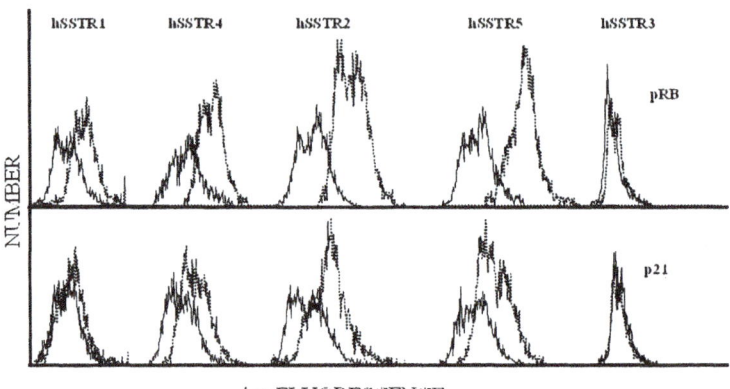

FIGURE 4. $G_1$ cell cycle arrest signaled via hSSTRs 1,2,4 and 5 in CHO-K1 cells is associated with induction of pRB (top panel) and p21 (bottom panel). Cells were labeled with anti pRB antibody and FITC conjugated second antibody. Solid lines represent the untreated cells and dotted lines D-Trp$^8$ SST-14-treated cells. Results are representative of four individual experiments.

The radionuclide-tagged SST analogs are effective in labeling of cell surface SSTRs thereby enabling preoperative localization of tumor mass. An entirely different approach was developed by Schally's group to target SSTR positive tumors with vapreotide-tagged chemotoxic agents like doxorubicin [79-83]. The present generation of SST agonists tagged with radionuclides as well as SST analogs like AN-238 (doxorubicin-conjugated vapreotide) bind with greater affinity to SSTRs 2 and 5 than to SSTR3 [81,84,85]. Although intracellular accumulation of radionuclide-tagged SST analogs can occur through SSTRs 2,4 and 5, the efficacy of radiation-induced tumor cell damage in such cases will not be optimal since these subtypes do not induce p53 and Bax. To date SSTR3-selective agonists carrying radionuclide or chemotoxin have not been developed. Many tumors harbor mutated p53 gene and these, even if SSTR3 positive, will show blunted response to radiation-induced killing by radionuclides and chemotoxins internalized through SSTR3-mediated endocytosis. Another factor that needs to be resolved is how the cytotoxic action of SST transduced through SSTR3 is influenced by the cytostatic effect manifested by SST through the other SSTR subtypes. It is now well established that heteromeric association between SSTR subtypes can alter the intracellular effect of SST agonists. For instance, agonist-induced heterodimerization of SSTR1 and SSTR5 increases their negative coupling to adenylylcyclase [86] and of SSTR2 and SSTR5 to Ras [87]. By contrast,

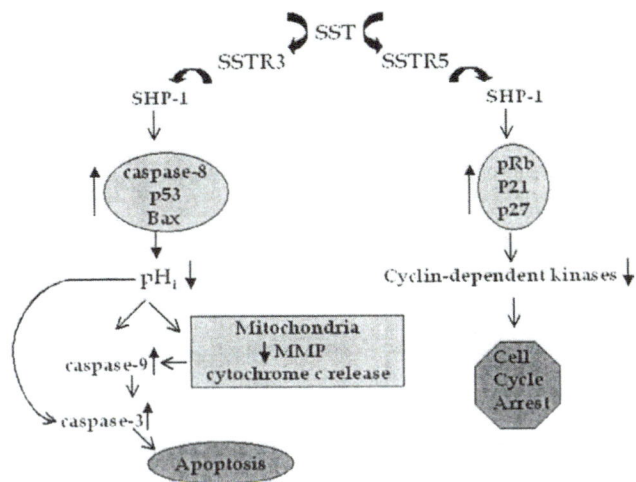

FIGURE 5. *Subtype selective hSSTR-mediated antiproliferative signaling mechanisms. SHP-1-dependent distinct antiproliferative actions of SST arise from SSTR subtype-selective regulation of cytostatic and cytotoxic signaling cascades. Disruption of pH homeostasis leading to the reduction of intracellular pH (pHᵢ) caused by the induction of caspase-8, p53 and Bax promotes acidification-dependent activation of effector caspases both directly and through activation of caspase-9 through proapoptotic changes in the mitochondria (reduction in mitochondrial membrane potential, MMP and release of cytochrome c into the cytoplasm) leading to apoptosis. By contrast, hSSTR5-mediated cytostatic action promotes cell cycle arrest due to the suppression of cyclin-dependent kinases through induction of Rb, p21 and p27.*

physical association between SSTR2 and SSTR3 blunts SSTR3-mediated signaling [88]. A clear understanding of the complexities involved require a precise knowledge of the relative expression pattern of SSTR subtypes at a protein level in a given tumor, development of truly subtype-selective SST agonists and the influence of modulation SSTR-mediated signal transduction by the physical interaction between different SSTR subtypes.

## ACKNOWLEDGEMENTS

I wish to thank Drs. K. Sharma, M. Thangaraju, and Mr. G. Martino for their valuable contribution to the studies cited in this review. I am especially indebted to the late Dr. Y.C. Patel for his contribution, collaboration and encouragement for over two decades. Our work cited in this review was supported by grants from the Canadian Institutes of Health (formerly Canadian Medical Research Council) and National Cancer Institute of Canada.

# REFERENCES

1.  Liebow, C, Reilly, C, Serrano, M, and Schally, AV, *Somatostatin analogues inhibit growth of pancreatic cancer by stimulating tyrosine phosphatase.* Proceedings of the National Academy of Sciences of the United States of America, 1989. **86**: 2003-2007.

2.  Schally, AV, *Oncological applications of somatostatin analogues.* Cancer Res, 1988. **48**: 6977-6985.

3.  Lamberts, SW, Krenning, EP, and Reubi, JC, *The role of somatostatin and its analogs in the diagnosis and treatment of tumors.* Endocr Rev, 1991. **12**: 450-482.

4.  Hofland, LJ and Lamberts, SW, *The pathophysiological consequences of somatostatin receptor internalization and resistance.* Endocr Rev, 2003. **24**: 28-47.

5.  Patel, YC and Srikant, CB, *Somatostatin receptors.* Trends Endocrinol Metabol, 1997. **8**: 398-405.

6.  Patel, YC and Srikant, CB, *Somatostatin and its receptors.* Adv Mol Cell Endocrinol, 1999. **3**: 43-73.

7.  Patel, YC, *Somatostatin and its receptor family.* Front Neuroendocrinol, 1999. **20**: 157-198.

8.  White, RE, Schonbrunn, A, and Armstrong, DL, *Somatostatin stimulates Ca(2+)-activated K+ channels through protein dephosphorylation.* Nature, 1991. **351**: 570-573.

9.  Renstrom, E, Ding, WG, Bokvist, K, and Rorsman, P, *Neurotransmitter-induced inhibition of exocytosis in insulin-secreting beta cells by activation of calcineurin.* Neuron, 1996. **17**: 513-522.

10. Liebow, C, Lee, MT, and Schally, A, *Antitumor effects of somatostatin mediated by the stimulation of tyrosine phosphatase.* Metabolism, 1990. **39**: 163-166.

11. Pan, MG, Florio, T, and Stork, PJ, *G protein activation of a hormone-stimulated phosphatase in human tumor cells.* Science, 1992. **256**: 1215-1217.

12. Rivard, N, Lebel, D, Laine, J, and Morisset, J, *Regulation of pancreatic tyrosine kinase and phosphatase activities by cholecystokinin and somatostatin.* American Journal of Physiology, 1994. **266**: G1130-1138.

13. Reardon, DB, Wood, SL, Brautigan, DL, Bell, GI, Dent, P, and Sturgill, TW, *Activation of a protein tyrosine phosphatase and inactivation of Raf-1 by somatostatin.* Biochemical Journal, 1996. **314**: 401-404.

14. Tahiri-Jouti, N, Cambillau, C, Viguerie, N, Vidal, C, Buscail, L, Saint Laurent, N, Vaysse, N, and Susini, C, *Characterization of a membrane tyrosine phosphatase in AR42J cells: regulation by somatostatin.* Am J Physiol, 1992. **262**: G1007-1014.

15. Vantus, T, Csermely, P, Teplan, I, and Keri, G, *The tumor-selective somatostatin analog, TT2-32 induces a biphasic activation of phosphotyrosine phosphatase activity in human colon tumor cell line, SW620.* Tumour Biol, 1995. **16**: 261-267.

16. Hierowski, MT, Liebow, C, du Sapin, K, and Schally, AV, *Stimulation by somatostatin of dephosphorylation of membrane proteins in pancreatic cancer MIA PaCa-2 cell line.* FEBS Lett, 1985. **179**: 252-256.

17. Srikant, CB and Shen, SH, *Octapeptide Somatostatin Analog Sms 201-995 Induces Translocation of Intracellular Ptp1c to Membranes in Mcf-7 Human Breast Adenocarcinoma Cells.* Endocrinology, 1996. **137**: 3461-3468.

18. Bousquet, C, Delesque, N, Lopez, F, Saint-Laurent, N, Esteve, JP, Bedecs, K, Buscail, L, Vaysse, N, and Susini, C, *sst2 somatostatin receptor mediates negative regulation of insulin receptor signaling through the tyrosine phosphatase SHP-1.* J Biol Chem, 1998. **273**: 7099-7106.

19.    Guillermet, J, Saint-Laurent, N, Rochaix, P, Cuvillier, O, Levade, T, Schally, AV, Pradayrol, L, Buscail, L, Susini, C, and Bousquet, C, *Somatostatin receptor subtype 2 sensitizes human pancreatic cancer cells to death ligand-induced apoptosis.* Proc Natl Acad Sci U S A, 2003. **100**: 155-160.

20.    Liu, D, Martino, G, Thangaraju, M, Sharma, M, Halwani, F, Shen, SH, Patel, YC, and Srikant, CB, *Caspase-8-mediated intracellular acidification precedes mitochondrial dysfunction in somatostatin-induced apoptosis.* J Biol Chem, 2000. **275**: 9244-9250.

21.    Lopez, F, Esteve, JP, Buscail, L, Delesque, N, Saint-Laurent, N, Theveniau, M, Nahmias, C, Vaysse, N, and Susini, C, *The tyrosine phosphatase SHP-1 associates with the sst2 somatostatin receptor and is an essential component of sst2-mediated inhibitory growth signaling.* Journal of Biological Chemistry, 1997. **272**: 24448-24454.

22.    Pages, P, Benali, N, Saint-Laurent, N, Esteve, JP, Schally, AV, Tkaczuk, J, Vaysse, N, Susini, C, and Buscail, L, *sst2 somatostatin receptor mediates cell cycle arrest and induction of p27(Kip1). Evidence for the role of SHP-1.* J Biol Chem, 1999. **274**: 15186-15193.

23.    Benali, N, Cordelier, P, Calise, D, Pages, P, Rochaix, P, Nagy, A, Esteve, JP, Pour, PM, Schally, AV, Vaysse, N, Susini, C, and Buscail, L, *Inhibition of growth and metastatic progression of pancreatic carcinoma in hamster after somatostatin receptor subtype 2 (sst2) gene expression and administration of cytotoxic somatostatin analog AN-238.* Proc Natl Acad Sci U S A, 2000. **97**: 9180-9185.

24.    Weckbecker, G, Raulf, F, Stolz, B, and Bruns, C, *Somatostatin analogs for diagnosis and treatment of cancer.* Pharmacology & Therapeutics, 1993. **60**: 245-264.

25.    Weckbecker, G, Tolcsvai, L, Stolz, B, Pollak, M, and Bruns, C, *Somatostatin analogue octreotide enhances the antineoplastic effects of tamoxifen and ovariectomy on 7,12-dimethylbenz(alpha)anthracene-induced rat mammary carcinomas.* Cancer Res, 1994. **54**: 6334-6337.

26.    Hoelting, T, Duh, QY, Clark, OH, and Herfarth, C, *Somatostatin analog octreotide inhibits the growth of differentiated thyroid cancer cells in vitro, but not in vivo.* Journal of Clinical Endocrinology & Metabolism, 1996. **81**: 2638-2641.

27.    Buscail, L, Esteve, JP, Saint-Laurent, N, Bertrand, V, Reisine, T, AM, OC, Bell, GI, Schally, AV, Vaysse, N, and Susini, C, *Inhibition of cell proliferation by the somatostatin analogue RC-160 is mediated by somatostatin receptor subtypes SSTR2 and SSTR5 through different mechanisms.* Proc Natl Acad Sci U S A, 1995. **92**: 1580-1584.

28.    Cattaneo, MG, Amoroso, D, Gussoni, G, Sanguini, AM, and Vicentini, LM, *A somatostatin analogue inhibits MAP kinase activation and cell proliferation in human neuroblastoma and in human small cell lung carcinoma cell lines.* FEBS Letters, 1996. **397**: 164-168.

29.    Cheung, NW and Boyages, SC, *Somatostatin-14 and its analog octreotide exert a cytostatic effect on GH3 rat pituitary tumor cell proliferation via a transient G0/G1 cell cycle block.* Endocrinology, 1995. **136**: 4174-4181.

30.    Florio, T, Scorizello, A, Fattore, M, D'Alto, V, Salzano, S, Rossi, G, Berlingieri, MT, Fusco, A, and Schettini, G, *Somatostatin inhibits PC Cl3 thyroid cell proliferation through the modulation of phosphotyrosine activity. Impairment of the somatostatinergic effects by stable expression of E1A viral oncogene.* J Biol Chem, 1996. **271**: 6129-6136.

31.    Medina, DL, Toro, MJ, and Santisteban, P, *Somatostatin interferes with thyrotropin-induced G1-S transition mediated by cAMP-dependent protein kinase and phosphatidylinositol 3- kinase. Involvement of RhoA and cyclin E x cyclin-dependent kinase 2 complexes.* J Biol Chem, 2000. **275**: 15549-15556.

32.    Medina, DL, Velasco, JA, and Santisteban, P, *Somatostatin is expressed in FRTL-5 thyroid cells and prevents thyrotropin-mediated down-regulation of the cyclin-dependent kinase inhibitor p27kip1.* Endocrinology, 1999. **140**: 87-95.

33.    Sharma, K and Srikant, CB, *Induction of wild type p53, Bax and a cation-insensitive acidic endonuclease during somatostatin signaled apoptosis in MCF-7 human breast cancer cells.* Int J Canc, 1998. **76**: 259-266.

154

34.    Srikant, CB, *Cell cycle dependent induction of apoptosis by somatostatin analog SMS 201-995 in AtT-20 mouse pituitary cells.* Biochemical & Biophysical Research Communications, 1995. **209**: 400-406.

35.    Thangaraju, M, Sharma, K, Leber, B, Andrews, DW, Shen, SH, and Srikant, CB, *Regulation of acidification and apoptosis by SHP-1 and Bcl-2.* J Biol Chem, 1999. **274**: 29549-29557.

36.    Thangaraju, M, Sharma, K, Liu, D, Shen, SH, and Srikant, CB, *Interdependent regulation of intracellular acidification and SHP-1 in apoptosis.* Cancer Res, 1999. **59**: 1649-1654.

37.    Thangaraju, M, Halwani, F, and Srikant, CB, *cAMP inhibits acidification-dependent apoptosis by attenuating the reduction in pHi distal to caspase-8.* Mol Cell Biochem, 2003: submitted.

38.    Candi, E, Melino, G, De Laurenzi, V, Piacentini, V, Guerreri, P, Spinedi, A, and Knight, RA, *Tamoxifen and somatostatin affect tumors by inducing apoptosis.* Cancer Lett, 1995. **96**: 141-145.

39.    Melen-Mucha, G, Winczyk, K, and Pawlikowski, M, *Somatostatin analogue octreotide and melatonin inhibit bromodeoxyuridine incorporation into cell nuclei and enhance apoptosis in the transplantable murine colon 38 cancer.* Anticancer Res, 1998. **18**: 3615-3619.

40.    Florio, T, Thellung, S, and Schettini, G, *Intracellular transducing mechanisms coupled to brain somatostatin receptors.* Pharmacol Res, 1996. **33**: 297-305.

41.    Yoshitomi, H, Fujii, Y, Miyazaki, M, Nakajima, N, Inagaki, N, and Seino, S, *Involvement of MAP kinase and c-fos signaling in the inhibition of cell growth by somatostatin.* American Journal of Physiology, 1997. **272**: E769-E774.

42.    Florio, T, Thellung, S, Arena, S, Corsaro, A, Spaziante, R, Gussoni, G, Acuto, G, Giusti, M, Giordano, G, and Schettini, G, *Somatostatin and its analog lanreotide inhibit the proliferation of dispersed human non-functioning pituitary adenoma cells in vitro.* Eur J Endocrinol, 1999. **141**: 396-408.

43.    Sharma, K and Srikant, CB, *G protein coupled receptor signaled apoptosis is associated with activation of a cation insensitive acidic endonuclease and intracellular acidification.* Biochem Biophys Res Commun, 1998. **242**: 134-140.

44.    Sharma, K, Patel, YC, and Srikant, CB, *C-Terminal region of human somatostatin receptor 5 is required for induction of Rb and $G_1$ cell cycle arrest.* Mol Endocrinol, 1999. **13**: 82-90.

45.    Thangaraju, M, Sharma, K, Shen, S-H, Leber, B, Andrews, DW, and Srikant, CB, *Functional inactivation of Bcl-2 by SHP-1 restores organelle-specific pH-dependent and -independent apoptotic signaling.* J Biol Chem, 2003.

46.    Zhao, H, Tourian Jr., L, Thangaraju, M, Leber, B, Andrews, DW, Shen, S-H, and Srikant, CB, *Involvement of c-jun-NH2 terminal kinase in the functional inactivation of Bcl-2 by SHP-1.* EMBO J, 2003: manuscript in preparation.

47.    Douziech, N, Calvo, E, Coulombe, Z, Muradia, G, Bastien, J, Aubin, RA, Lajas, A, and Morisset, J, *Inhibitory and stimulatory effects of somatostatin on two human pancreatic cancer cell lines: a primary role for tyrosine phosphatase SHP-1.* Endocrinology, 1999. **140**: 765-777.

48.    Ain, KB and Taylor, KD, *Somatostatin analogs affect proliferation of human thyroid carcinoma cell lines in vitro.* Journal of Clinical Endocrinology & Metabolism, 1994. **78**: 1097-1102.

49.    Charland, S, Boucher, MJ, Houde, M, and Rivard, N, *Somatostatin inhibits Akt phosphorylation and cell cycle entry, but not p42/p44 mitogen-activated protein (MAP) kinase activation in normal and tumoral pancreatic acinar cells.* Endocrinology, 2001. **142**: 121-128.

50.    Brevini, TA, Bianchi, R, and Motta, M, *Direct inhibitory effect of somatostatin on the growth of the human prostatic cancer cell line LNCaP: possible mechanism of action.* Journal of Clinical Endocrinology & Metabolism, 1993. **77**: 626-631.

51.    Patel, YC, Greenwood, M, Panetta, R, Hukovic, N, Grigorakis, S, Robertson, LA, and Srikant, CB, *Molecular biology of somatostatin receptor subtypes. [Review] [31 refs].* Metabolism: Clinical & Experimental, 1996. **45**: 31-38.

52. Patel, YC, Greenwood, MT, Panetta, R, Demchyshyn, L, Niznik, H, and Srikant, CB, *The somatostatin receptor family. [Review] [125 refs]*. Life Sciences, 1995. **57**: 1249-1265.

53. Panetta, R, Greenwood, MT, Warszynska, A, Demchyshyn, LL, Day, R, Niznik, HB, Srikant, CB, and Patel, YC, *Molecular cloning, functional characterization, and chromosomal localization of a human somatostatin receptor (somatostatin receptor type 5) with preferential affinity for somatostatin-28*. Molecular Pharmacology, 1994. **45**: 417-427.

54. Bruno, JF, Xu, Y, Song, J, and Berelowitz, M, *Molecular cloning and functional expression of a brain-specific somatostatin receptor*. Proceedings of the National Academy of Sciences of the United States of America, 1992. **89**: 11151-11155.

55. Xu, Y, Song, J, Bruno, JF, and Berelowitz, M, *Molecular cloning and sequencing of a human somatostatin receptor, hSSTR4*. Biochemical & Biophysical Research Communications, 1993. **193**: 648-652.

56. Demchyshyn, LL, Srikant, CB, Sunahara, RK, Kent, G, Seeman, P, Van Tol, HH, Panetta, R, Patel, YC, and Niznik, HB, *Cloning and expression of a human somatostatin-14-selective receptor variant (somatostatin receptor 4) located on chromosome 20*. Mol Pharmacol, 1993. **43**: 894-901.

57. Corness, JD, Demchyshyn, LL, Seeman, P, Van Tol, HH, Srikant, CB, Kent, G, Patel, YC, and Niznik, HB, *A human somatostatin receptor (SSTR3), located on chromosome 22, displays preferential affinity for somatostatin-14 like peptides*. FEBS Lett, 1993. **321**: 279-284.

58. Luthin, DR, Eppler, CM, and Linden, J, *Identification and quantification of Gi-type GTP-binding proteins that copurify with a pituitary somatostatin receptor*. Journal of Biological Chemistry, 1993. **268**: 5990-5996.

59. Gu, YZ and Schonbrunn, A, *Coupling specificity between somatostatin receptor sst2A and G proteins: isolation of the receptor-G protein complex with a receptor antibody*. Mol Endocrinol, 1997. **11**: 527-537.

60. Law, SF, Yasuda, K, Bell, GI, and Reisine, T, *Gi alpha 3 and G(o) alpha selectively associate with the cloned somatostatin receptor subtype SSTR2*. Journal of Biological Chemistry, 1993. **268**: 10721-10727.

61. Law, SF, Zaina, S, Sweet, R, Yasuda, K, Bell, GI, Stadel, J, and Reisine, T, *Gi alpha 1 selectively couples somatostatin receptor subtype 3 to adenylyl cyclase: identification of the functional domains of this alpha subunit necessary for mediating the inhibition by somatostatin of cAMP formation*. Molecular Pharmacology, 1994. **45**: 587-590.

62. Florio, T, Rim, C, Hershberger, RE, Loda, M, and Stork, PJ, *The somatostatin receptor SSTR1 is coupled to phosphotyrosine phosphatase activity in CHO-K1 cells*. Mol Endocrinol, 1994. **8**: 1289-1297.

63. Bito, H, Mori, M, Sakanaka, C, Takano, T, Honda, Z, Gotoh, Y, Nishida, E, and Shimizu, T, *Functional coupling of SSTR4, a major hippocampal somatostatin receptor, to adenylate cyclase inhibition, arachidonate release and activation of the mitogen-activated protein kinase cascade*. Journal of Biological Chemistry, 1994. **269**: 12722-12730.

64. Cordelier, P, Esteve, JP, Bousquet, C, Delesque, N, O'Carroll, AM, Schally, AV, Vaysse, N, Susini, C, and Buscail, L, *Characterization of the antiproliferative signal mediated by the somatostatin receptor subtype sst5*. Proc Natl Acad Sci U S A, 1997. **94**: 9343-9348.

65. Reisine, T and Bell, GI, *Molecular biology of somatostatin receptors*. Endocr Rev, 1995. **16**: 427-442.

66. Sharma, K, Patel, YC, and Srikant, CB, *Subtype-selective induction of wild-type p53 and apoptosis, but not cell cycle arrest, by human somatostatin receptor 3*. Mol Endocrinol, 1996. **10**: 1688-1696.

67. Rocheville, M, Kumar, U, Semaan, L, Sasi, R, Srikant, CB, Khare, S, Chan, M, and Patel, YC. *Apoptotic signaling by somatostatin receptor type 3 (SSTR3) requires molecular signals in the receptor C-tail. in ENDO 2000, 82nd Annu Mtg of the Endocrine Society*. 2000. Toronto.

156

68.     Sakanaka, C, Ferby, I, Waga, I, Bito, H, and Shimizu, T, *On the mechanism of cytosolic phospholipase A2 activation in CHO cells carrying somatostatin receptor: wortmannin-sensitive pathway to activate mitogen-activated protein kinase.* Biochem Biophys Res Commun, 1994. **205**: 18-23.

69.     Shimizu, T, Mori, M, Bito, H, Sakanaka, C, Tabuchi, S, Aihara, M, and Kume, K, *Platelet-Activating Factor and Somatostatin Activate Mitogen-Activated Protein Kinase (Map Kinase) and Arachidonate Release.* Journal of Lipid Mediators & Cell Signalling, 1996. **14**: 103-108.

70.     Florio, T, Yao, H, Carey, KD, Dillon, TJ, and Stork, PJ, *Somatostatin activation of mitogen-activated protein kinase via somatostatin receptor 1 (SSTR1).* Mol Endocrinol, 1999. **13**: 24-37.

71.     Florio, T, Thellung, S, Arena, S, Corsaro, A, Bajetto, A, Schettini, G, and Stork, PJ, *Somatostatin receptor 1 (SSTR1)-mediated inhibition of cell proliferation correlates with the activation of the MAP kinase cascade: role of the phosphotyrosine phosphatase SHP-2.* J Physiol Paris, 2000. **94**: 239-250.

72.     Sellers, LA, Alderton, F, Carruthers, AM, Schindler, M, and Humphrey, PPA, *Receptor Isoforms Mediate Opposing Proliferative Effects through Gbeta gamma - Activated p38 or Akt Pathways.* Mol Cell Biol, 2000. **20**: 5974-5985.

73.     Rauly, I, Saint-Laurent, N, Delesque, N, Buscail, L, Esteve, JP, Vaysse, N, and Susini, C, *Induction of a negative autocrine loop by expression of sst2 somatostatin receptor in NIH 3T3 cells.* Journal of Clinical Investigation, 1996. **97**: 1874-1883.

74.     Hukovic, N, Panetta, R, Kumar, U, and Patel, YC, *Agonist-dependent regulation of cloned human somatostatin receptor types 1-5 (hSSTR1-5): subtype selective internalization or upregulation.* Endocrinology, 1996. **137**: 4046-4049.

75.     Krenning, EP, Kwekkeboom, DJ, Reubi, JC, Van Hagen, PM, van Eijck, CH, Oei, HY, and Lamberts, SW, *111In-octreotide scintigraphy in oncology. [Review] [16 refs].* Metabolism: Clinical & Experimental, 1992. **41**: 83-86.

76.     Smith-Jones, PM, Stolz, B, Bruns, C, Albert, R, Reist, HW, Fridrich, R, and Macke, HR, *Gallium-67/gallium-68-[DFO]-octreotide--a potential radiopharmaceutical for PET imaging of somatostatin receptor-positive tumors: synthesis and radiolabeling in vitro and preliminary in vivo studies.* J Nucl Med, 1994. **35**: 317-325.

77.     de Jong, M, Breeman, WA, Bernard, BF, Rolleman, EJ, Hofland, LJ, Visser, TJ, Setyono-Han, B, Bakker, WH, van der Pluijm, ME, and Krenning, EP, *Evaluation in vitro and in rats of 161Tb-DTPA-octreotide, a somatostatin analogue with potential for intraoperative scanning and radiotherapy.* Eur J Nucl Med, 1995. **22**: 608-616.

78.     Hofland, LJ, Vankoetsveld, PM, Waaijers, M, and Lamberts, SWJ, *Internalisation of Isotope-Coupled Somatostatin Analogues.* Digestion, 1996. **57**: 2-6.

79.     Szepeshazi, K, Schally, AV, Halmos, G, Armatis, P, Hebert, F, Sun, B, Feil, A, Kiaris, H, and Nagy, A, *Targeted cytotoxic somatostatin analogue AN-238 inhibits somatostatin receptor-positive experimental colon cancers independently of their p53 status.* Cancer Res, 2002. **62**: 781-788.

80.     Szepeshazi, K, Schally, AV, Halmos, G, Sun, B, Hebert, F, Csernus, B, and Nagy, A, *Targeting of cytotoxic somatostatin analog AN-238 to somatostatin receptor subtypes 5 and/or 3 in experimental pancreatic cancers.* Clin Cancer Res, 2001. **7**: 2854-2861.

81.     Plonowski, A, Schally, AV, Nagy, A, Sun, B, and Szepeshazi, K, *Inhibition of PC-3 human androgen-independent prostate cancer and its metastases by cytotoxic somatostatin analogue AN-238.* Cancer Res, 1999. **59**: 1947-1953.

82.     Koppan, M, Nagy, A, Schally, AV, Arencibia, JM, Plonowski, A, and Halmos, G, *Targeted cytotoxic analogue of somatostatin AN-238 inhibits growth of androgen-independent Dunning R-3327-AT-1 prostate cancer in rats at nontoxic doses.* Cancer Res, 1998. **58**: 4132-4137.

83.     Nagy, A, Schally, AV, Halmos, G, Armatis, P, Cai, RZ, Csernus, V, Kovacs, M, Koppan, M, Szepeshazi, K, and Kahan, Z, *Synthesis and biological evaluation of cytotoxic analogs of somatostatin containing doxorubicin or its intensely potent derivative, 2-pyrrolinodoxorubicin.* Proc Natl Acad Sci U S A, 1998. **95**: 1794-1799.

84.     Reubi, JC, Schar, JC, Waser, B, Wenger, S, Heppeler, A, Schmitt, JS, and Macke, HR, *Affinity profiles for human somatostatin receptor subtypes SST1-SST5 of somatostatin*

*radiotracers selected for scintigraphic and radiotherapeutic use.* Eur J Nucl Med, 2000. **27**: 273-282.

85.     Plonowski, A, Schally, AV, Nagy, A, Kiaris, H, Hebert, F, and Halmos, G, *Inhibition of metastatic renal cell carcinomas expressing somatostatin receptors by a targeted cytotoxic analogue of somatostatin AN-238.* Cancer Res, 2000. **60**: 2996-3001.

86.     Rocheville, M, Lange, DC, Kumar, U, Patel, SC, Patel, RC, and Patel, YC, *Receptors for dopamine and somatostatin: formation of hetero-oligomers with enhanced functional activity [see comments].* Science, 2000. **288**: 154-157.

87.     Cattaneo, MG, Taylor, JE, Culler, MD, Nisoli, E, and Vicentini, LM, *Selective stimulation of somatostatin receptor subtypes: differential effects on Ras/MAP kinase pathway and cell proliferation in human neuroblastoma cells.* FEBS Lett, 2000. **481**: 271-276.

88.     Pfeiffer, M, Koch, T, Schroder, H, Klutzny, M, Kirscht, S, Kreienkamp, H-J, Hollt, V, and Schulz, S, *Homo- and Heterodimerization of Somatostatin Receptor Subtypes. INACTIVATION OF sst3 RECEPTOR FUNCTION BY HETERODIMERIZATION WITH sst2A.* J Biol Chem, 2001. 276: 14027-14036.

# 10
# SOMATOSTATIN RECEPTOR SIGNALING VIA PROTEIN TYROSINE PHOSPHATASES

Hicham Lahlou, Julie Guillermet, Fabienne Vernejoul, Stéphane Pyronnet, Corinne Bousquet, Louis Buscail and Christiane Susini
*INSERM U531, CHU Rangueil, IFR 31, 31403 Toulouse, France*

## INTRODUCTION

Protein phosphorylation, which is controlled by the opposing dynamic activities of protein tyrosine kinases and protein tyrosine phosphatases (PTPs), is a key cellular control mechanism for regulation of cell growth, metabolism, differentiation and development. During the past few years, molecular cloning has established the existence of a structurally diverse family of intracellular and transmembrane PTPs, which share sequence similarity over a stretch of 240-250 amino acids, the PTP catalytic domain. This domain is characterized by an 11-residue sequence motif (I/V)HCxAGxxR(S/T)G containing the critical catalytic cysteine residue involved in the formation of thiophosphate intermediate. Crystal structures of PTP and enzymological studies have conducted to the development of PTP mutants. Mutation of the catalytic cysteine (C->S) or the aspartic residue in the WPD loop (D->A) generates catalytically inactive enzymes, which yet retain their ability to bind substrates and are often used for PTP functional studies. The cytoplasmic PTPs possess one PTP domain whereas the transmembrane PTPs have frequently two PTP domains and variable extracellular domains. Depending on the subclass, PTPs may recognize cell surface proteins, matrix components or soluble ligands. Specificity of PTPs arises from both PTP catalytic domain and additional sequences attached to either the N or C terminus PTP domain, which regulate enzyme activity and/or PTP targeting to specific subcellular location and substrates[1].

PTPs are tightly regulated enzymes which can deliver both positive and negative signals in transduction cascades. Most growth factors, cytokines, and hormone receptor-mediated signaling pathways involve protein tyrosine phosphorylation and dephosphorylation. However, which PTP modulates each event is far from being elucidated.

Somatostatin (SST) inhibits a wide range of biological actions including cell growth and hormone secretion. Its actions are mediated by a family of G protein-coupled somatostatin receptors (SSTR) designed sst1-sst5. Each receptor subtype is coupled to multiple signal transduction pathways [1]. This review will focus on recent studies in which insight has been gained into the role of specific PTPs in SSTR subtype signaling.

**Role of PTP in somatostatin receptor signaling**

First evidence that SST can activate a PTP came from studies of Schally's group. They demonstrated that in membranes isolated from human pancreatic cancer cells, SST stimulated a PTP, which dephosphorylated tyrosine-phosphorylated epidermal growth factor (EGF) receptors and abrogated EGF mitogenic action[2, 3]. On the same cellular models, Stork and co-workers have shown the implication of a pertussis toxin-sensitive $G_{i/o}$ protein in SST-dependent PTP activation[4]. *In vivo*, intravenous injection of SST analog SMS also stimulated a particulate rat pancreatic PTP[5]. Stimulation of PTPs by SST has been subsequently reported in various models, *in vitro* and *in vivo*. Indeed, in S49 lymphoma cells, which express endogenous sst2, SST stimulates a membrane $G\alpha_{i/o}$ protein-regulated PTP, which is recovered in a high molecular weight complex[6]. Activation of PTP has also been involved in SST-induced repression of AP-1 transcriptional activity and consequent inhibition of pituitary GH3 cell proliferation[7]. In PC C13 normal thyroid cells, SST activated a PTP to inhibit cell proliferation[8]. Biochemical studies of SST-sensitive PTP activities in rat pancreatic acini, which highly express the sst2 receptor subtype, demonstrated that SST analogs induce a rapid stimulation of a membrane PTP with an apparent molecular mass of 70 kDa[9]. In addition, a 66 kDa protein immunoreactive to anti-SHP-1 antibodies was copurified with pancreatic membrane SSTRs suggesting that SHP-1 is associated with SSTRs at the membrane level and that SHP-1 is one PTP candidate for sst2-mediated early signaling[10]. More recently, Florio et al (2003) reported that SST and analogs inhibit human GH-secreting pituitary adenoma cell proliferation in vitro, via the activation of tyrosine phosphatases whereas SST-mediated inhibition of hormone release involves the inhibition of voltage-dependent calcium channels and adenylyl cyclase activities[11].

Stable or transient expression of each of the five cloned SSTRs in heterologous Chinese hamster ovary (CHO)-DG44, CHO-K1 and NIH 3T3 cells was used to demonstrate that SST activation of sst1-sst4 receptors promoted a rapid stimulation of a membrane PTP activity, which was sensitive to pertussis toxin treatment[6, 12-14]. Concerning sst5, no stimulation of any membrane PTP activity was observed in NIH 3T3 or CHO-K1 cells expressing sst5 after a 10-15 min cell treatment with SST[6, 13]. However, an increase in PTP activity was observed in CHO-K1 cells expressing sst5 after a 24 h treatment with SST and this activity was involved in sst5-mediated cell cycle arrest[15].

**SHP-1 is involved in somatostatin receptor signaling**

SHP-1 is a nontransmembrane PTP that belongs, like SHP-2, to the SH2 domain-containing PTP family (SHP). SHP-1 is predominantly expressed in hematopoietic cells but also substantially in a variety of non-hematopoietic cells, especially in normal and malignant epithelial cells. SHP-1 associates *in vivo*, *via* its SH2 domains, with activated growth factor tyrosine kinase receptors, cytokine receptors and with antigen receptors (reviewed in[16]). It is well demonstrated that SHPs are essentially inactive under basal conditions, but their catalytic activity dramatically increases upon SH2 domain binding to tyrosine phosphorylated molecules. Mechanisms for SHP enzymatic activation by pTyr peptides have been provided by SHP-2 crystal structure[17]. Under basal conditions, SHP N-terminal SH2 domains bind the phosphatase catalytic domain and directly block its active site. Binding of SHP SH2 domains to phosphotyrosylated peptides disrupts such an inhibitory intramolecular interaction and thereby promotes SHP catalytic activation. SHP-1 has been mainly reported as a negative regulator. SHP-1 negative effects on receptor-mediated signaling is mediated by dephosphorylation of the tyrosine phosphorylated receptor itself or that of receptor-associated tyrosine phosphorylated mediators, which results in terminating signaling (reviewed in[16]). The critical role of SHP-1 as a negative regulator of hematopoietic cell signal transduction is consistent with the multiple defects observed in hematopoietic cells from *motheaten* mice, which exhibit homozygous mutations in the SHP-1 gene[18].

Since one of the roles of SHP-1 is to antagonize growth factor signaling, one might therefore speculate that SHP-1 might be activated by factors that negatively regulate cell growth such as SST. In CHO cells stably co-expressing sst2 and SHP-1, we reported that SHP-1 constitutively associated with sst2 as shown by co-immunoprecipitation studies using specific antibodies. Activation of sst2 by SST resulted in a rapid and transient recruitment of SHP-1 by sst2 (peaking at 30 s) followed by an increase in SHP-1 activity (Fig. 1). Such a rapid and transient activation of SHP-1 by SST was also reported in PANC-1 cells, which express the sst2 receptor subtype[19]. Stimulation of SHP-1 activity by SST was abolished by cell pretreatment with pertussis toxin. $G_{\alpha i3}$ was specifically immunoprecipitated by anti-sst2 and anti-SHP-1 antibodies and SST induced a rapid dissociation of $G_{\alpha i3}$ from sst2 suggesting that $G_{\alpha i3}$ is involved in the SHP-1-sst2 complexes. Finally, SST inhibited cell proliferation and this effect was suppressed in cells co-expressing sst2 and the catalytic inactive SHP-1 (C453S). All these data identify SHP-1 as one PTP associated with sst2 and demonstrate that this enzyme may be an initial key transducer of the sst2 antimitogenic signaling.

SHP-1 effects result from its interaction with a broad spectrum of signaling targets of the sst2 receptor. One of SHP-1 substrate is the insulin receptor. As known for ligand-activated receptor tyrosine kinases, activated insulin receptor

162

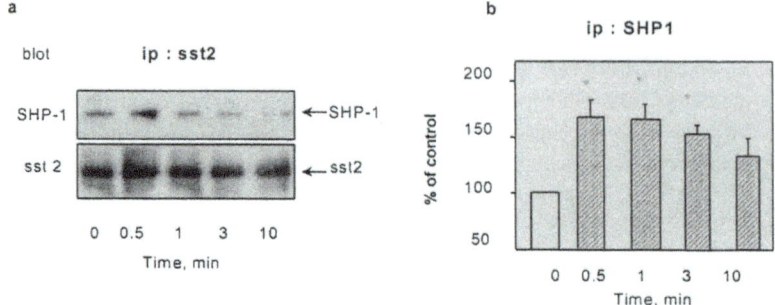

*Figure 1. SHP-1-sst2 association (a) and SHP-1 activity (b) following SMS somatostatin analog treatment of CHO cells co-expressing sst2 and SHP-1. a.Sst2 immunoprecipitates were analyzed by SDS-PAGE and sequentially immunoblotted with anti-SHP-1 and anti-sst2 antibodies. b. SHP-1 immunoprecipitates were assayed for PTP activity in the presence of [33]-P labeled poly (Glu, Tyr). Reprinted with permission of Lopez et al., (1997).*

is rapidly dephosphorylated and this is related to the insulin-induced recruitment of SHP-1 to insulin receptor and of consequent SHP-1 activation. In CHO cells expressing sst2 (CHO/sst2), SST analog, RC-160, inhibits the insulin-induced mitogenic signal by accelerating and amplifying SHP-1 inhibitory action on the insulin signaling pathway[20]. SHP-1 is required for sst2-mediated $G_1$ cell cycle arrest through activation of the Ras/Rap1/Braf/ERK pathway leading to the induction of the cyclin-dependent kinase inhibitor p27[Kip1]. In mouse pancreatic acini, RC-160 stimulates ERK pathway and reverts mitogen-induced down-regulation of p27[Kip1] but these effects do not occur in acini from *viable motheaten* mice expressing an inactive SHP-1[21, 22]. Neuronal nitric oxide synthase (nNOS) is another SHP-1 substrate, which is critical for sst2-induced cell growth arrest. In CHO/sst2 cells stimulated by serum, RC-160 induces a rapid recruitment of nNOS by SHP-1, which leads to nNOS tyrosine dephosphorylation and activation. The resulting NO induces an up-regulation of p27[Kip1] [23]. These findings provide evidence that SHP-1 is required for SST signaling cascade leading to p27[Kip1] up-regulation.

The role of SHP-1 in sst2 signaling is emphasized by data obtained with heterologous expression of sst2 in NIH 3T3 cells. In these cells, expression of sst2 induces an up-regulation of the SST gene and an increase in SST secretion, which is associated with a constitutive SHP-1 activation, and with an enhanced SHP-1 protein expression. This results in an autocrine inhibition of cell proliferation. Blocking SHP-1 activity with orthovanadate or SHP-1 protein expression with SHP-1 antisense oligonucleotides decreases sst2-mediated autocrine inhibition of cell proliferation[24].

sst2 expression is lost in 90 % of human pancreatic adenocarcinomas and most of pancreatic cancer derived cell lines[25]. Heterologous expression of sst2 in non-expressing human pancreatic cancer BxPC-3 cells and in hamster pancreatic cancer PC-1.0 cells also triggers an increase of SST secretion that induces an autocrine inhibition of cell proliferation. Furthermore, sst2 expression reverses the malignant properties of these pancreatic cancer cells and decreases cell invasiveness by a mechanism dependent on SHP-1-mediated E-cadherin tyrosine dephosphorylation[26, 27].

In addition to its effects in the negative regulation of activated signaling cascades, SHP-1 is implicated in SST-mediated cellular apoptosis signaling[26, 28]. In human breast cancer cells, which express several SSTR subtypes, Srikant and coworkers demonstrated that SST induces SHP-1 translocation from the cytosolic to the membrane fraction and this probably accounts for an increase in SHP-1 activity[29]. Using the catalytic inactive SHP-1 mutant, they reported that in these cells, cytotoxic action of SST, which involves caspase-8 activation leading to intracellular acidification, is also SHP-1-mediated[30]. Such an involvement of SHP-1 in cytotoxic activity of SST was also demonstrated in pancreatic tumor cells BxPC3 expressing sst2[30].

**Role of SHP-2 in somatostatin receptor signaling**

SHP-2, another member of the non-transmembrane SH2-containing PTP family, is ubiquitously expressed and seems to play a positive role in different cell signaling. SHP-2 binds directly, *via* its SH2 domains, to activated growth factor receptors. Cytokine receptors and immunoreceptors. SHP-2 can be also recruited by diverse signaling proteins and cell adhesion molecules[31]. SHP-2 plays a critical role for transducing various signals that lead to cell growth, differentiation or migration[16, 31].

However, several studies argue in favor of the role of SHP-2 in SST signaling. Florio et al (1999) reported in CHO-K1 cells over-expressing sst1 a rapid SHP-2 stimulation by SST, starting within 1 min and reaching a maximum at 10 min. Furthermore, they demonstrated that activation of SHP-2 is involved in SST-induced activation of the Ras/Raf-1/MAP kinase pathway, and consequent increased expression of cyclin-dependent inhibitor p21[cip/Waf1] and cell cycle arrest[32]. In thyroid PC C13 cells, they also showed that SST stimulates SHP-2 activity and inhibits cell proliferation. However, SHP2 activity does not seem to be sufficient to inhibit PC C13 cell growth[33]. SHP-2 is also involved in sst2 signaling. Co-expression of sst2 with a catalytically inactive SHP-2 abrogates somatostatin-induced PTP stimulation in NIH 3T3 cells transformed with Ha-Ras[G12V] [34]. We recently identified the Src tyrosine kinase and SHP-2 as sst2-associated molecules acting upstream of SHP-1. In CHO/sst2, Src and SHP-2 are both associated with sst2. SHP-2 directly associates with phosphorylated tyrosine 228 and 312, which are located in sst2 ITIMs (immunoreceptor tyrosine-based inhibitory motifs), one in the third intracellular loop, L-C-$Y_{228}$-L-F-I, and the second in the C-terminal tail, I-L-$Y_{312}$-A-F-L). Upon sst2 activation by SST, Src becomes activated by a $G_{\beta\gamma}$-dependent mechanism. In turn, Src activation leads to sst2 hyper-phosphorylation and SHP-2 activation. Then, dephosphorylation of signal relay molecules allows SHP-1 recruitment and activation[35].

**Role of PTPη in SST signaling**

The rat tyrosine phosphatase r-PTPη is a transmembrane PTP whose expression is negatively regulated by neoplastic cell transformation. A dramatic reduction in DEP-1/HPTPη (the human homolog of r-PTPη) expression has been shown in human thyroid carcinomas. Reexpression of the r-PTPη gene in highly malignant rat thyroid cells transformed by retroviruses carrying the v-mos and v-ras-Ki oncogenes suppresses their malignant phenotype. PTPη causes G1 growth arrest through the induction of $p27^{kip1}$[36].

The role of PTPη in SST signaling has been demonstrated by Florio's group. They reported that SST inhibited PC Cl3 thyroid cell proliferation through the activation of a membrane phosphotyrosine phosphatase[8]. Conversely, in PC Cl3 cells stably expressing the v-mos oncogene (PC mos), blockade of both SST-mediated inhibition of cell proliferation and stimulation of PTP was associated with a reduction in the expression of PTPη, suggesting that PTPη may participate in SST-induced antiproliferative signal. To directly prove the involvement of r-PTPη in SST action, they stably transfected this PTP in PC mos cells. This new cell line (PC mos/PTPη) recovered the ability of SST to inhibit cell proliferation and to stimulate PTP activity. The activation of r-PTPη by SST caused an inhibition of insulin-induced ERK1/2 activation downstream of MAPK kinase with the subsequent blockade of the phosphorylation, ubiquitination, and proteasome degradation of $p27^{kip1}$[33].

# CONCLUSION

Many studies have implicated tyrosine phosphatases in the down-regulation of growth factor signaling pathways. It is becoming evident that some of them, including SHP-1, SHP-2 and PTPη, have a role in SSTR signaling. However, there are many issues which remain to be clarified. These include the identification of SST-mediated pathways in which these PTPs participate, as well as their physiological substrates and binding partners and the precise molecular mechanisms, which permit their interaction with sst receptors and/or associated molecules. In addition, the respective role of these PTP for each receptor subtype pathway awaits further study. Genetic approaches including gene targeting, RNA interference (RNAi) and new technologies including proteomic analysis and cell imaging for the detection and analysis of signalling complexes, will provide powerful tools to elucidate these important questions.

## REFERENCES

1.  Patel YC. Molecular pharmacology of somatostatin receptor subtypes. J Endocrinol Invest 1997; 20:348-67.
2.  Hierowski MT, Liebow C, du Sapin K, Schally AV. Stimulation by somatostatin of dephosphorylation of membrane proteins in pancreatic cancer MIA PaCa-2 cell line. FEBS Lett 1985; 179:252-6.
3.  Liebow C, Reilly C, Serrano M, Schally AV. Somatostatin analogues inhibit growth of pancreatic cancer by stimulating tyrosine phosphatase. Proc Natl Acad Sci U S A 1989; 86:2003-7.
4.  Pan MG, Florio T, Stork PJ. G protein activation of a hormone-stimulated phosphatase in human tumor cells. Science 1992; 256:1215-7.
5.  Rivard N, Lebel D, Laine J, Morisset J. Regulation of pancreatic tyrosine kinase and phosphatase activities by cholecystokinin and somatostatin. Am J Physiol 1994; 266:G1130-8.
6.  Dent P, Wang Y, Gu YZ, et al. S49 cells endogenously express subtype 2 somatostatin receptors which couple to increase protein tyrosine phosphatase activity in membranes and down-regulate Raf-1 activity in situ. Cell Signal 1997; 9:539-49.
7.  Todisco A, Seva C, Takeuchi Y, Dickinson CJ, Yamada T. Somatostatin inhibits AP-1 function via multiple protein phosphatases. Am J Physiol 1995; 269:G160-6.
8.  Florio T, Scorizello A, Fattore M, et al. Somatostatin inhibits PC Cl3 thyroid cell proliferation through the modulation of phosphotyrosine activity. Impairment of the somatostatinergic effects by stable expression of E1A viral oncogene. J Biol Chem 1996; 271:6129-36.
9.  Colas B, Cambillau C, Buscail L, et al. Stimulation of a membrane tyrosine phosphatase activity by somatostatin analogues in rat pancreatic acinar cells. Eur J Biochem 1992; 207:1017-24.
10. Zeggari M, Esteve JP, Rauly I, et al. Co-purification of a protein tyrosine phosphatase with activated somatostatin receptors from rat pancreatic acinar membranes. Biochemical Journal 1994; 303:441-8.
11. Florio T, Thellung S, Corsaro A, et al. Characterization of the intracellular mechanisms mediating somatostatin and lanreotide inhibition of DNA synthesis and growth hormone release from dispersed human GH-secreting pituitary adenoma cells in vitro. Clin Endocrinol (Oxf) 2003; 59:115-28.
12. Florio T, Rim C, Hershberger RE, Loda M, Stork PJ. The somatostatin receptor SSTR1 is coupled to phosphotyrosine phosphatase activity in CHO-K1 cells. Molecular Endocrinology 1994; 8:1289-97.
13. Buscail L, Esteve JP, Saint-Laurent N, et al. Inhibition of cell proliferation by the somatostatin analogue RC-160 is mediated by somatostatin receptor subtypes SSTR2 and SSTR5 through different mechanisms. Proceedings of the National Academy of Sciences of the United States of America 1995; 92:1580-4.
14. Reardon DB, Wood SL, Brautigan DL, Bell GI, Dent P, Sturgill TW. Activation of a protein tyrosine phosphatase and inactivation of Raf-1 by somatostatin. Biochemical Journal 1996; 314:401-4.
15. Sharma K, Patel YC, Srikant CB. C-terminal region of human somatostatin receptor 5 is required for induction of Rb and G1 cell cycle arrest. Mol Endocrinol 1999; 13:82-90.
16. Neel BG, Tonks NK. Protein tyrosine phosphatases in signal transduction. Curr Opin Cell Biol 1997; 9:193-204.
17. Hof P, Pluskey S, Dhe-Paganon S, Eck MJ, Shoelson SE. Crystal structure of the tyrosine phosphatase SHP-2. Cell 1998; 92:441-50.

18.    Shultz LD, Schweitzer PA, Rajan TV, et al. Mutations at the murine motheaten locus are within the hematopoietic cell protein-tyrosine phosphatase (Hcph) gene. Cell 1993; 73:1445-54.

19.    Douziech N, Calvo E, Coulombe Z, et al. Inhibitory and stimulatory effects of somatostatin on two human pancreatic cancer cell lines: a primary role for tyrosine phosphatase SHP-1. Endocrinology 1999; 140:765-77.

20.    Bousquet C, Delesque N, Lopez F, et al. sst2 somatostatin receptor mediates negative regulation of insulin receptor signaling through the tyrosine phosphatase Shp-1. J Biol Chem 1998; 273:7099-106.

21.    Pages P, Benali N, Saint-Laurent N, et al. sst2 somatostatin receptor mediates cell cycle arrest and induction of p27(Kip1). Evidence for the role of Shp-1. J Biol Chem 1999; 274:15186-93.

22.    Lahlou H, Saint-Laurent N, Esteve JP, et al. sst2 Somatostatin receptor inhibits cell proliferation through Ras-, Rap1-, and B-Raf-dependent ERK2 activation. J Biol Chem 2003; 278:39356-71.

23.    Lopez F, Ferjoux G, Cordelier P, et al. Neuronal nitric oxide synthase is a SHP-1 substrate involved in sst2 somatostatin receptor growth inhibitory signaling. Faseb J 2001; 17:17.

24.    Rauly I, Saint-Laurent N, Delesque N, et al. Induction of a negative autocrine loop by expression of sst2 somatostatin receptor in NIH 3T3 cells. Journal of Clinical Investigation 1996; 97:1874-83.

25.    Buscail L, Saint-Laurent N, Chastre E, et al. Loss of sst2 somatostatin receptor gene expression in human pancreatic and colorectal cancer. Cancer Research 1996; 56:1823-7.

26.    Rochaix P, Delesque N, Esteve JP, et al. Gene therapy for pancreatic carcinoma: local and distant antitumor effects after somatostatin receptor sst2 gene transfer [see comments]. Hum Gene Ther 1999; 10:995-1008.

27.    Benali N, Cordelier P, Calise D, et al. Inhibition of growth and metastatic progression of pancreatic carcinoma in hamster after somatostatin receptor subtype 2 (sst2) gene expression and administration of cytotoxic somatostatin analog AN-238. Proc Natl Acad Sci U S A 2000; 97:9180-5.

28.    Thangaraju M, Sharma K, Leber B, Andrews DW, Shen SH, Srikant CB. Regulation of acidification and apoptosis by SHP-1 and Bcl-2. J Biol Chem 1999; 274:29549-57.

29.    Srikant CB, Shen SH. Octapeptide somatostatin analog SMS 201-995 induces translocation of intracellular PTP1C to membranes in MCF-7 human breast adenocarcinoma cells. Endocrinology 1996; 137:3461-8.

30.    Guillermet J, Saint-Laurent N, Rochaix P, et al. Somatostatin receptor subtype 2 sensitizes human pancreatic cancer cells to death ligand-induced apoptosis. Proc Natl Acad Sci U S A 2003; 100:155-60.

31.    Stein-Gerlach M, Wallasch C, Ullrich A. SHP-2, SH2-containing protein tyrosine phosphatase-2. Int J Biochem Cell Biol 1998; 30:559-66.

32.    Florio T, Yao H, Carey KD, Dillon TJ, Stork PJ. Somatostatin activation of mitogen-activated protein kinase via somatostatin receptor 1 (SSTR1). Mol Endocrinol 1999; 13:24-37.

33.    Florio T, Arena S, Thellung S, et al. The activation of the phosphotyrosine phosphatase eta (r-PTP eta) is responsible for the somatostatin inhibition of PC Cl3 thyroid cell proliferation. Mol Endocrinol 2001; 15:1838-52.

34.    Reardon DB, Dent P, Wood SL, Kong T, Sturgill TW. Activation in vitro of somatostatin receptor subtypes 2, 3, or 4 stimulates protein tyrosine phosphatase activity in membranes from transfected Ras-transformed NIH 3T3 cells: coexpression with catalytically inactive SHP-2 blocks responsiveness. Mol Endocrinol 1997; 11:1062-9.

35.    Ferjoux G, Lopez F, Esteve JP, et al. Critical role of Src and SHP-2 in sst2 somatostatin receptor-mediated activation of SHP-1 and inhibition of cell proliferation. Mol Biol Cell 2003.

36.    Trapasso F, Iuliano R, Boccia A, et al. Rat protein tyrosine phosphatase eta suppresses the neoplastic phenotype of retrovirally transformed thyroid cells through the stabilization of p27(Kip1). Mol Cell Biol 2000; 20:9236-46.

# 11
# EXPRESSION AND FUNCTION OF SOMATOSTATIN AND ITS RECEPTORS IN IMMUNE CELLS

David E. Elliott, M.D., Ph.D.
University of Iowa, College of Medicine, Department of Internal Medicine, Iowa City, Iowa, USA

## INTRODUCTION

A successful immune response requires tightly coordinated reactions by multiple cell types. Cytokines elaborated by immune cells provide critical signals that orchestrate the response. Somatostatin (SST), in addition to acting as a hormone and neurotransmitter, functions as a cytokine regulating immune and inflammatory responses. This chapter will focus on the immunoregulatory function of SST and SST receptor usage by immune and inflammatory cells.

### The SST Iimmunoregulatory Circuit in Murine Schistosomiasis

Our laboratory has studied the SST immunoregulatory circuit using the murine schistosomiasis model of chronic granulomatous inflammation. *Schistosoma* are tropical parasitic worms that infect one in every 30 people worldwide (1). Adult *Schistosoma mansoni* lay eggs that lodge in the host's intestines and liver where they release antigenic glycolipids and proteins. Granulomas form around the eggs under the control of T cells. Mice are capable of developing schistosomiasis mansoni that mimics human disease (2), making it a superb model for the study of granulomatous inflammation. Murine schistosome egg granulomas are T helper cell type 2 (Th2) responses comprising 50% eosinophils, 30% macrophages, 10% T cells and small numbers of B cells, mast cells and fibroblasts. T cells isolated from schistosome granulomas and stimulated with parasite egg antigen secrete predominantly IL4, IL5 and IL10. Also present are Th1, IFNγ-producing T cells (3) but they are tightly regulated. Murine schistosome granulomas express a SST immunoregulatory circuit.

We found that murine granuloma macrophages make and release SST-14 (Figure 1) (4,5). Local macrophage SST-14 production may help limit granuloma size. Mice treated *in vivo* with octreotide (OCT) develop granulomas 60% the size that form in sham-treated mice (6). Murine splenic macrophages also make SST-14. Spleen cells from schistosome-infected mice express mRNA for authentic preprosomatostatin as determined by RT-PCR and DNA sequencing (5). However, spleno-cytes from uninfected mice do not express SST mRNA. Granulomas do not form in the spleens of schistosome-infected mice. Therefore, SST-producing cells may

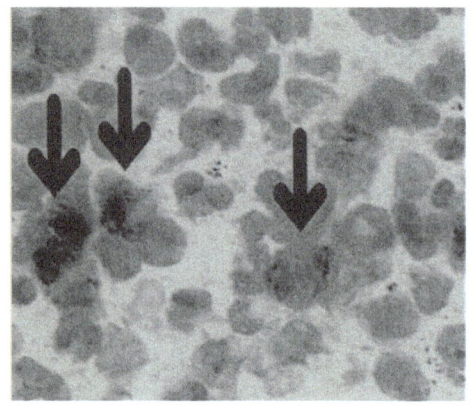

**Figure 1.** Granuloma macrophages make SST (arrows) shown by immunohistochemistry using monoclonal anti-SST (CURES607).

migrate to the spleen or mediators that are produced by hepatic and intestinal granulomas may induce splenic SST production in schistosome-infected mice. The later mechanism is supported by induction of SST transcription in splenocytes of normal mice by brief (4hr) treatment with LPS, IFNγ, TNFα, IL-10, PgE2 or dibutyryl cAMP (5). Splenocytes from uninfected Rag1 and Scid mutant mice that lack T and B cells express SST mRNA in response to IL10 (5). Immunohistochemistry demonstrated that it was splenic macro-phages from mice that make SST in response to cytokine stimulation (7). SST transcripts also are made by well characterized macrophage cell lines such as P388D1, J774A.1 and RAW267.4.

The immunoregulatory effect of SST-14 and OCT is substantial. Both suppress antigen-stimulated IFNγ production by granuloma and splenic T cells of schistosome-infected mice (6,8). Although Th2-type granulomas form in schistosomiasis, these lesions do contain many T cells capable of IFNγ production (3). Addition of nanomolar SST-14 or OCT to granuloma and spleen cell cultures stimulated with low doses of schistosome egg antigen suppresses IFNγ release by T cells. This effect requires only $10^{-10}$M SST. SST does not suppress IFNγ production by granuloma or splenic T cells stimulated with high doses of schistosome egg antigen.

IFNγ is a pivotal cytokine that regulates the activity of many cell types including B cells, T cells, natural killer (NK) cells, and macrophages (9-11). IFNγ activates macrophages and NK cells, promotes antigen presentation, and augments Th1 responses. Excessive IFNγ release can promote inflammation and tissue injury. Insufficient IFNγ release can permit microbial invasion. Exposure to IFNγ induces murine B cells to switch immunoglobulin class synthesis to complement-fixing IgG2a. Thus, IgG2a provides a useful measure of the physiologic effects of IFNγ. Schistosome infected mice treated with OCT *in vivo* develop granulomas that do not secrete IgG2a. SST inhibits

IgG2a production from antigen-stimulated splenic and granuloma B cells by blocking T cell IFNγ release (12).

Schistosome granuloma T helper cells specifically bind SST with high affinity (6). Therefore, granuloma T cells display cell surface receptors for SST. Five SST receptors (SSTR1-5) have been cloned and sequenced (13-21). We developed sensitive and specific RT-PCR assays to determine if granuloma cells express mRNA for these SST receptors (8,22) and found that murine granuloma cells express mRNA only for SSTR2. Furthermore, thymocytes and splenocytes from infected and normal animals also express SSTR2 mRNA.

In addition, unfractionated splenocytes (containing immune and other cell types) express mRNA for SSTR3 and SSTR4. Thus, these other receptors may have contributed to SST-14-mediated inhibition of splenic T cell IFNγ release. However, we found that specific SSTR2-blocking rabbit antiserum (23) prevented SST-14 and OCT-mediated inhibition of splenic T cell IFNγ production (8). This antiserum also prevents SST-14 or OCT inhibition of granuloma T cell IFNγ release. Thus, SST-14-mediated inhibition of murine T cell IFNγ production requires functional SSTR2. Because SSTR2 is the only SST receptor expressed by granuloma cells at the mRNA level, it is likely that the anti-inflammatory effects of SST-14 or OCT are mediated by SSTR2.

Two isoforms (A & B) of SSTR2 exist that result from alternate RNA splicing (24-26). These variants differ in the C-terminal cytoplasmic tails and may activate alternative signal transduction pathways producing different effects (27). Murine schistosome granuloma and spleen cells transcribe mRNA for both the SSTR2A and SSTR2B isoforms (22). We developed three quantitative, sensitive and specific competitive RT-PCR assays to determine the abundance of each SSTR2 isoform in murine granulomatous inflammation. We found that SSTR2A accounts for 99% of the SSTR2 transcripts in granuloma cells (8). Similar results were obtained for spleen cells.

We used these competitive PCR assays to determine if SSTR2 mRNA levels changed due to cytokine exposure. Exposure to cytokines or inflammatory mediators such as SST-14 ($10^{-6}$M), OCT ($10^{-6}$M), LPS (30 μg/ml), PgE$_2$ ($1 \times 10^{-6}$M), dibutyryl cAMP ($1 \times 10^{-4}$ M), rTNFα (30 ng/ml), rTGFβ1 (250 pg/ml), anti-CD3 (145-2C11, 1μg/ml), rIL-4 (200 U/ml), rIL-10 (30 ng/ml), rIL-12 (5ng/ml) or rIFNγ (200 U/ml) did not appear to alter the steady state mRNA level of SSTR2A or SSTR2B in murine granuloma cells or splenocytes (8). These factors also did not significantly alter SSTR2, SSTR2A or SSTR2B mRNA levels in cloned Th1 or Th2 cell lines. This suggests that murine T cells may use a constitutive SSTR2 promoter (28).

While SST-14 binds to granuloma CD4+ T cells, other inflammatory cell types also express SSTR2 mRNA. T cells comprise 10% of the schistosome granuloma cell population but account for 50% of the total granuloma cell

SSTR2 mRNA content as measured by competitive RT-PCR (29). This implies that T cells are the predominant but not exclusive employers of $SSTR_2$ at sites of inflammation in the mouse. We found that B cell and macrophage cell lines also express SSTR2 mRNA as determined by RT-PCR. B cells and macrophages serve as antigen-presenting cells and help regulate T cell responses. It is possible that SST-14 regulates T cell IFNγ release by acting on these accessory cells rather than acting directly on T cells. To determine if SST-14 acts directly on T cells, we examined the well defined D1.1 Th1 cell line (30) that expresses SSTR2. We expanded these cells to pure culture, free of any other cell type, and stimulated the D1.1 cells with biotinylated anti-CD3 (145-2C11, ATCC) and anti-CD4 (GK 1.5, ATCC) crosslinked with streptavidin. This signals through the T cell antigen receptor in the absence of antigen-presenting cells. Both SST-14 and OCT at $10^{-7}$M inhibited TCR-stimulated IFNγ production by D1.1 Th1 cells (8). Thus, although SST-14 may regulate other inflammatory cell types, SST-14 can directly regulate T cell IFNγ production through interaction with SSTR2. We are currently studying SSTR2 deficient mice using schistosomiasis and other models. Mice that lack SSTR2 have augmented IFNγ responses (unpublished observations). This shows that endogenous SST-14 signals through SSTR2 and regulates the murine Th1/Th2 response profile.

SST and substance P (SP) are components of the Th1/Th2 cytokine circuit. SST inhibits while SP augments T cell IFNγ release (31). Mice that lack the SP receptor NK1 develop small schistosome egg granulomas with impaired IFNγ and IgG2a production (32). Thus, SST-14 and SP have opposing effects. Furthermore, SP inhibits the production of SST-14 by macrophages. As little as $10^{-9}$M SP prevents IFNγ, IL10, and LPS-induced SST transcription by splenic macrophages from normal mice (7). Substance P also causes a 90% inhibition of ongoing SST mRNA expression by granuloma macrophages from schistosome-infected mice. Hence, SP both directly stimulates T cell IFNγ release and prevents production of SST that would otherwise inhibit IFNγ secretion (Figure 2).

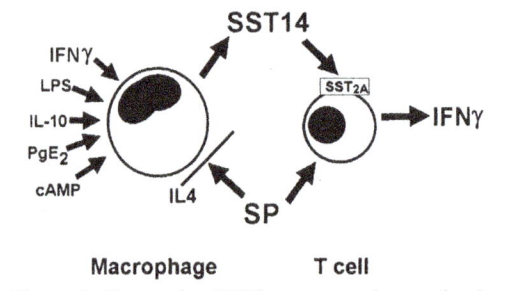

**Figure 2.** The murine SST immunoregulatory circuit. Inflammatory mediators induce SST production by murine macrophages. SST inhibits (gray arrow) IFNγ production by T cells by signaling through $SST_2$. Substance P inhibits macrophage SST production and promotes T cell IFNγ release. IL4 blocks SP from inhibiting SST synthesis.

Another important cytokine is IL4. IL4 is made by Th2 cells and is critical for most of the characteristics of a Th2 response. Strong Th2 responses prevent development of IFNγ-dominant Th1 responses. Conversely, strong Th1 responses prevent development of IL4-dominant Th2 responses. Remarkably, IL4 prevents SP from inhibiting SST production(7). Thus, IL4 allows macrophage SST production, that inhibits T cell IFNγ production that would otherwise suppress IL4 responses. SST and SP help regulate T helper cell circuitry in inflammation.

## Other Examples of Immunoregulation by SST in the Mouse

SST and SST agonists can suppress proliferation of murine lymphocytes. SST at $10^{-9}$ to $10^{-7}$M inhibits proliferation of murine spleen, mesenteric lymph node, and Peyer's patch cells stimulated with the T cell mitogen, concanavalin A (ConA) (33,34). The SST agonist BIM 23014 (BIM) at $10^{-8}$ to $10^{-5}$M also produced a 25 to 40% decrease in ConA-stimulated proliferation of murine Peyer's patch lymphocytes (35). Treatment with SST-14 or BIM *in vivo* for 7 days prior to sacrifice, suppressed the ConA-stimulated proliferation of Peyer's patch lymphocytes by 65% and 72% respectively as compared to control mice. Addition of SST or BIM to the cell cultures further suppressed proliferation. In that study the *in vivo ex*posure to SST-14 or BIM also inhibited NK activity from spleen or Peyer's patch lymphocytes (35).

SST influences cytokine release. We found that SST inhibits IFNγ release from Th1 cells. Levite reported that SST ($10^{-8}$M) caused an antigen-stimulated Th1-type cell line that normally secretes IFNγ to release IL4 and a Th2-type line that normally secretes IL4 to release IFNγ (36). This suggests that exposure to SST can cause murine T helper cells to switch phenotype. We study murine schistosomiasis which causes a predominate Th2-type response with a majority of IL4-secreting T cells. Instead of increasing IFNγ production, we find that SST inhibits antigen-induced IFNγ release in this model. We find no effect on granuloma or splenic cell IL4 production. Other murine models need to be investigated to determine if SST can deviate established T helper cell phenotypes.

SST may influence B cell function. As noted above, SST-14 suppresses IgG2a production by splenic and granuloma B cells of schistosome-infected mice (12). This effect is due to suppression of IFNγ that is required for immunoglobulin class switching to IgG2a. Stanisz et. al. found that SST-14 suppressed IgA and IgM production by splenocytes cultured in the presence of the T cell mitogen, conconavalin A (37). SST also may have direct activity on murine B cells that express SST receptors. The murine IgA-secreting myeloma MOPC-315 specifically binds SST-14 (38) and SST-14 ($10^{-12}$ to $10^{-9}$M) inhibited proliferation of MOPC-315 (2-30%) at 24 hours. Low dose ($10^{-12}$M) appeared to enhance (by 40%) and higher doses ($10^{-8}$ to $10^{-6}$M) inhibit (by 60%) IgA synthesis by MOPC-315 (38).

## The SST Immunoregulatory Circuit in Rats

SST also is present in the thymus and spleen of rats. Using a sensitive radioimmunoassay, Kronheim et. al. in 1976 demonstrated immunoreactive SST in rat thymus and spleen (39). Northern analysis and RNA protection assays showed SST mRNA in rat thymus (40,41) and spleen (41). In the later study, radioimmunoassay confirmed synthesis of SST peptide. The thymus was a richer source of SST mRNA and SST peptide than was spleen (39,41).

As in the mouse, investigations have shown that SST can suppress rat lymphocyte proliferation. In one study, addition SST-14 ($10^{-8}$M) inhibited by 50% the ConA-stimulated proliferation of rat thymocytes as measured by thymidine incorporation and centromere separation. Response was bimodal with higher doses ($10^{-6}$ M) not effective.(42).

Endogenously produced SST may inhibit lymphocyte proliferation in culture. A stable antisense oligonucleotide complementary to the translation start site of SST inhibited endogenous SST synthesis and augmented spontaneous splenocyte proliferation 15 fold (43). Although addition of SST agonist (RC 160) abolished the antisense-stimulated proliferation, this required a high dose ($10^{-5}$M) to be effective. The antisense used in this study contained a potential CpG sequence absent in the control oligonucleotide. CpG sequences can directly stimulate lymphocyte proliferation (44) so other antisense constructs or SST receptor antagonists should be tested in this system.

There is some controversy over which SST receptors are employed by rat immune cells. Sedqi et. al. showed by RT-PCR that rat (Sprague Dawley) thymocytes express mRNA for SSTR2 but not SSTR1 or SSTR3. Stimulation of the thymocytes with phytohemagglutinin (PHA, 5 μg/ml) and interleukin-1 (1ng/ml) induced expression of SSTR1 but not SSTR3. Similar to our findings in the mouse, rat thymocyte SSTR2 mRNA levels did not change with mitogen stimulation (45). However, ten Bokum et. al. found no expression of SSTR2 by spleen, thymus lymph node and peripheral blood lymphocytes from several strains of rats assayed by RT-PCR. Instead, SSTR3 appeared to be the dominant transcript with variable expression of SSTR4 (46). They found that splenocytes specifically bound [$^{125}$I-Tyr$^0$]-SST-28 but not [$^{125}$I-Tyr$^3$]-OCT and that OCT administered as Sandostatin-LAR (10 & 30mg/kg) did not affect adjuvant induced arthritis (46). However, in a study publish in abstract form, Rees, et. al. showed that BIM, which binds similarly to OCT, did suppress adjuvant-induced arthritis (47).

Octreotide and BIM also suppressed inflammation in the Sprague-Dawley rat acute carrageenin-induced aseptic inflammation model. In this model 2% carrageenin in normal saline is injected subcutaneously into an area that was injected with air one day previously. Seven hours after injection of carrageenin the inflammation was evaluated and consisted mostly of PMN. Immunoreactive SST was present in macrophages, fibroblasts, and endothelial cells. Addition of OCT or BIM to the carrageenin mixture decreased lesion volume by up to 40% and cell number by up to 48%. Treatment with BIM also suppressed expression of immunoreactive TNFα and substance P in the lesions (48).

**The SST Immunoregulatory Circuit in Humans**

As in mouse and rat, SST also appears to be made in human lymphoid organs and sites of inflammation. Normal thymic tissue removed from children undergoing cardiovascular surgery contains mRNA for SST (49). Although peptidic nerves may account for some SST mRNA, cultured thymic epithelial cells transcribed this message (49). SST is also made at some sites of inflammation. The number of dendritic cells (specialized antigen-presenting cells) containing SST is greatly increased in the inflamed epidermis of patients with psoriasis as compared to skin from uninvolved areas or healthy controls (50).

SST can inhibit mitogen-stimulated proliferation of human lymphocytes. SST ($10^{-10}$M) inhibited proliferation of PHA-stimulated peripheral blood mononuclear cells (PBMC) but this inhibition of proliferation was not seen in higher concentrations ($>10^{-7}$M) of SST-14 (51,52). SST at $10^{-10}$M also inhibited proliferation of human Molt-4 T lymphoblasts (51,53).

SST can regulate other lymphocyte functions. SST and OCT at high doses ($10^{-6}$M) inhibited IFNγ release by human PBMC stimulated with PHA or ConA (54). SST ($10^{-8}$M) also inhibited tumor necrosis factor and lymphotoxin production by PBMC (55). SST at $10^{-12}$ to $10^{-6}$M inhibited interleukin 2 receptor display (a measure of lymphocyte activation) by PHA-stimulated PBMC and lamina propria lymphocytes (LPMC). IL2R display by LPMC was more sensitive to inhibition than was that of PBMC (52). But not everything is inhibited. SST ($10^{-8}$M) promoted the adhesion of human peripheral blood T cells to fibronectin-coated dishes (56).

Human lymphocytes have receptors for SST. In 1981 Bhathena and colleagues demonstrated that human peripheral blood lymphocytes and monocytes specifically bound SST (57). LPMC also specifically bind SST-14 (52). Autoradiography with radiolabeled octreotide showed that lymphoid follicle germinal centers in human tonsil, ileum, colon, and appendix contained cells bearing SST receptors. Cells in primary follicles without germinal centers did not bind octreotide (58).

Several human lymphocyte tumor lines express SST receptors as determined by radioligand binding. Molt-4 (human T cell line), MT-2 (adult T cell leukemia), Isk (EBV transformed B cell line), and other B cell lines (ER, RPMI-8866) specifically bound [$^{125}$I-Tyr] SST-14 (59). That study did not find binding by Jurkat T cells but Sreedharan et. al. found that Jurkat T cells and U266 IgE producing myeloma cells did bind [$^{125}$I-Tyr] SST-14. Low abundance of high affinity binding sites ($K_d$ = 3-5 pM) and a high abundance of low affinity binding sites ($K_d$ = 66-100 nM) were detected by both cell types. (60).

Like the mouse, human immune cells seem to employ SSTR2. In a study by Tsutsumi et. al., SSTR2 mRNA was expressed by peripheral blood lymphocytes from normal volunteers. Peripheral blood lymphocytes from 2 of 5 leukemia patients had increased SSTR2/$\beta$-actin mRNA ratio as compared to normals. SSTR2 transcripts were upregulated in normal PBMC stimulated for 48hours with PHA. Virally (EBV)-transformed cells also had increased SSTR2/$\beta$-actin mRNA levels (61). RT-PCR analysis of flash frozen human thymic samples showed transcripts for SSTR1, SSTR2A, and SSTR3 (49). SSTR4 and SSTR5 were absent. Cultured thymic epithelial cells (TEC) expressed transcripts for SSTR1 and SSTR2A. The authors did not detect SST receptor mRNA in isolated thymocytes but this may be due to the culture conditions as freshly isolated thymocytes do express receptors for SST (62).

Immunohistochemistry using anti-SSTR2A (R2-88) antiserum demonstrated the presence of SSTR2A on cells in the germinal centers of intestinal lymphoid follicles (63). The R2-88 anti-serum localized SSTR2A expression to macrophages in the synovium of patients with rheumatoid arthritis (64) and granulomas of patients with sarcoidosis (65).

Transcripts for SSTR2 are also detected in many human lymphocyte cell lines. Tsutsumi et. al. found that Jurkat (T), MT-2 (T), HPB-ALL (T), Sommer T cells, BT-1 (B), RPMI 8075 B cells, Odà myeloma cells, and K562 granulocytic leukemia cells all express mRNA for SSTR2. None of the cell lines expressed SSTR1. Most cell lines co-expressed SSTR3, SSTR4 or SSTR5. (61). Cardoso et. al. confirmed that Jurkat T cells express SSTR3 but did not find SSTR2 mRNA (66). Others found SSTR2 transcripts in HSB-2 and MT-1 T cells and JY, and TMM B cells, but 14 other lines did not express SSTR2 mRNA and no other SST receptor subtype mRNA was transcribed (67). Lymphoid cell lines grown *in vitro* can accumulate mutations over time that may account for some of the variation seen in these studies.

**SST Receptor Imaging in Inflammation and Lymphoma**
Octreotide scanning, commonly used to locate carcinoid tumors, also identifies areas of inflammation. Scintigraphy using [111]In-pentetreotide or [111]In-octreotide showed localized uptake in areas of granulomatous inflammation in patients with sarcoidosis, pulmonary tuberculosis, aspergillosis, and Wegener's granulomatosis (68,69). Scintigraphy also identified inflamed joints in patients with rheumatoid arthritis (70) and orbital inflammation in patients with Grave's disease (71).

In a study of 46 patients with sarcoidosis, scintigraphy using [111]In-pentetreotide was more sensitive than standard imaging to document disease in thoracic and extrapulmonary sites. However imaging did not seem to provide prognostic information about disease severity or course (72).

Scintigraphy can distinguish lymphoma. In 4 patients examined using [111]In-octreotide *in vivo*, radioligand scanning identified the involvement of lymph nodes (73). The same study evaluated lesions from 31 patients with lymphoma. Autoradiography (*ex vivo*) showed that 25 of 29 B cell non-Hodgkin's lymphomas bound [125]I-Tyr³ octreotide. A T cell lymphoma and Hodgkin's lymphoma with nodular sclerosis were also SST receptor positive (73).

Neuroendocrine tumors that adsorb labeled octreotide express transcripts for SSTR2 (74). Octreotide uptake in inflammation also may be due to SSTR2 expression. Immunohistochemistry with rabbit anti-SSTR2A antiserum demonstrated receptors on macrophages at sites of inflammation obtained from patients with rheumatoid arthritis, sarcoidosis, giant cell arteritis, and Wegener's granulomatosis (64,65). Furthermore, germinal centers in human intestinal lymphoid follicles and Peyer's patch that bind [125]I-[Tyr³]-octreotide also label with antibody specific for SSTR2A (63). Thus, SSTR2A appears to be highly expressed at sites of inflammation and immune cell activation.

**Therapeutic Use of SST and SST Agonists in Inflammation**
SST and its agonists decrease inflammatory responses *in vivo*. We showed that mice treated with octreotide develop smaller granulomas around schistosome eggs (6). Others showed that treatment with SST (0.2 μg BID) delayed onset of diabetes in the NOD-scid/scid transfer model of murine IDDM (75). Intra-articular injection of SST reduced established fibrin-induced arthritis in rabbits (76)., Octreotide and BIM 23014 suppressed the carrageenan-induced aseptic inflammatory reaction in rats (48). Octreotide treatment decreased colonic mucosal platelet activating factor, and leukotrtriene $B_4$ content and inhibited mucosal damage in rats with acetic acid-induced acute colitis (77). And administration of octreotide decreased colonic proinflammatory cytokines TNFα, IL1β and IFNγ in the rat trinitrobenzene sulfonic acid (TNBS) model of colitis (78).

SST has been used clinically as well. Repeated intra-articular injection of SST-14 (750 μg) reduced joint pain, synovial membrane thickness, erythrocyte sedimentation rate, and C-reactive protein in patients with rheumatoid arthritis (79,80). In one study published in abstract form, open-label vapreotide was used in 21 patients with Crohn's disease. The mean Crohn's activity index dropped from 269 ± 65 to 164 ± 89 (p<0.001) with 9 patients achieving remission (81). SST or it's agonists also have been used in psoraisis and result in a 30 to 80% clearance rate (82-86). In a small placebo-controlled trial, lanreotide (40mg q 2 weeks), significantly improved the inflammatory ocular manifestations of thyroid disease (87).

# CONCLUSION

SST is a true cytokine. Inflammatory mediators such as IFNγ induce SST production by murine macrophages. SST is made by activated macrophages at sites of inflammation. SST and its agonists inhibit inflammatory reactions in mice, rats, rabbits, and humans. Thus, SST functions in a feedback circuit to regulate inflammatory reactions.

SST inhibits IFNγ release by Th1 cells. IFNγ is a pivotal pro-inflammatory cytokine that regulates macrophage, T, B, and NK cell function. IFNγ promotes Th1 and inhibits Th2 cell development. Therefore, SST influences the critical balance between Th1 and Th2 responses. In this regard SST is opposed by substance P that promotes IFNγ release and prevents SST production.

In mice, and probably in humans, SST signals inflammatory cells through the SSTR2A receptor. SSTR2 couples to protein phosphotyrosine phosphatases (PTPase) and inhibitory G-proteins (88,89). In co-transfected CHO cells, SSTR2A associates with SHP-1 and functional SHP-1 is required for the inhibition of proliferation by SSTR2 (90). This observation has application to the immune system.

Antigen-induced T cell stimulation results from a protein phosphotyrosine kinase cascade (91). Initial signaling is highly dependent on tyrosine phosphokinase activity and also is regulated by PTPase. CD45 is a PTPase that positively regulates TCR signaling by dephosphorylating an inhibitory site on Lck. Other PTPases serve to down regulate T cell activation. An important inhibitory PTPase is SHP-1 that dephosphorylates Zap-70 to down-modulate T cell receptor signaling (92). SHP-1 also inhibits other cascades such as the MAPK pathway (93). The PTPases help regulate the threshold for T cell responses.

It is intriguing to speculate that SST signals through SSTR2A to activate PTPases and increase the threshold for T cell responses. It may have similar action on macrophages, B cells, and NK cells. Alternatively, SST/SSTR2A may influence other pathways that more directly regulate IFNγ production.

SST-14 is an immunoregulatory peptide made by cells of the nervous, neuroendocrine, and immune systems. This chapter has focused on SST production by macrophages and action on lymphocytes. Additionally, SST provides an avenue for cross-regulation by the nervous and endocrine systems. SST released from nerves or endocrine cells could alter local immune responses. At the same time, SST made by inflammatory macrophages could modulate local nerve and endocrine cell responses.

# ACKNOWLEDGEMENTS

Grants from the National Institute of Health (DK02428) and the Crohn's and Colitis Foundation support research by the author.

# REFERENCES

1. Elliott, D.E. Schistosomiasis: pathophysiology, diagnosis, and treatment. Gastroenterology Clinics of North America 1996 25(3):599-625.
2. Elliott DE. Methods used to study immunoregulation of schistosome egg granulomas. Methods: A Companion to Methods in Enzymology 1996;9:255-67.
3. Rakasz E, Blum AM, Metwali A, Elliott DE , Li J, Ballas ZK, Qadir K, Lynch R, Weinstock JV. Localization and regulation of IFN-gamma production within the granulomas of murine schistosomiasis in IL-4-deficient and control mice. J.Immunol. 1998;160(10):4994-9.
4. Weinstock JV, Blum AM, Malloy T. Macrophages within the granulomas of murine Schistosoma mansoni are a source of a somatostatin 1-14-like molecule. Cellular Immunology 1990;131(2):381-90.
5. Elliott DE, Blum AM, Li J, Metwali A, Weinstock JV. Preprosomatostatin messenger RNA is expressed by inflammatory cells and induced by inflammatory mediators and cytokines. J.Immunol. 1998;160(8):3997-4003.
6. Blum AM, Metwali A, Mathew RC, Cook G, Elliott D, Weinstock JV. Granuloma T lymphocytes in murine schistosomiasis mansoni have somatostatin receptors and respond to somatostatin with decreased IFN-gamma secretion. J.Immunol. 1992;149(11):3621-6.
7. Blum AM, Elliott DE, Metwali A, Li J, Qadir K, Weinstock JV. Substance P regulates somatostatin expression in inflammation. J.Immunol. 1998;161(11):6316-22.
8. Elliott DE, Li J, Blum AM, Metwali A, Patel YC, Weinstock JV. SSTR2A is the dominant somatostatin receptor subtype expressed by inflammatory cells, is widely expressed and directly regulates T cell IFN-gamma release. Eur.J.Immunol. 1999;29(8):2454-63.
9. Farrar MA, Schreiber RD. The molecular cell biology of interferon-gamma and its receptor. Annual Review of Immunology 1993;11:571-611.
10. Romagnani S. Lymphokine production by human T cells in disease states. Annual Review of Immunology 1994;12:227-57.
11. Swain SL, Bradley LM, Croft M, Tonkonogy S, Atkins G, Weinberg AD, Duncan DD, Hedrick SM, Dutton RW, Huston G. Helper T-cell subsets: phenotype, function and the role of lymphokines in regulating their development. Immunological Reviews 1991;123:115-44.
12. Blum AM, Metwali A, Mathew RC, Elliott D , Weinstock JV. Substance P and somatostatin can modulate the amount of IgG2a secreted in response to schistosome egg antigens in murine schistosomiasis mansoni. J.Immunol. 1993;151(12):6994-7004.
13. Bruno JF, Xu Y, Song J, Berelowitz M. Molecular cloning and functional expression of a brain-specific somatostatin receptor. Proc.Natl.Acad.Sci.USA 1992;89(23):11151-5.
14. Corness JD, Demchyshyn LL, Seeman P, Van Tol HH, Srikant CB, Kent G, Patel YC, Niznik HB. A human somatostatin receptor (SSTR3), located on chromosome 22, displays preferential affinity for somatostatin-14 like peptides. FEBS Lett. 1993;321(2-3):279-84.
15. Demchyshyn LL, Srikant CB, Sunahara RK, Kent G, Seeman P, Van Tol HH, Panetta R, Patel YC, Niznik HB. Cloning and expression of a human somatostatin-14-selective receptor variant (somatostatin receptor 4) located on chromosome 20. Mol Pharmacol 1993;43(6):894-901.
16. O'Carroll AM, Lolait SJ, Konig M, Mahan LC. Molecular cloning and expression of a pituitary somatostatin receptor with preferential affinity for somatostatin-28. Mol Pharmacol 1992;42(6):939-46.
17. Panetta R, Greenwood MT, Warszynska A, Demchyshyn LL, Day R, Niznik, HB, Srikant CB, Patel YC. Molecular cloning, functional characterization, and chromosomal

localization of a human somatostatin receptor (somatostatin receptor type 5) with preferential affinity for somatostatin-28. Mol Pharmacol 1994;45(3):417-27.

18. Rohrer L, Raulf F, Bruns C, Buettner R, Hofstaedter F, Schule R. Cloning and characterization of a fourth human somatostatin receptor. Proc.Natl.Acad.Sci.USA 1993;90(9):4196-200.

19. Yamada Y, Reisine T, Law SF, Ihara Y, Kubota A, Kagimoto S, Seino M, Seino Y, Bell GI, Seino S. Somatostatin receptors, an expanding gene family: cloning and functional characterization of human SSTR3, a protein coupled to adenylyl cyclase. Mol.Endocrinol. 1992;6(12):2136-42.

20. Yamada Y, Post SR, Wang K, Tager HS, Bell GI, Seino S. Cloning and functional characterization of a family of human and mouse somatostatin receptors expressed in brain, gastrointestinal tract, and kidney. Proc.Natl.Acad.Sci.USA 1992;89(1):251-5.

21. Yasuda K, Rens-Domiano S, Breder CD, Law SF, Saper CB, Reisine T, Bell GI. Cloning of a novel somatostatin receptor, SSTR3, coupled to adenylylcyclase. J.Biol.Chem. 1992;267(28):20422-8.

22. Elliott DE, Metwali A, Blum AM, Sandor M , Lynch R, Weinstock JV. T lymphocytes isolated from the hepatic granulomas of schistosome-infected mice express somatostatin receptor subtype II (SSTR2) messenger RNA. J.Immunol. 1994;153(3):1180-6.

23. Patel YC, Panetta R, Escher E, Greenwood M, Srikant CB. Expression of multiple somatostatin receptor genes in AtT-20 cells. Evidence for a novel somatostatin-28 selective receptor subtype. J.Biol.Chem. 1994;269(2):1506-9.

24. Patel YC, Greenwood M, Kent G, Panetta R , Srikant CB. Multiple gene transcripts of the somatostatin receptor SSTR2: tissue selective distribution and cAMP regulation. Biochem.Biophys.Res.Commun. 1993;192(1):288-94.

25. Reisine T, Kong H, Raynor K, Yano H, Takeda J, Yasuda K, Bell GI. Splice variant of the somatostatin receptor 2 subtype, somatostatin receptor 2B, couples to adenylyl cyclase. Mol Pharmacol 1993;44(5):1016-20.

26. Vanetti M, Kouba M, Wang X, Vogt G, Hollt V. Cloning and expression of a novel mouse somatostatin receptor (SSTR2B). FEBS Lett. 1992;311(3):290-4.

27. Alderton F, Fan TP, Schindler M, Humphrey PP. Rat somatostatin sst2(a) and sst2(b) receptor isoforms mediate opposite effects on cell proliferation. British Journal of Pharmacology 1998;125(8):1630-3.

28. Kraus J, Woltje M, Schonwetter N, Hollt V. Alternative promoter usage and tissue specific expression of the mouse somatostatin receptor 2 gene. FEBS Lett. 1998;428(3):165-70.

29. Weinstock JV, Elliott D. The substance P and somatostatin interferon-g immunoregulatory circuit. Annals of the New York Academy of Sciences 1998;840:532-9.

30. Boom WH, Liano D, Abbas AK. Heterogeneity of helper/inducer T lymphocytes. II. Effects of interleukin 4- and interleukin 2-producing T cell clones on resting B lymphocytes. Journal of Experimental Medicine 1988;167(4):1350-63.

31. Blum AM, Metwali A, Cook G, Mathew RC, Elliott D, Weinstock JV. Substance P modulates antigen-induced, IFN-gamma production in murine Schistosomiasis mansoni. J.Immunol. 1993;151(1):225-33.

32. Blum AM, Metwali A, Kim-Miller M, Li J, Qadir K, Elliott DE, Lu B, Fabry Z, Gerard N, Weinstock JV. The substance P receptor is necessary for a normal granulomatous response in murine schistosomiasis mansoni. J.Immunol. 1999;162(10):6080-5.

33. Stanisz AM, Befus D, Bienenstock J. Differential effects of vasoactive intestinal peptide, substance P, and somatostatin on immunoglobulin synthesis and proliferations by lymphocytes from Peyer's patches, mesenteric lymph nodes, and spleen. J.Immunol. 1986;136(1):152-6.

34. Krco CJ, Gores A, Go VL. Gastrointestinal regulatory peptides modulate in vitro immune reactions of mouse lymphoid cells. Clinical Immunology & Immunopathology 1986;39(2):308-18.

35. Agro A, Padol I, Stanisz AM. Immunomodulatory activities of the somatostatin analogue BIM 23014c: effects on murine lymphocyte proliferation and natural killer activity. Regulatory Peptides 1991;32(2):129-39.

36. Levite M. Neuropeptides, by direct interaction with T cells, induce cytokine secretion and break the commitment to a distinct T helper phenotype. Proc.Natl.Acad.Sci.USA 1998;95(21):12544-9.

37. Stanisz AM, Scicchitano R, Payan DG, Bienenstock J. In vitro studies of immunoregulation by substance P and somatostatin. Annals of the New York Academy of Sciences 1987;496:217-25.

38. Scicchitano R, Dazin P, Bienenstock J, Payan DG, Stanisz AM. The murine IgA-secreting plasmacytoma MOPC-315 expresses somatostatin receptors. J.Immunol. 1988;141(3):937-41.

39. Kronheim S, Berelowitz M, Pimstone BL. A radioimmunoassay for growth hormone release-inhibiting hormone: method and quantitative tissue distribution. Clinical Endocrinology 1976;5(6):619-30.

40. Fuller PJ, Verity K. Somatostatin gene expression in the thymus gland. J.Immunol. 1989;143(3):1015-7.

41. Aguila MC, Dees WL, Haensly WE, McCann SM. Evidence that somatostatin is localized and synthesized in lymphoid organs. Proc.Natl.Acad.Sci.USA 1991;88(24):11485-9.

42. Mascardo RN, Barton RW, Sherline P. Somatostatin has an antiproliferative effect on concanavalin A-activated rat thymocytes. Clinical Immunology & Immunopathology 1984;33(1):131-8.

43. Aguila MC, Rodriguez AM, Aguila-Mansilla HN, Lee WT. Somatostatin antisense oligodeoxynucleotide-mediated stimulation of lymphocyte proliferation in culture. Endocrinology 1996;137(5):1585-90.

44. Krieg AM, Yi AK, Matson S, Waldschmidt TJ, Bishop GA, Teasdale R, Koretzky GA, Klinman DM. CpG motifs in bacterial DNA trigger direct B-cell activation. Nature 1995;374(6522):546-9.

45. Sedqi M, Roy S, Mohanraj D, Ramakrishnan S, Loh HH. Activation of rat thymocytes selectively upregulates the expression of somatostatin receptor subtype-1. Biochem.Mol.Biol.Int. 1996;38(1):103-12.

46. ten Bokum AM, Lichtenauer-Kaligis EG, Melief MJ, van Koetsveld PM, Bruns C, van Hagen PM, Hofland LJ, Lamberts SW, Hazenberg MP. Somatostatin receptor subtype expression in cells of the rat immune system during adjuvant arthritis. Journal of Endocrinology 1999;161(1):167-75.

47. Rees RG, Eckland DJA, Lightman SL, Brewerton DA. The effects of somatostatin analogue, BM23014, on adjuvant arthritis in rats. British Journal of Rheumatology 1989;28(Supplement 2):40

48. Karalis K, Mastorakos G, Chrousos GP, Tolis G. Somatostatin analogues suppress the inflammatory reaction in vivo. Journal of Clinical Investigation 1994;93(5):2000-6.

49. Ferone D, van Hagen PM, van Koetsveld PM, Zuijderwijk J, Mooy DM, Lichtenauer-Kaligis EG, Colao A, Bogers AJ, Lombardi G, Lamberts SW, et al. In vitro characterization of somatostatin receptors in the human thymus and effects of somatostatin and octreotide on cultured thymic epithelial cells. Endocrinology 1999;140(1):373-80.

50. Talme T, Schultzberg M, Sundqvist KG, Marcusson JA. Colocalization of somatostatin- and HLA-DR-like immunoreactivity in dendritic cells of psoriatic skin [see comments]. Acta Dermato-Venereologica 1997;77(5):338-42.

51. Payan DG, Hess CA, Goetzl EJ. Inhibition by somatostatin of the proliferation of T-lymphocytes and Molt-4 lymphoblasts. Cellular Immunology 1984;84(2):433-8.

52. Fais S, Annibale B, Boirivant M, Santoro A, Pallone F, Delle FG. Effects of somatostatin on human intestinal lamina propria lymphocytes. Modulation of lymphocyte activation. Journal of Neuroimmunology 1991;31(3):211-9.

53. Tang SC, Braunsteiner H, Wiedermann CJ. Regulation of human T lymphoblast growth by sensory neuropeptides: augmentation of cholecystokinin-induced inhibition of Molt-4 proliferation by somatostatin and vasoactive intestinal peptide in vitro. Immunology Letters 1992;34(3):237-42.

54. Yousefi S, Ghazinouri A, Vaziri N, Tilles J, Carandang G, Cesario T. The effect of somatostatin on the production of human interferons by mononuclear cells. Proceedings of the Society for Experimental Biology & Medicine 1990;194(2):114-8.

55. Yousefi S, Vaziri N, Carandang G, Le W, Yamamoto R, Granger G, Ocariz J, Cesario T. The paradoxical effects of somatostatin on the bioactivity and production of cytotoxins derived from human peripheral blood mononuclear cells. British Journal of Cancer 1991;64(2):243-6.

56. Levite M, Cahalon L, Hershkoviz R, Steinman L, Lider O. Neuropeptides, via specific receptors, regulate T cell adhesion to fibronectin. J.Immunol. 1998;160(2):993-1000.

57. Bhathena SJ, Louie J, Schechter GP, Redman RS, Wahl L, Recant L. Identification of human mononuclear leukocytes bearing receptors for somatostatin and glucagon. Diabetes 1981;30(2):127-31.

58. Reubi JC, Horisberger U, Waser B, Gebbers JO, Laissue J. Preferential location of somatostatin receptors in germinal centers of human gut lymphoid tissue. Gastroenterology 1992;103(4):1207-14.

59. Nakamura H, Koike T, Hiruma K, Sato T, Tomioka H, Yoshida S. Identification of lymphoid cell lines bearing receptors for somatostatin. Immunology 1987;62(4):655-8.

60. Sreedharan SP, Kodama KT, Peterson KE, Goetzl EJ. Distinct subsets of somatostatin receptors on cultured human lymphocytes. J.Biol.Chem. 1989;264(2):949-52.

61. Tsutsumi A, Takano H, Ichikawa K, Kobayashi S, Koike T. Expression of somatostatin receptor subtype 2 mRNA in human lymphoid cells. Cellular Immunology 1997;181(1):44-9.

62. Ferone D, van Hagen PM, Colao A, Annunziato L, Lamberts SW, Hofland, LJ. Somatostatin receptors in the thymus. Annals of Medicine 1999;31 Suppl 2:28-33.

63. Reubi JC, Laissue JA, Waser B, Steffen DL, Hipkin RW, Schonbrunn A. Immunohistochemical detection of somatostatin sst2a receptors in the lymphatic, smooth muscular, and peripheral nervous systems of the human gastrointestinal tract: facts and artifacts. Journal of Clinical Endocrinology & Metabolism 1999;84(8):2942-50.

64. ten Bokum AM, Melief MJ, Schonbrunn A, van der Ham F, Lindeman J, Hofland LJ, Lamberts SW, van Hagen PM . Immunohistochemical localization of somatostatin receptor sst2A in human rheumatoid synovium. Journal of Rheumatology 1999;26(3):532-5.

65. ten Bokum AM, Hofland LJ, de Jong G, Bouma J, Melief MJ, Kwekkeboom, DJ, Schonbrunn A, Mooy CM, Laman JD, et al. Immunohistochemical localization of somatostatin receptor sst2A in sarcoid granulomas. European Journal of Clinical Investigation 1999;29(7):630-6.

66. Cardoso A, el Ghamrawy C, Gautron JP, Horvat B, Gautier N, Enjalbert, Krantic S. Somatostatin increases mitogen-induced IL-2 secretion and proliferation of human Jurkat T cells via sst3 receptor isotype. Journal of Cellular Biochemistry 1998;68(1):62-73.

67. van Hagen PM, Hofland LJ, ten Bokum AM, Lichtenauer-Kaligis EG, Kwekkeboom DJ, Ferone D, Lamberts SW. Neuropeptides and their receptors in the immune system. Annals of Medicine 1999;31 Suppl 2:15-22.

68. Vanhagen PM, Krenning EP, Reubi JC, Kwekkeboom DJ, Bakker WH, Mulder, AH, Laissue I, Hoogstede HC, Lamberts SW. Somatostatin analogue scintigraphy in granulomatous diseases. European Journal of Nuclear Medicine 1994;21(6):497-502.

69. Ozturk E, Gunalp B, Ozguven M, Ozkan S, Sipit T, Narin Y, Bayhan H. The visualization of granulomatous disease with somatostatin receptor scintigraphy. Clinical Nuclear Medicine 1994;19(2):129-32.

70. Vanhagen PM, Markusse HM, Lamberts SW, Kwekkeboom DJ, Reubi JC, Krenning EP. Somatostatin receptor imaging. The presence of somatostatin receptors in rheumatoid arthritis. Arthritis & Rheumatism 1994;37(10):1521-7.

71. Postema PT, Kwekkeboom DJ, van Hagen PM, Krenning EP. Somatostatin-receptor scintigraphy in Graves' orbitopathy. European Journal of Nuclear Medicine 1996;23(6):615-7.

72. Kwekkeboom DJ, Krenning EP, Kho GS, Breeman WA, van Hagen PM. Somatostatin receptor imaging in patients with sarcoidosis. European Journal of Nuclear Medicine 1998;25(9):1284-92.

73. Reubi JC, Waser B, van Hagen M, Lamberts SW, Krenning EP, Gebbers JO, Laissue JA. In vitro and in vivo detection of somatostatin receptors in human malignant lymphomas. International Journal of Cancer 1992;50(6):895-900.

74. John M, Meyerhof W, Richter D, Waser B, Schaer JC, Scherubl H, Boese-Landgraf J, Neuhaus P, Ziske C, Molling K, et al. Positive somatostatin receptor scintigraphy correlates with the presence of somatostatin receptor subtype 2. Gut 1996;38(1):33-9.
75. Bowman MA, Campbell L, Darrow BL, Ellis TM, Suresh A, Atkinson MA. Immunological and metabolic effects of prophylactic insulin therapy in the NOD-scid/scid adoptive transfer model of IDDM. Diabetes 1996;45(2):205-8.
76. Matucci-Cerinic M, Borrelli F, Generini S, Cantelmo A, Marcucci I, Martelli F, Romagnoli P, Bacci S, Conz A, Marinelli P. Somatostatin-induced modulation of inflammation in experimental arthritis. Arthritis & Rheumatism 1995;38(11):1687-93.
77. Eliakim R, Karmeli F, Okon E, Rachmilewitz D. Octreotide effectively decreases mucosal damage in experimental colitis. Gut 1993;34(2):264-9.
78. Lamrani A, Tulliez M, Chauvelot-Moachon L, Chaussade S, Mauprivez C, Hagnere AM, Vidon N. Effects of octreotide treatment on early TNF-alpha production and localization in experimental chronic colitis. Alimentary Pharmacology & Therapeutics 1999;13(5):583-94.
79. Coari G, Di Franco M, Iagnocco A, Di Novi MR, Mauceri MT, Ciocci A. Intra-articular somatostatin 14 reduces synovial thickness in rheumatoid arthritis: an ultrasonographic study. International Journal of Clinical Pharmacology Research 1995;15(1):27-32.
80. Fioravanti A, Govoni M, La Montagna G, Perpignano G, Tirri G, Trotta, Bogliolo A, Ciocci A, Mauceri MT, Marcolongo R. Somatostatin 14 and joint inflammation: evidence for intraarticular efficacy of prolonged administration in rheumatoid arthritis. Drugs Under Experimental & Clinical Research 1995;21(3):97-103.
81. Cortot A, Dupas JL, Colombel JF, Canva JY, Rigaud D, Gehenot M, Deschamps V, and Bonfils S. Vapreotide (Octastatine, RC 160) in the treatment of acute Crohn's disease (CD). Gut 1994;35F285
82. Weber G, Klughardt G, Neidhardt M, Galle K, Frey H, Geiger A. Treatment of psoriasis with somatostatin. Archives of Dermatological Research 1982;272(1-2):31-6.
83. Guilhou JJ, Boulanger A, Barneon G, Vic P, Meynadier J, Tardieu JC, Clot J. Somatostatin treatment of psoriasis. Archives of Dermatological Research 1982;274(3-4):249-57.
84. Matucci-Cerinic M, Lotti T, Cappugi P, Boddi V, Fattorini L, Panconesi E. Somatostatin treatment of psoriatic arthritis. International Journal of Dermatology 1988;27(1):56-8.
85. Venier A, De Simone C, Forni L, Ghirlanda G, Uccioli L, Serri F, Frati L. Treatment of severe psoriasis with somatostatin: four years of experience. Archives of Dermatological Research 1988;280 Suppl:S51-S54
86. Camisa C, O'Dorisio TM, Maceyko RF, Schacht GE, Mekhjian HS, Howe BA. Treatment of psoriasis with chronic subcutaneous administration of somatostatin analog 201-995 (sandostatin). I. An open-label pilot study. Cleveland Clinic Journal of Medicine 1990;57(1):71-6.
87. Krassas GE, Kaltsas T, Dumas A, Pontikides N, Tolis G. Lanreotide in the treatment of patients with thyroid eye disease. European Journal of Endocrinology 1997;136(4):416-22.
88. Buscail L, Delesque N, Esteve JP, Saint-Laurent N, Prats H, Clerc P, Robberecht P, Bell GI, Liebow C, Schally AV. Stimulation of tyrosine phosphatase and inhibition of cell proliferation by somatostatin analogues: mediation by human somatostatin receptor subtypes SSTR1 and SSTR2. Proc.Natl.Acad.Sci.USA 1994;91(6):2315-9.
89. Dent P, Wang Y, Gu YZ, Wood SL, Reardon DB, Mangues R, Pellicer A, Schonbrunn A, Sturgill TW. S49 cells endogenously express subtype 2 somatostatin receptors which couple to increase protein tyrosine phosphatase activity in membranes and down-regulate Raf-1 activity in situ. Cellular Signalling 1997;9(7):539-49.
90. Lopez F, Esteve JP, Buscail L, Delesque N, Saint-Laurent N, Theveniau, Nahmias C, Vaysse N, Susini C. The tyrosine phosphatase SHP-1 associates with the sst2 somatostatin receptor and is an essential component of sst2-mediated inhibitory growth signaling. J.Biol.Chem. 1997;272(39):24448-54.
91. Alberola-Ila J, Takaki S, Kerner JD, Perlmutter RM. Differential signaling by lymphocyte antigen receptors. Annual Review of Immunology 1997;15:125-54.

92. Plas DR, Johnson R, Pingel JT, Matthews RJ, Dalton M, Roy G, Chan, AC, Thomas ML. Direct regulation of ZAP-70 by SHP-1 in T cell antigen receptor signaling. Science 1996;272(5265):1173-6.
93. Pani G, Fischer KD, Mlinaric-Rascan I, Siminovitch KA. Signaling capacity of the T cell antigen receptor is negatively regulated by the PTP1C tyrosine phosphatase. Journal of Experimental Medicine 1996;184(3):839-52.

# 12
# PHYSIOLOGY OF SOMATOSTATIN RECEPTORS: FROM GENETICS TO MOLECULAR ANALYSIS*

Hans-Jürgen Kreienkamp, Chong Wee Liew, Dietmar Bächner, Marie-Germaine Mameza, Michaela Soltau, Arne Quitsch, Marcus Christenn, Wolf Wente and Dietmar Richter
*Institut für Zellbiochemie und Klinische Neurobiologie, Universität Hamburg, Martinistrasse 52, Hamburg 20246, Germany*

## INTRODUCTION

The molecular cloning of somatostatin (SST) receptors in the early 1990s brought about an unexpected multiplicity of receptor subtypes (1,2); whereas only two different types of ligand binding sites could be discriminated in different tissues by conventional pharmacological radioligand binding techniques, it became clear that five different receptor (SSTR1-5) genes are expressed in a highly tissue-specific and developmentally regulated manner (3). A similar number of different G-protein coupled receptors has been observed before also for other hormones such as dopamine (4) or acetylcholine (5). In contrast to these other receptor systems, however, SSTRs can not easily be differentiated in a functional way by analysis of their intracellular coupling to G-proteins, as all five subtypes can be coupled to activation of pertussis-toxin sensitive G-proteins of the Gi/o class (6). Consequently, activation of SSTRs leads to a reduction of cellular cAMP levels (6) the activation of potassium conductances (7) and inhibition of voltage gated calcium channels (8). These effects have been observed not only in heterologous expression systems but also in more physiological situations involving endogenously expressed receptors (e.g. 9). However, not all SSTR-mediated actions may be easily explained by this coupling pattern; inhibition of cell growth by SST-analogs for example requires the activation of a tyrosine phosphatase which is a rather uncommon effector for Gi-coupled receptors (10); in hypothalamic neurons, activation of one receptor subtype (SSTR2) attenuates glutamate-induced depolarization, while another subtype (SSTR1) augments the response to glutamate (11,12). Thus in native systems receptor signal transduction may differ from that observed in heterologous expression system, probably due to the presence of additional factors which affect the functional state of the receptor. We attempted to shed light on these phenomena by the identification of SSTR-interacting proteins; these efforts are summarized in the last part of this chapter.

Meanwhile, it has become an important and difficult task to assign physiological functions to individual receptor subtypes. Several approaches have been taken to solve this question; an important step was the analysis of the cellular and subcellular distribution of receptor subtypes first by *in situ* hybridization (3; 13-15) and later on by immunohistochemical analysis (e.g. 16-18). In addition the availability of cloned receptors has made possible the development of subtype specific receptor agonists and antagonists; both these aspects are reviewed in chapters XYZ. A third approach has been the generation of mice deficient for individual subtype genes. It should be noted at this stage however, that the analysis of SSTR-ko mice is still largely in progress, and valid conclusions can be drawn only with respect to the regulatory effects of SST in some endocrine systems. A major problem in the analysis of traditional knockout mice is certainly the presence of compensatory effects, i.e. the presence of a regulatory process which detects the absence of the physiological influence of the gene of interest at early developmental stages. Detailed studies of the mRNA levels of SSTR1-5 in adult SSTR2-ko mice, however, yielded little evidence of compensatory regulation (D. Hoyer, Basel, personal communication) suggesting that once the expression pattern of the SSTR genes has been defined during early development little if any cross regulations occur at later stages.

This chapter reviews recent progress in the studies of SSTR knockout mice and the attempts to assign physiological functions to individual SSTR subtypes. In a separate effort we will summarize experiments devoted to the identification of SST-like transmitter systems in *Drosophila melanogaster*. The analysis of receptor function in simpler organisms such as Drosophila with a well-defined genetic background and possibly with a less receptor subtype complexity may be an alternative approach in assigning a function to individual receptor subtypes.

## GENETIC ANALYSIS: THE ROLE OF SSTR2 AND SSTR1 IN THE REGULATION OF ENDOCRINE AND EXOCRINE SECRETION

The function of SSTR2 has first been analyzed by Zheng et al. (19) who deleted the mouse SSTR2 gene by homologous recombination; this strain of mice has been analyzed by several laboratories with respect to the regulation of growth hormone, glucagon and gastric acid release (see below). The data of Zheng et al. (19) show that SST is involved in the feedback regulation of growth hormone at the level of the hypothalamus.

An analysis of the endocrine pancreas in SSTR2-deficient mice was performed by Strowski et al. (20) in isolated pancreatic cells; here it was shown that SST is still able to regulate the release of insulin by beta cells of pancreatic islets while the release of glucagon was no longer subject to control by SST agonists in SSTR2-ko mice, thus assigning a clear and specific function to SSTR2. These data are in agreement with the specific expression of SSTR2 in alpha cells pancreatic islets that was observed by immunocytochemical analysis and *in situ* hybridization. In addition, these results are also in agreement with pharmacological studies with SSTR5- and SSTR2-specific agonists which show that SSTR5 is the receptor subtype which regulates the release of insulin. The presence of SSTR5 in insulin-containing beta cells had also been shown earlier by immunocytochemical analysis (21).

In the exocrine pancreas, it has long been known that SST negatively influences the release of gastric acid; SSTR2 deficient mice showed that the rate of gastric acid release was increased almost tenfold when compared to wild type littermates, probably due to lack of regulation by SST via this receptor subtype (22). Again these data are in agreement with previous studies by Rossowski and Coy (23) who used SSTR2 and SSTR5 specific agonists to show that SSTR2 (and not SSTR5) is the major regulator of gastric acid release.

SSTR1 deficient mice were similarly generated by homologous recombination in ES cells (24). As observed before with the SSTR2-deficient mice, SSTR1-ko mice are viable and are born at the expected Mendelian frequencies. There are no apparent differences between knockout animals and their heterozygous and wild-type littermates, suggesting that the SSTR1 gene is not essential for normal development.

We investigated if the lack of SSTR1 affects the regulation of growth hormone secretion by SST; whereas all SSTR subtypes are expressed in the pituitary (2), SSTR2 has been suggested to be the major regulator of growth hormone secretion, an assignment which is largely based on pharmacological data using subtype specific agonists. In isolated pituitary cells derived from wild-type or SSTR1-ko animals, the SSTR2 specific agonist octreotide but not the SSTR1 specific agonist CH-275 efficiently suppressed growth hormone release stimulated by forskolin treatment. This pattern was changed, however, when the basal rate of growth hormone release was analyzed. In wild-type animals the SSTR1-specific drug clearly reduced hormone release similarly to octreotide and the (non-specific) natural agonist SST-14. Whereas octreotide and - to a lesser degree - SST-14 were active in this assay in SSTR1-deficient animals, CH-275 had no effect. These data confirm the predominance of SSTR2-mediated effects on growth hormone secretion but also show that SSTR1 can mediate the inhibitory effect of SST-14 in some situations. This is in agreement with an earlier observation by Roosterman et al. (25) who demonstrated that SSTR1 may be coupled to the inhibition of voltage-activated calcium channels in endocrine cells.

## SSTR FUNCTIONS IN THE CENTRAL NERVOUS SYSTEM

Due to the very complex temporal and spatial expression pattern of SSTR subtypes in the central nervous system it has been difficult to attribute specific physiological functions to individual SSTRs in the CNS. SST-14 has been shown to influence a broad spectrum of physiological processes including locomotor activity and cognitive functions. Recent progress in this field has been achieved by Viollet et al. (26) by a careful behavioural analysis of the SSTR2-ko mice initially described by Zheng et al. (19).

A first major finding of this study is the almost complete loss of binding sites for radiolabelled SST-14 in SSTR2-ko animals, with the notable exception of the hippocampal CA1 region where SST-14 binding remains rather unaffected. Thus the ko-model confirms earlier observations that the distribution of SST-binding sites in the rodent brain is largely overlapping with immunoreactivity for the SSTR2 (16), suggesting that SSTR2 is by far the most abundant receptor subtype. The residual binding sites in the CA1 region cannot be easily attributed to one receptor subtype, but strong pharmacological evidence points to SSTR4 as the receptor being present in this part of the brain. This is in agreement with the presence of SSTR4 mRNA and immunoreactivity in the hippocampus. It is however somewhat surprising how small the contribution of SSTR1 and 3 to total SST-binding in the brain must be, as mRNAs coding for these receptors were found to be also rather abundant in the brain when analyzed by *in situ* hybridization experiments (27,28) The fifth receptor subtype, SSTR5 shows only very limited expression in the brain (29) and would therefore not be expected to account for a large proportion of binding sites.

In a behavioural analysis, SSTR2-deficient mice exhibited a reduced locomotor activity in the so-called open-field test in which mice are placed into an open box and their movements are monitored. This observation is not due to motor defects of these mice, as validated e.g. by the uncompromised ability of SSTR2-ko mice to walk on a rotating horizontal pole. The hypoactive behaviour rather appears to be related to an increase in anxiety and an alteration of the physiological stress response, leading to an increased fear to explore novel areas. In agreement with this, hypoactivity of SSTR2-ko mice is only observed in novel environments (such as the open-field), but not after habituation of the animals for example in their home cage. Increased anxiety is also supported in other tests like the elevated-plus maze, where SSTR2-deficient mice do not explore novel areas as much as their wild-type counterparts. Thus the action of SST elicited through SSTR2 receptor appears to reduce stress and/or stress related anxiety in mice. Interestingly, SSTR2-ko mice exhibit a strong increase in pituitary ACTH release which may be related to the alteration of the stress response in these animals.

Based on the transient expression of SSTR1 in the brain, a role for this receptor subtype in neuronal development has been postulated (30). As this role does not become evident in the SSTR1-knockout model, it can be speculated that compensatory effects such as upregulation of other receptor systems might mask an effect of the loss of the SSTR1 gene. Recent attempts in this laboratory to generate double-knockout mice deficient in SSTR1 and 2 revealed - besides the low number of littermates - no obvious phenotype; thus physiological relevant information has to await more detailed behavioural analysis of this knockout strain.

To eliminate possible compensatory effects due to the concomitant expression of one or the other SST receptor subtype during development we started to generate a transgenic mouse in which the SSTR1 is temporally targeted deleted in neurons. In brief, the generation of the transgenic mice included several steps including (i) the design of the construct with flanking regions upstream (9 kb) and downstream (2 kb) of the SSTR 1 gene; (ii) sequencing of the construct by the linker-scanning technique. Once the targeting construct was generated, extensive analyses were carried out to verify the important features of the construct including (iii) detailed sequencing and restriction analysis to diagnose the construct and its selection markers; as well as test deletion of the 'floxed' region in E. coli. The final targeting vector contained loxP-sites for the inducible deletion of the SSTR1 gene, and Frt-sites for the deletion of the neomycin cassette (C.W. Liew, unpublished)

Targeting construct DNA of large quantity and high quality was prepared by modifying the culture conditions and purification steps by combining the methods of Qiagen column and CsCl gradient purification. After transfection and homologous recombination in ES cells, five cell lines were established with the desired homologous recombination event. Chimeric and subsequently heterozygous mice were generated from these lines. Homozygous mutant mice were obtained by crossbreeding and are presently used for further breeding with the appropriate 'inducer' mice. As inducer mice we used a mouse line where expression of the cre recombinase is driven by the CaM kinase promoter, which restricts expression to the forebrain, beginning at postnatal day 3. Crossing this strain into our SSTR1/lox line will (i) enable us to test the correct induction of SSTR1 gene deletion, and (ii) provide us with the first set of mice where the SSTR1 gene is deleted postnatally. The generation of these transgenic mice are underway (C. W. Liew, unpublished).

## IDENTIFICATION OF A SSTR-LIKE GENE IN MAMMALS

The SSTR 1-like receptor SLC-1 was identified by its homology to members of the somatostatin receptor family (31) yet it was neither activated by SST-14 nor SST-28. Only recently the nature of this receptor has been elucidated by a 'reverse physiology' (also called 'reverse pharmacology') strategy (32) using rat brain extracts and a sensitive screening system in which SLC-1 was functionally coexpressed with a G-protein-gated inwardly rectifying potassium (GIRK) channel in *Xenopus* oocytes. Upon several purification steps, followed by mass spectrometric analysis and peptide sequencing the ligand was identified as melanin-concentrating hormone (MCH) indicating that the SLC-1 is a MCH receptor (33; reviewed in ref. 34).

*Xenopus* oocytes expressing the MCH receptor responded to nanomolar concentrations of synthetic MCH not only by the activation of GIRK-mediated currents but also by the induction of Ca-ion dependent chloride currents mediated by phospholipase C indicating that this receptor can couple either to the $G_i$ or $G_q$ mediated signal transduction pathway suggesting that MCH may serve for a number of distinct brain functions. The main function of MCH/MCH-R in higher vertebrates seems to be the regulation of appetite behaviour and energy metabolism as indicated by null mutations of MCH or MCH-R in transgenic mice, resulting in a lean phenotype and altered energy metabolism [35,36].

## SSTR-LIKE GENES IN A SIMPLE MODEL ORGANISM

A better understanding of gene functions may be gained from the study of simple model organisms such as the invertebrate *Drosophila melanogaster* where elaborate genetic systems are available; in addition, due to the limited complexity for example of the *Drosophila* nervous system a phenotypic trait of receptor deficient animals may be more obvious. With respect to SST (and the closely related opioid) receptors, however, it has been unclear if similar receptors exist in invertebrates (37). In addition, the issue has been complicated for neuropeptide receptors in general by the lack of a clear assignment between the numerous insect neuropeptides, and those neuropeptide receptors that have been known for some time, e.g. NPY-like and tachykinin-like receptors in *Drosophila melanogaster*. We have addressed this issue by searching for receptors similar to SSTRs using a strategy based on reverse transcriptase PCR with degenerate oligonucleotides. By this method we identified a receptor cDNA in *Drosophila* with strong similarity to SST, opioid, and also galanin receptors (38). When functionally expressed in *Xenopus* oocytes, this novel receptor was however not activated by any of the known mammalian peptide ligands. This prompted us to search for the native ligand of the new receptor by the reverse-physiological approach (32). For this purpose the functional coexpression of the receptor with the GIRK channel in *Xenopus* oocytes was used to screen for bioactive substances in a peptide extract prepared from *Drosophila* heads. An active peptide was detected in this extract, purified to homogeneity and identified by mass spectrometry as the *Drosophila* homolog of the allatostatin peptides which are widely found in invertebrates ranging from crabs to insects, but which have so far not been detected in vertebrates. The novel receptor was accordingly named AlstR for **al**lato**st**atin **r**eceptor (38).

Numerous peptides have been described as allatostatins based on their ability to inhibit juvenile hormone synthesis from the retrocerebral *corpora allata* complex of insects. They can be grouped into the allatostatin subfamilies A, B, and C. The ligand that is recognized by AlstR belongs to the subfamily A. Peptides of the subfamily A vary in sequence and length but all carry a conserved pentapeptide YXFGL-amide at the C-terminus. The conserved pentapeptide is sufficient to trigger AlstR signalling provided it is amidated (32,39). Members of the allatostatin subfamily B are characterized by the C-terminal motif XWXXXXXXW-amide, whereas the allatostatin peptides C are pyro-Glu or N-Gln decapentapeptides.

By combining *Drosophila* genome database search and RT-PCR-based cDNA isolation, we succeeded in cloning two other allatostatin receptors activated by members of the allatostatin C subfamily. The receptors were termed Drostar1 and 2 for **Dro**sophila allato**sta**tin C **r**eceptor1 and 2 (40). By reverse pharmacology, the physiological ligands were identified as pyro-Glu-allatostatin C and Gln-allatostatin C. Only the allatostatin C peptides activated the two Drostars whereas no cross-activation of the two receptors occurred when complemented with allatostatin A peptides; conversely, AlstR receptors are not activated by allatostatin C.

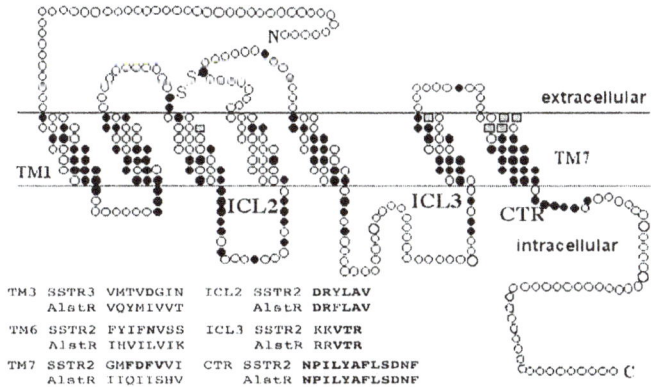

**Fig. 1. Conserved regions in SSTR-like receptors.** *A model of SSTRs/AlstR is shown; residues conserved between SSTRs, opioid and galanin receptors as well as Drosophila AlstR are indicated in black. A comparison of the sequences involved in agonist binding in transmembrane (TM) regions 3, 6 and 7, and in G-protein specificity in intracellular loops (ICL) 2 and 3 and in the carboxy-terminal region (CTR) is shown at the bottom. Residues in SSTRs which have been shown to be involved in agonist binding or selectivity are indicated by squares and bold print; sequences conserved between SSTRs and AlstR in intracellular segments are also indicated by bold print.*

A sequence comparison between AlstRs, Drostars SSTRs, opioid and galanin receptors shows that similarity between the mammalian and the *Drosophila* receptors is highest in the transmembrane regions facing the intracellular side of the membrane known to interact with G-proteins; this is in line with our functional expression studies showing that, similar to its mammalian counterparts, the *Drosophila* AlstR and Drostar couple to the activation of GIRK channels in *Xenopus* oocytes, but not to the endogenous phospholipase C system. Thus, *Drosophila* allatostatin receptors and mammalian SSTRs may activate the same mammalian effector system, presumably via coupling to G-proteins of the $G_i/G_o$-class. Amino acid residues which dictate specificity in G-protein recognition have been identified close to the membrane in intracellular loops 2 and 3 and the carboxy terminus of GPCRs (41), and these are the regions where amino acid similarity is high between the two receptor classes (Fig. 1).

On the other hand, sequences are clearly distinct between allatostatin and SST receptors within those parts of the receptor molecule facing the ligand-binding region at the extracellular side of the membrane. This is particularly true for the aspartate residue in the third transmembrane domain ($Asp_{124}$ in SSTR3) and the sequences between the sixth and seventh transmembrane regions, which determine the selectivity of the ligand for the SSTRs. As shown earlier the aspartate residue of SSTR is part of a narrow and selective pocket forming an ion pair with the positively charged amino acid residues lysine at position 4 or 9 of SST and by portions of TM6 and 7 facing the extracellular side of the plasma membrane (42,43). In contrast, in AlstRs the aspartate residue is replaced by isoleucine (38,39), in Drostars by threonine (40). That these amino acid residues may also play a role in ligand binding was shown by site directed mutagenesis of Drostar1 (40). When converting threonine ($Thr_{140}$) of Drostar1 into an aspartate, the recombinant Drostar1 receptor showed a 3-fold reduced affinity for pyroGlu-allatostatin C; no activity was observed with SST-14. A Drostar1 chimeric mutant carrying the second binding region of SSTR2 from the middle of TM6 to the middle of TM7 was neither activated by pyroGlu-allatostatin C nor SST-14. Clearly, there is no cross-reactivity between the allatostatin and the SST ligand/receptor-systems even in the chimeric mutants. Thus, the data suggest that during evolution of this group of receptors a common scaffold of a receptor protein coupled to inhibitory G-proteins has been maintained, whereas the extracellular parts facing the ligand are rather variable in order to accommodate different peptide ligands. Both SSTRs and *Drosophila* allatostatin receptors presumably use part of TM3 and the extracellular portions of TM6 and 7 to accommodate their agonists, as is underlined by our mutagenesis study.

Although there are no sequence similarities between SST and allatostatin peptides they share some functions and expression patterns. For instance, similar to the inhibitory function of SST, allatostatin also block the release of insect hormones. In *Drosophila*, allatostatin immunoreactivity is found in interneurons of the central nervous system, a feature shared with the mammalian SST peptide (44) which may reflect a general importance of inhibitory neuropeptides in interneurons.

Taken together, although there do exist some similarities between mammalian SST and insect allatostatin receptors, the insect equivalent to the mammalian somatostatin peptides appears to be absent. This assumption gains support by our inability to identify SST- or opioid-precursor like sequences in the *Drosophila* genome database although exon-intron boundaries within a predicted peptide may complicate such a search. As a conclusion, in insects a number of GPCRs – although structurally related to vertebrate receptors - may use an entirely different set of peptides to activate their receptors. For a functional annotation of the many receptor sequences generated in genome projects such as those for *Drosophila melanogaster* or *Caenorhabditis elegans*, it may therefore in many cases become necessary to identify the physiological ligands of novel receptors by the reverse pharmacology technique.

**Identification of SSTR-Interacting Proteins**

Signal transduction by seven TM receptors has been classically viewed as being performed solely by heterotrimeric G-proteins (45). Thus the assignment of a receptor such as the five SSTRs to a particular set of G-proteins (i.e. the Gi/Go-coupled pertussin toxin-sensitive G-proteins) would predict that in most or all cellular systems this receptor would have the same effect on second messenger systems. However, this dogma has been challenged recently by the observation that additional intracellular proteins may physically interact with GPCRs and influence the outcome of receptor signalling. Thus it was shown by Hall et al. (46) that activation of the ß$_2$-adrenergic receptor leads to an interaction of the receptor C-terminus with the PDZ (postsynaptic density-95/discs large/ZO-1) domain of the regulatory factor of a Na$^+$/H$^+$ exchanger protein, leading to activation of exchange activity. PDZ-domains are known as protein interaction domains consisting of about 90 amino acid-long repeated sequences that bind in a sequence-specific manner to the C-terminus of another protein provided it terminates in a peptide motif -X-S/T-X-V/L with a free carboxyl group. Although hydrophobic C-terminal residues seem to be preferred by many PDZ-containing proteins exceptions from this rule have also been found (reviewed by Sheng and Sala, 2001; ref. 47).

*SSTR1*. When inspecting the C-terminus of mammalian SSTRs we realized that most of them contain a typical recognition motif for PDZ-containing proteins (Table 1). To identify potential candidates interacting with the various SST receptor subtypes we used the respective C-terminus as a bait in a yeast two-hybrid system. When the C-terminus of SSTR1 was used as a bait the human Skb1Hs was identified as interacting protein which is homologous to the yeast protein Skb1 known to down-regulate mitosis in *Schizosaccharomyces pombe* via binding to the Shk1 protein kinase; the latter is a homolog to the mammalian p21$^{cdc42/Rac}$-activated protein kinases (47). SSTR1 and Skb1Hs were co-localized at the cell surface when co-expressed in human embryonic kidney cells; these cells showed a strong increase in SST binding compared to cells expressing the receptor alone suggesting that Skb1Hs may act like a chaperone by correctly targeting the receptor to the cell surface.

**Table 1. C-terminus of SSTR1-5 and their interacting partners**

| C-terminus | Interacting proteins | Function |
|---|---|---|
| SSTR1-GTCASR<u>ISTL</u> | Shk1 kinase bdg. protein | chaperone |
| SSTR2-LLNGDL**QTSI** | SSTRIP/Shank | scaffolding |
| SSTR3-GDKASTL**SHL** | MUPP | transport |
| SSTR4-RVPFTK**TTTF** | PSD-95 | scaffolding |
| SSTR5-ANGLMQ**TSRI** | PIST | transport |

*The PDZ-recognition motif at the C-terminus of SSTR1-5 is marked in bold letters and essential amino acid residues are underlined. Except for Shk1 kinase binding protein all the other interacting proteins contain a PDZ (postsynaptic density-95/discs large/ZO-1) domain. Possible functions of the interacting proteins are listed. Shank, Src homology 3 domain and ankyrin repeat-containing; MUPP, multiple PDZ domain protein; PSD-95, postsynaptic density-95 kD protein; PIST, PDZ protein interacting specifically with TC10 (TC10, small GTPase of the rho family).*

*SSTR2.* Using the intracellular C-terminus of the SSTR2 as a bait, we detetcted strong interactions with the PDZ domain of a novel human protein which we termed <u>s</u>omato<u>s</u>tatin <u>r</u>eceptor <u>i</u>nteracting <u>p</u>rotein or SSTRIP. Cloning of the full-length human SSTRIP protein revealed a multidomain structure consisting of a set of seven ankyrin repeats at the N-terminus followed by an SH3 (<u>S</u>rc <u>h</u>omology 3) domain, the PDZ domain mediating the interaction with SSTR2, a set of several proline rich domains and a SAM region (<u>S</u>terile <u>a</u>lpha <u>m</u>otif) at its C-terminus (ref. 48; Fig. 2).

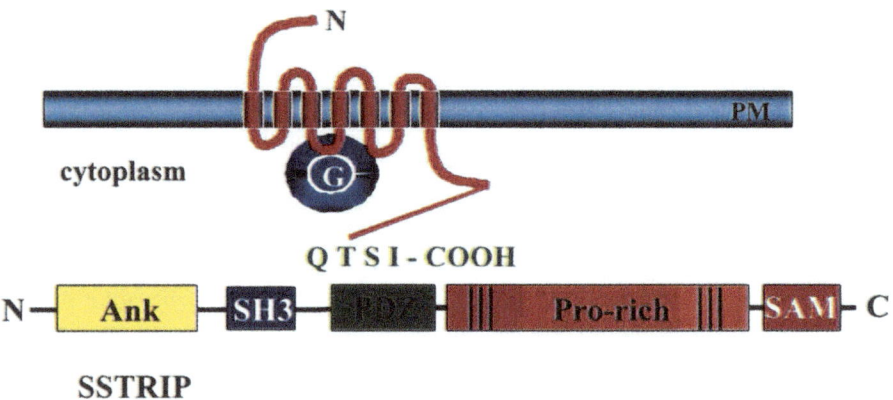

***Fig. 2. The C-terminus of SSTR2 interacts with the PDZ-domain of SSTRIP.*** *Ank, ankyrin repeat; SH3, Src homology 3; Pro-rich, Proline-rich domain; SAM, sterile alpha-motif; G, G-protein; Q, Glutamine; T, Threonine; S, Serine; I, Isoleucine; PM, plasma membrane.*

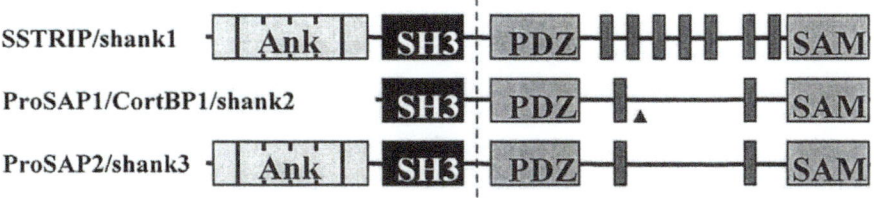

*Fig. 3. Domain organization of SSTRIP/ProSAP/Shank proteins. The different protein interaction domains of the three family members are indicated. Note that different splice variants have been described for all isoforms, which may lead to the expression of N-terminally (vertical line) or C-terminally (arrowhead for ProSAP1/Shank2) truncated proteins. Ank, ankyrin repeats; small grey boxes, proline-rich domains.*

The domain structure of SSTRIP (or its rat ortholog Shank1 or synamon; 49,50) is more or less conserved in two other proteins termed cortactin binding protein1 (CortBP1), ProSAP1 or Shank2, and ProSAP2 or Shank3 (49, 51-54; see Fig. 3). In addition, the sequence of the PDZ domain is almost identical in all three proteins, which is in line with our observation that the PDZ domain of CortBP1 exhibits a similar high affinity to the SSTR2 C-terminus as the SSTRIP PDZ domain (55).

Interaction of SSTR2 with SSTRIP/CortBP1 provides a link for the receptor to the actin-based cytoskeleton; CortBP1 interacts with the actin binding protein cortactin via one of its proline rich regions and the SH3-domain of cortactin (Fig. 4; ref. 55). Interestingly, cortactin is a major substrate of the src tyrosine kinase in v-src transformed fibroblasts; tyrosine phosphorylation or overexpression of cortactin (due to amplification of the corresponding genomic region in breast cancer cells) has been implicated in the progression of tumours (56). We pursued the idea that Tyr-phosphorylated cortactin, due to its indirect interaction with SSTR2, might be a preferred substrate of a SSTR2-induced phosphatase activity; however we could so far not detect any experimental evidence for this hypothesis in MCF7 breast adenocarcinoma cells (M. Soltau, unpublished).

As the PDZ domains of the variants of this multidomain SSTRIP protein family are structurally very similar, it appears likely that they are responsible for anchoring SSTR2 in many different localizations in neurons including presynaptic as well as somatodendritic, postsynaptic compartments. However, the strong enrichment of proteins of this family in the postsynaptic density fraction from rat brain suggests an important structural function which certainly goes far beyond an anchoring function for SSTRs. For instance, the interaction of the PDZ domain with the C-terminus of the so-called GKAP proteins (Fig. 4; ref. 50), in addition to the strong enrichment of all family members in the postsynaptic density of excitatory synapses in the brain, identifies the SSTRIP protein family as major structural components of the postsynaptic apparatus in glutamatergic synapses. GKAPs provide a link to the NMDA-receptor complex in the synapse through an intermediate interaction with the SAP-90/PSD-95 family of proteins (57,58). A further interaction of a second proline rich region

196

of SSTRIP (Fig. 4) with the *enabled/VASP homology* (EVH) domain of homer/Vesl proteins links SSTRIP/Shanks to a second branch of glutamatergic signalling in excitatory synapses, i.e. the metabotropic glutamate receptors (59), suggesting that the SSTRIP/Shank proteins represent a major master scaffold protein family within the architecture of the postsynaptic density. This notion gets further support by our findings that for instance, the N-terminal ankyrin repeats of Shank1 and 3 specifically interact with the cytoskeletal protein α-fodrin (Fig. 4; ref. 60,61). Mapping of the interacting domains of α-fodrin revealed that the highly conserved spectrin repeat 21 is sufficient to bind to the ankyrin repeats. Both interacting partners are coexpressed widely in the rat brain and are colocalized in synapses of hippocampal cultures. Our data suggest that the complexes consisting of α-fodrin and either Shank1 or Shank3 may represent a dynamic substructure of the synapse responsible for alterations of the functional architecture that occur during the organization and reorganization of spines and synapses in the central nervous system.

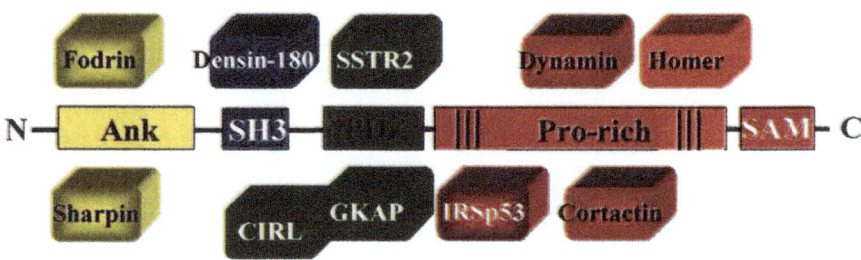

**Fig. 4. Interacting partners of the multidomain protein SSTRIP.** *SSTRIP domain and its respective interacting partners are marked by the same colour. Note that IRSp53 and homer bind to different proline-rich regions (59,62-64). GKAP, guanylate kinase associated protein (50); CIRL, calcium independent recptor for alpha-latrotoxin (67). For further details see ref. 68.*

As another interacting partner of the SSTRIP/Shank protein family we recently identified IRSp53 (insulin receptor substrate protein of 53 kDa; ref. 62-64) which itself is a multidomain protein. The SH3 domain of IRSp53 specifically binds to a proline-rich motif of SSTRIP different to that known for binding homer (Fig.4; ref. 59). IRSp53 known as a target of the small GTPase cdc42 has been implicated in the formation of filopodia which is induced by cdc42 (65,66). The interaction of IRSp53 with SSTRIP requires activated cdc42; our data show that SSTRIP interferes with the induction of filopodia by cdc42/IRSp53, presumably by displacing other effector proteins from the SH3 domain of IRSp53. Consequently, overexpression of IRSp53 in cultivated neurons induces elongated dendritic spines or filopodia, and coexpression of SSTRIP abrogates this effect. Thus, SSTRIP itself steps in as an effector of the cdc42/IRSp53-signalling pathway, leading to enhanced maturation of dendritic spines and the assembly of receptor-associated signalling complexes (Soltau et al., Moll. Cell Neurosci. 21 (4):575-83, 2002).

As interacting partner for the SH3 domain of SSTRIP the postsynaptic density protein Densin-180 was identified in a yeast two-hybrid assay (Fig. 4). This interaction was verified in vitro and in vivo by co-immunoprecipitation experiments using either transfected HEK cells or brain lysates. Densin-180 is thought to be a transmembrane adhesion molecule participating in building the synaptic cleft. SSTRIP-Densin-180 interaction may suggest that SSTRIP is involved in localizing Densin-180 in the PSD (A. Quitsch, unpublished).

In summary, these data clearly demonstrate that SSTRIP is a member of a highly important synaptic scaffolding protein family of the brain that provides multiple links between synaptic receptors and the cytoskeleton. *SSTR3*. Using the C-terminus of human SSTR3 in a yeast two-hybrid screen as bait we identified the **mu**ltiple **P**DZ domain **p**rotein1 (MUPP1) as SSTR3 interacting partner (C. W. Liew, unpublished). MUPP1 contains 13 PDZ domains (69-71), appears to be concentrated at tight junctions in epithelial cells and myelinating Schwann cells (69,70). Mapping of the SSTR3-MUPP1 interaction revealed that the 10th PDZ domain of MUPP1 is involved. Preliminary data indicate that in the mouse brain MUPP1 forms large macromolecular junctional complexes containing proteins such as PSD-95, the epithelial tight junction protein ZO-1 and MAGI-3 (MAGUK with inverted orientation; MAGUK, **m**embrane-**a**ssociated **gu**anylate **k**inase), as well as other signalling proteins including the CaM-kinase alpha subunit. Overexpression of MUPP together with SSTR3 potentiates the cytotoxic effect achieved by overexpression of SSTR3 alone; thus, we currently investigate if the cytotoxic effect of SSTR3 is due to disruption of MUPP-mediated formation of cell adhesion complexes.

*SSTR4*. By using a slightly modified purification strategy based on an immobilized C-terminal peptide, we could show that the C-terminus of the rat SSTR4 binds to the second PDZ domain of PSD-95; both proteins can efficiently be co-immunoprecipitated from rat brain lysates, indicating that interaction occurs in vivo as well (M. Christenn, unpublished).

*SSTR5*. In the case of SSTR5, we identified several interacting proteins one of which is PIST (**P**DZ protein **i**nteracting **s**pecifically with **T**C10 [TC10, small GTPase of the rho family]), a protein that is found in the Golgi apparatus (72). PIST was found to be colocalized with SSTR5 in the Golgi apparatus of transfected cells, and we are currently analyzing if PIST plays a role in the intracellular transport of SSTR5 to the cell surface (W. Wente, unpublished).

Taken together, the GPCR-mediated signal transduction process depends on a number of factors including multidomain proteins that interact specifically with the C-terminus of the receptor. Most of these interacting proteins are members of a rather diverse PDZ-containing protein family that recognize a consensus sequence at the C-terminus of GPCR. The best-studied example is the SSTR receptor family where most of the receptor subtypes terminate in a PDZ consensus sequence - a prerequisite for binding to a PDZ-protein. Other interacting proteins may contain a zinc-finger domain as it is the case for MIZIP (ref. 73; melanin-concentrating hormone receptor1 interacting zinc-finger protein) which binds to the C-terminus of MCH-R1 (33) or an S-adenosyl methionine binding-domain as in Skb1Hs - the interacting partner for SSTR1 (48). The so far identified GPCR-interacting factors have been reported as chaperon, transport or scaffold proteins. They often form larger protein complexes via their multidomain structure as it has been shown for the SSTRIP/Shank proteins. The function of these receptor-multidomain protein complexes may reside, for instance, in the tissue-specific modulation of the SST-mediated signalling process.

## ACKNOWLEDGEMENT

This work was supported by Deutsche Forschungsgemeinschaft grants SFB 545/B7 (to H.-J.K. and D.R.), GRK 255/2 (to D.R.), BA 2160/1-1 (to D.B.), and by European Commission grant QLG3-CT-1999-00908 (to D.R.).

## REFERENCES

1.  Patel, Y.C. (1999). Somatostatin and its receptor family. Front Neuroendocrinol 20, 157-98.
2.  Darlison, M. G. & Richter, D. (1999). Multiple genes for neuropeptides and their receptors: Co-evolution and physiology. Trends in Neurosciences 22, 81-88.
3.  Wulfsen, I., Meyerhof, W., Fehr, S., and Richter, D. (1993). Expression patterns of rat somatostatin receptor genes in pre- and postnatal brain and pituitary. J Neurochem 61, 1549-52.
4.  Civelli, O., Bunzow, J. R., and Grandy, D.K. (1993). Molecular diversity of the dopamine receptors. Annu Rev Pharmacol Toxicol 33, 281-307.
5.  Wess, J., (1996). Molecular biology of muscarinic acetylcholine receptors. Crit Rev Neurobiol 10, 69-99.
6.  Patel, Y.C., Greenwood, M.T., Warszynska, A., Panetta, R., and Srikant, C.B. (1994). All five cloned human somatostatin receptors (hSSTR1-5) are functionally coupled to adenylyl cyclase. Biochem Biophys Res Commun 198, 605-12.
7.  Kreienkamp, H.-J., Hönck, H.-H., and Richter, D. (1997). Coupling of rat somatostatin receptor subtypes to a G-protein gated inwardly rectifying potassium channel (GIRK1). FEBS Lett 419, 92-4.
8.  Glassmeier, G., Hopfner, M., Riecken, E.O., Mann, B., Buhr, H., Neuhaus, P., Meyerhof, W., and Scherubl, H. (1998). Inhibition of L-type calcium channels by octreotide in isolated human neuroendocrine tumor cells of the gut. Biochem Biophys Res Commun 250, 511-5.
9.  Sodickson, D.L., and Bean, B.P. (1998). Neurotransmitter activation of inwardly rectifying potassium current in dissociated hippocampal CA3 neurons: interactions among multiple receptors. J Neurosci 18, 8153-62.
10. Pan, M.G., Florio, T., and Stork, P.J. (1992). G protein activation of a hormone-stimulated phosphatase in human tumor cells. Science 256, 1215-7.
11. Viollet, C., Lanneau, C., Faivre-Bauman, A., Zhang, J., Djordjijevic, D., Loudes, C., Gardette, R., Kordon, C., and Epelbaum, J. (1997). Distinct patterns of expression and physiological effects of sst1 and sst2 receptor subtypes in mouse hypothalamic neurons and astrocytes in culture. J Neurochem 68, 2273-80.
12. Lanneau, C., Viollet, C., Faivre-Bauman, A., Loudes, C., Kordon, C., Epelbaum, J., and Gardette, R. (1998). Somatostatin receptor subtypes sst1 and sst2 elicit opposite effects on the response to glutamate of mouse hypothalamic neurones: an electrophysiological and single cell RT-PCR study. Eur J Neurosci 10, 204-12.

13. Breder, C.D., Yamada, Y., Yasuda, K., Seino, S., Saper, C.B., and Bell, G.I. (1992). Differential expression of somatostatin receptor subtypes in brain. J Neurosci 12, 3920-34.

14. Senaris, R.M., Humphrey, P.P., and Emson, P.C. (1994). Distribution of somatostatin receptors 1, 2 and 3 mRNA in rat brain and pituitary. Eur J Neurosci 6, 1883-96.

15. Beaudet, A., Greenspun, D., Raelson, J., and Tannenbaum, G.S. (1995). Patterns of expression of SSTR1 and SSTR2 somatostatin receptor subtypes in the hypothalamus of the adult rat: relationship to neuroendocrine function. Neuroscience 65, 551-61.

16. Dournaud, P., Gu, Y.Z., Schonbrunn, A., Mazella, J., Tannenbaum, G.S., and Beaudet, A. (1996). Localization of the somatostatin receptor SST2A in rat brain using a specific anti-peptide antibody. J Neurosci 16, 4468-78.

17. Schulz, S., Schreff, M., Schmidt, H., Handel, M., Przewlocki, R., and Höllt, V. (1998). Immunocytochemical localization of somatostatin receptor sst2A in the rat spinal cord and dorsal root ganglia. Eur J Neurosci 10, 3700-8.

18. Schulz, S., Schmidt, H., Handel, M., Schreff, M., and Höllt, V. (1998). Differential distribution of alternatively spliced somatostatin receptor 2 isoforms (sst2A and sst2B) in rat spinal cord. Neurosci Lett 257, 37-40.

19. Zheng, H., Bailey, A., Jiang, M.H., Honda, K., Chen, H.Y., Trumbauer, M.E., Van der Ploeg, L.H., Schaeffer, J.M., Leng, G., and Smith, R.G. (1997). Somatostatin receptor subtype 2 knockout mice are refractory to growth hormone-negative feedback on arcuate neurons. Mol Endocrinol 11, 1709-17.

20. Strowski, M.Z., Parmar, R.M., Blake, A.D., and Schaeffer, J.M. (2000). Somatostatin inhibits insulin and glucagon secretion via two receptors subtypes: an in vitro study of pancreatic islets from somatostatin receptor 2 knockout mice. Endocrinology 141, 111-7.

21. Mitra, S.W., Mezey, E., Hunyady, B., Chamberlain, L., Hayes, E., Foor, F., Wang, Y., Schonbrunn, A., and Schaeffer, J.M. (1999). Colocalization of somatostatin receptor sst5 and insulin in rat pancreatic beta-cells. Endocrinology 140, 3790-6.

22. Martinez, V., Curi, A.P., Torkian, B., Schaeffer, J.M., Wilkinson, H.A., Walsh, J.H., and Tache, Y. (1998). High basal gastric acid secretion in somatostatin receptor subtype 2 knockout mice. Gastroenterology 114, 1125-32.

23. Rossowski, W.J., and Coy, D.H. (1998) New potent somatostatin receptor 2 antagonist inhibits SRIF, and SRIF-mediated effects on gastric acid and pancreatic secretion in rats. 25th Meeting of the Federation of European Biochemical Societies (FEBS), Abstracts, p.133.

24. Kreienkamp, H.-J., Akgün, E., Baumeister, H., Meyerhof, W., and Richter, D. (1999). Somatostatin receptor subtype 1 modulates basal inhibition of growth hormone release in somatotrophs. FEBS Lett 462, 464-6.

25. Roosterman, D., Glassmeier, G., Baumeister, H., Scherubl, H., and Meyerhof, W. (1998). A somatostatin receptor 1 selective ligand inhibits $Ca^{2+}$ currents in rat insulinoma 1046-38 cells. FEBS Lett 425, 137-40.

26. Viollet, C., Vaillend, C., Videau, C., Bluet-Pajot, M.T., Ungerer, A., L'Heritier, A., Kopp, C., Potier, B., Billard, J., Schaeffer, J., Smith, R.G., Rohrer, S.P., Wilkinson, H., Zheng, H., and Epelbaum, J. (2000). Involvement of sst2 somatostatin receptor in locomotor, exploratory activity and emotional reactivity in mice. Eur J Neurosci 12, 3761-70.

27. Meyerhof, W., Paust, H.-J., Schönrock, C., and Richter, D. (1991). Cloning of a cDNA encoding a novel putative G-protein-coupled receptor expressed in specific rat brain regions. DNA and Cell Biology 10, 689-94.

28. Meyerhof, W., Wulfsen, I., Schönrock, C., Fehr, S., and Richter, D. (1992). Molecular cloning of a somatostatin-28 receptor and comparison of its expression pattern with that of a somatostatin-14 receptor in rat brain. Proc Natl Acad Sci USA 89, 10267-71.

29. Stroh, T., Kreienkamp, H.-J., and Beaudet, A. (1999). Immunohistochemical distribution of the somatostatin receptor subtype 5 in the adult rat brain: predominant expression in the basal forebrain. J Comp Neurol 412, 69-82.

30. Hartmann, D., Fehr, S., Meyerhof, W., and Richter, D. (1995). Distribution of somatostatin receptor subtype 1 mRNA in the developing cerebral hemispheres of the rat. Dev Neurosci 17, 246-55.

31. Kolakowski, L.F., Jung, B.P., Ngyen, T., Johnson, M.P., Lynch, K.R., Cheng, R., Heng, H.H.Q., George, S.R. and O'Dowd, B.F. (1996). Characterization of a human gene related to genes encoding somatostatin receptors. FEBS Lett. 398, 253-258.

32. Reinscheid, R.K., Nothacker, H.P., and Civelli, O. (1999). Orphan receptors and the concept of reverse physiology: discovery of the novel neuropeptide orphanin FQ/nociceptin. Results Probl Cell Differ 26, 193-214.

33. Bächner, D., Kreienkamp, H., Weise, C., Buck, F., and Richter, D. (1999). Identification of melanin concentrating hormone (MCH) as the natural ligand for the orphan somatostatin-like receptor 1 (SLC-1). FEBS Lett. 457, 522-524.

34. Saito, Y., Nothacker, H., and Civelli, O. (2000). Melanin-concentrating hormone receptor: an orphan receptor fits the key Trends Endocrinol. Metab. 11, 299-303.

35. Shimada, M., Tritos, N.A., Lowell, B.B., Flier, J.S., and Maratos-Flier, E. (1998). Mice lacking melanin-concentrating hormone are hypophagic and lean. Nature 396, 670-674.

36. Marsh, D.J., Weingarth, D.T., Novi, D.E., Chen, H.Y., Trumbauer, M.E., Chen, A.S., Guan, X.M., Jiang, M.M., Feng, Y., Camacho, R.E., Shen, Z., Frazier, E.G., Yu, H., Metzger, J.M., Kuca, S.J., Shearman, L.P., Gopal-Truter, S., MacNeil, D.J., Strack, A.M., MacIntyre, D.E., Van der Ploeg, L.H., and Qian, S. (2002). Melanin-concentrating hormone 1 receptor-deficient mice are lean, hyperactive, and hyperphagic and have altered metabolism. Proc. Natl. Acad. Sci. U S A. 99, 3240-3245.

37. Li, X., Keith, D.E., Jr., and Evans, C.J. (1996). Mu opioid receptor-like sequences are present throughout vertebrate evolution. J Mol Evol 43, 179-84.

38. Birgül, N., Weise, C., Kreienkamp, H.-J., and Richter, D. (1999). Reverse physiology in Drosophila: identification of a novel allatostatin-like neuropeptide and its cognate receptor structurally related to the mammalian somatostatin/galanin/opioid receptor family. EMBO J 18, 5892-900.

39. Auerswald, L., Birgül, N., Gäde, G., Kreienkamp, H.-J., and Richter, D. (2001). Structural, functional and evolutionary characterization of novel members of the allatostatin receptor family from insects. Biochem Biophys Res Commun 282, 904-9.

40. Kreienkamp, H.-J., Larusson, H.J., Witte, I., Roeder, T., Birgül, N., Hönck, H.-H., Harder, S., Ellinghausen, G., Buck, F., and Richter, D. (2002). Functional annotation of two orphan G-protein coupled receptors, Drostar1 and 2, from Drosophila melanogaster and their ligands by reverse pharmacology. J Biol Chem 277, 39937-39943.

41. Strader, C.D., Fong, T.M., Tota, M.R., Underwood, D., and Dixon, R.A. (1994). Structure and function of G protein-coupled receptors. Annu Rev Biochem. 63, 101-32.

42. Kaupmann, K., Bruns, C., Raulf, F., Weber, H.P., Mattes, H., and Lübbert, H. (1995). Two amino acids, located in transmembrane domains VI and VII, determine the selectivity of the peptide agonist SMS 201-995 for the SSTR2 somatostatin receptor. EMBO J 14, 727-35.

43. Nehring, R.B., Meyerhof, W., and Richter, D. (1995). Aspartic acid residue 124 in the third transmembrane domain of the somatostatin receptor subtype 3 is essential for somatostatin-14 binding. DNA Cell Biol 14, 939-44.

44. Somogyi, P., Hodgson, A.J., Smith, A.D., Nunzi, M.G., Gorio, A., and Wu, J.Y. (1984). Different populations of GABAergic neurons in the visual cortex and hippocampus of cat contain somatostatin- or cholecystokinin- immunoreactive material. J Neurosci 4, 2590-603.

45. Gilman, A.G. (1987). G proteins: transducers of receptor-generated signals. Annu Rev Biochem 56, 615-49.

46. Hall, R.A., Premont, R.T., Chow, C.W., Blitzer, J.T., Pitcher, J.A., Claing, A., Stoffel, R.H., Barak, L.S., Shenolikar, S., Weinman, E.J., Grinstein, S., and Lefkowitz, R.J. (1998). The beta2-adrenergic receptor interacts with the $Na^+/H^+$-exchanger regulatory factor to control $Na^+/H^+$exchange. Nature 392, 626-30.

47. Sheng, M., and Sala, C. (2001). PDZ domains and the organization of supramolecular complexes. Annu. Rev. Neurosci. 24, 1-29.

48. Schwärzler, A., Kreienkamp, H.-J., and Richter, D. (2000). Interaction of the somatostatin receptor subtype 1 with the human homolog of the Shk1 kinase binding protein from yeast. J Biol Chem 275, 9557-62.

49. Zitzer, H., Hönck, H.-H., Bächner, D., Richter, D., and Kreienkamp, H.-J. (1999). Somatostatin receptor interacting protein defines a novel family of multidomain proteins present in human and rodent brain. J Biol Chem 274, 32997-33001.

50. Naisbitt, S., Kim, E., Tu, J.C., Xiao, B., Sala, C., Valtschanoff, J., Weinberg, R.J., Worley, P.F., and Sheng, M. (1999). Shank, a novel family of postsynaptic density proteins that binds to the NMDA receptor/PSD-95/GKAP complex and cortactin. Neuron 23, 569-82.

51. Yao, I., Hata, Y., Hirao, K., Deguchi, M., Ide, N., Takeuchi, M., and Takai, Y. (1999). Synamon, a novel neuronal protein interacting with synapse-associated protein 90/postsynaptic density-95-associated protein. J Biol Chem 274, 27463-6.

52. Du, Y., Weed, S.A., Xiong, W.C., Marshall, T.D., and Parsons, J.T. (1998). Identification of a novel cortactin SH3 domain-binding protein and its localization to growth cones of cultured neurons. Mol Cell Biol 18, 5838-51.

53. Boeckers, T.M., Winter, C., Smalla, K.H., Kreutz, M.R., Bockmann, J., Seidenbecher, C., Garner, C.C., and Gundelfinger, E.D. (1999). Proline-rich synapse-associated proteins ProSAP1 and ProSAP2 interact with synaptic proteins of the SAPAP/GKAP family. Biochem Biophys Res Commun 264, 247-52.

54. Boeckers, T.M., Kreutz, M.R., Winter, C., Zuschratter, W., Smalla K.H., Sanmarti-Vila, L., Wex, H., Langnaese, K., Bockmann, J., Garner, C.C., and Gundelfinger, E.D. (1999). Proline-rich synapse-associated protein-1/cortactin binding protein 1 (ProSAP1/CortBP1) is a PDZ-domain protein highly enriched in the postsynaptic density. J Neurosci 19, 6506-18.

55. Zitzer, H., Richter, D., and Kreienkamp, H.-J. (1999). Agonist-dependent interaction of the rat somatostatin receptor subtype 2 with cortactin-binding protein 1. J Biol Chem 274, 18153-6.

56. Wu, H., and Parsons, J.T. (1993). Cortactin, an 80/85-kilodalton pp60src substrate, is a filamentous actin-binding protein enriched in the cell cortex. J Cell Biol 120, 1417-26.

57. Schuuring, E., Verhoeven, E., Litvinov, S., and Michalides, R.J. (1993). The product of the EMS1 gene, amplified and overexpressed in human carcinomas, is homologous to a v-src substrate and is located in cell-substratum contact sites. Mol Cell Biol 13, 2891-98.

58. Kim, E., Naisbitt, S., Hsueh, Y.P., Rao, A., Rothschild, A., Craig, A.M., and Sheng, M. (1997). GKAP, a novel synaptic protein that interacts with the guanylate kinase-like domain of the PSD-95/SAP90 family of channel clustering molecules. J Cell Biol 136, 669-78.

59. Takeuchi, M., Hata, Y., Hirao, K., Toyoda, A., Irie, M., and Takai, Y. (1997). SAPAPs. A family of PSD-95/SAP90-associated proteins localized at postsynaptic density. J Biol Chem 272, 11943-51.

60. Böckers, T., Mameza, M.G., Kreutz, M.R., Bockmann, J., Richter, D., Gundelfinger, E.D., and Kreienkamp, H.-J. (2001). Synaptic scaffolding proteins in rat brain: Ankyrin repeats of the multidomain Shank protein family interact with the cytoskeletal protein alpha-fodrin. J. Biol. Chem. 276, 40104-40112.

61. Mameza, M.-G. (2003). Wechselwirkungen postsynaptischer Proteine aus Mensch und Ratte. Thesis. University of Hamburg, Germany pp 1-102.

62. Soltau, M., Richter. D., and Kreienkamp, H.-J. (2002). The insulin receptor substrate IRSp53 links the postsynaptic shank1/PSD-95 complex to the small G-protein cdc42. Molecular and Cellular Neuroscience 21, 575-583.

63. Soltau, M. (2003). Untersuchungen zur Funktion des humanen Proteins Shank1 in der Synaptogenese von Säugern. Thesis. University of Hamburg, Germany pp 1-98.

64. Bockmann, J., Kreutz, M.R., Gundelfinger, E.D., and Böckers, T.M. (2002). ProSAP/Shank postsynaptic density proteins interact with insulin receptor tyrosine kinase substrate IRSp53. J Neurochem, 83, 1013-1017.

65. Krugmann, S., Jordens, I., Gevaert, K., Driessens, M., Vandekerckhove, J., and Hall, A. (2001). Cdc42 induces filopodia by promoting the formation of an IRSp53:Mena complex. Curr Biol, 11, 1645-1655.

66. Govind, S., Kozma, R., Monfries, C., Lim, L., and Ahmed, S. (2001). Cdc42Hs facilitates cytoskeletal reorganization and neurite outgrowth by localizing the 58-kD insulin receptor substrate to filamentous actin. J Cell Biol, 152, 579-594.

67. Kreienkamp, H.-J., Zitzer, H., Gundelfinger, E.D., Richter, D., and Böckers, T.M. (2000). The calcium independent receptor for alpha-latrotoxin from human and rodent brain interacts with members of the ProSAP/SSTRIP/Shank family of multidomain proteins. J. Biol. Chem. 275, 32387-32390.

68. Sheng M., and Kim, M.J. (2002). Postsynaptic signaling and plasticity mechanisms. Science 298, 776-80.
69. Ullmer, C., Schmuck, K., Figge, A., and Lubbert, H. (1998). Cloning and characterization of MUPP1, a novel PDZ domain protein. FEBS Lett. 424, 63-68.
70. Hamazaki, Y., Itoh, M., Sasaki, H., Furuse, M., and Tsukita, S. (2002). Multi-PDZ domain protein 1 (MUPP1) is concentrated nat tight junctions through its possible interaction with claudin-1 and junction adhesion molecule. J. Biol. Chem. 277, 455-461.
71. Poliak, S., Matlis, S., Ullmer, C., Scherer, S.S., and Peles, E. (2002). Distinct claudins and associated PDZ proteins from different autotypic thight junctions in myelinating Schwann cells. J. Cell Biol. 159, 361-371.
72. Neudauer, C.L., Joberty, G., and Macara, I.G. (2001). PIST: a novel PDZ/coiled-coiled domain binding partner for the rho-family GTPase TC10. Biochem. Biophys. Res. Commun.280, 541-547.
73. Bächner, D., Kreienkamp, H.-J., and Richter, D. (2002). MIZIP, a highly conserved, vertebrate specific melanin-concentrating hormone receptor 1 interacting zinc-finger protein. FEBS Lett 526, 124-128.

# 13
# SOMATOSTATIN RECEPTOR IMAGING

Dik J. Kwekkeboom, Marion de Jong and Eric P. Krenning
*University Hospital Rotterdam, Rotterdam, the Netherlands*

## INTRODUCTION

In diagnosing or staging cancer, conventional imaging techniques like CT scanning or MRI usually seem adequate in detecting tumor localizations. Difficulties are encountered, however, when patients harbour small tumors, especially in the abdomen, a site notorious for the lack of sensitivity that is achieved with even the most refined imaging modalities. For the localization of for instance neuroendocrine gastroenteropancreatic tumors, like carcinoids, insulinomas, and gastrinomas, sensitivities of CT and MRI are at most 50%, depending on the size and localization of the lesions. Therefore, much effort has been put into developing other means to visualize such tumors. Because both neuroendocrine and non-neuroendocrine tumors possess peptide hormone and/or growth factor receptors, radiolabeled peptides that bind to such receptors may be of potential use to be applied in tumor imaging and staging. The first of such peptides to be successfully applied in vivo was the radiolabeled somatostatin (SST) analogue $[^{111}In-DTPA^0]$-octreotide. Experience with in vivo SST receptor imaging (SRI) has been gained in many institutions, and will be summarized in this chapter.

### SOMATOSTATIN AND SOMATOSTATIN RECEPTORS

SST receptors (SSTR) have been identified on many cells and tumors of neuroendocrine origin, like the somatotroph cells of the anterior pituitary, and pancreatic islet-cell tumors (1,2). Also cells and tumors not known as classically neuroendocrine of origin, such as activated lymphocytes, lymphomas and breast cancer, may possess these receptors (3,4). The SST analogue octreotide binds to SSTR subtypes sstr2, sstr3, and sstr5 on both tumorous and non-tumorous tissues (See Chapter 6).

Because of its relatively long effective half-life, $[^{111}In-DTPA^0]$-octreotide is a radiolabeled SST analogue which can be used to visualize SSTR-bearing tumors efficiently after 24 and 48 h, when interfering background radioactivity is minimized by renal clearance.

**Scanning Protocol**

The preferred dose of $[^{111}In\text{-}DTPA^0]$-octreotide is about 200 MBq. With such a dose it is possible to perform Single Photon Emission Tomography (SPET), which produces tomographic slices of the body. SPET may increase the sensitivity to detect SSTR-positive tissues and gives a better anatomical delineation than planar views. The acquisition of sufficient counts per view and also obtaining spot images with a sufficient counting time instead of performing whole-body scanning with a too low count density, are other important points that may make the difference between a successful localizing study and a disappointing one.

Planar images are obtained with a double head or large field of view gamma camera, equipped with medium-energy parallel-hole collimators. The pulse height analyzer windows are centered over both $^{111}$In photon peaks (172 kev and 245 keV) with a window width of 20%. The acquisition parameters for planar images (preferably spot views) are 300,000 preset counts or 15 min per view for the head and neck, and 500,000 counts or 15 min for the remainder of the body. If "whole-body" acquisition is used, scan speed should not exceed 3 cm/min. Using higher scan speeds, like 8 cm/min, will result in failure to recognize small SSTR positive lesions and lesions with a low density of these receptors (5). For SPECT images with a triple-head camera the acquisition parameters are: 40 steps of 3 degrees each, 64 x 64 matrix, and at least 30 sec per step (45 sec for SPECT of the head). SPECT analysis is performed with a Wiener or Metz filter on original data. The filtered data are reconstructed with a Ramp filter. Planar and SPECT studies are preferably performed 24 h after injection of the radiopharmaceutical. Planar studies after 24 and 48 h can be carried out with the same protocol. Repeat scintigraphy after 48 h is especially indicated when 24 h scintigraphy shows accumulation in the abdomen, which may also represent radioactive bowel content.

**Normal Scintigraphic Findings and Artifacts**

Normal scintigraphic features include visualization of the thyroid, spleen, liver, and kidneys, and in part of the patients of the pituitary (Figure 1, panels a-c). Also, the urinary bladder and the bowel (to a variable degree) are usually visualized. The visualization of the pituitary, thyroid, and spleen is due to receptor binding, whereas the uptake in the liver and kidneys is for the most part due to metabolism, although SSTRs have been demonstrated in renal tubular cells and vasa recta (6). There is a predominant renal clearance of the SST analogue, although hepatobiliary clearance into the bowel also occurs, necessitating the use of laxatives in order to facilitate the interpretation of abdominal images.

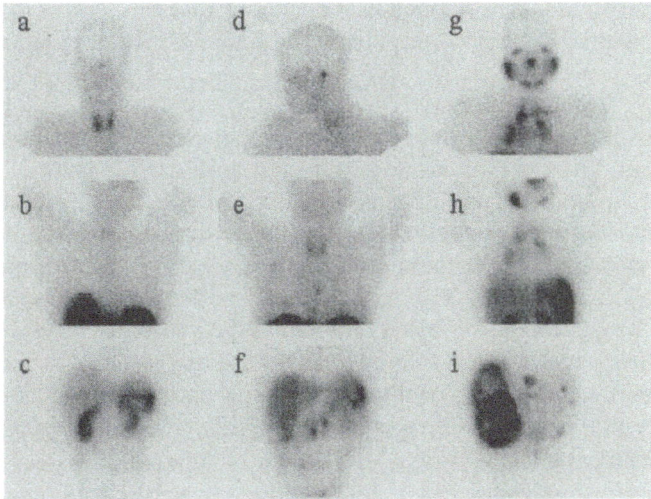

*Figure1. Anterior SRI planar images 24 h p.i. a-c: Normal visualization of the pituitary and thyroid (a,b) and liver, spleen and kidneys (c). d: Uptake in a tympanic paraganglioma. e,f: Multiple lesions in the chest and abdomen in a patient with a metastasized neuroendocrine tumor. g,h: Characteristic uptake in lacrimal and salivary glands, as well as bihilar and mediastinal pathology in a patient with sarcoidosis. i: High uptake in mutiple metstases, mainly in the liver, in a carcinoid patient.*

False-positive results of SRI have been reported. In virtually all cases the term "false-positive" is a misnomer because SSTR positive lesions that are not related to the pathology for which the investigation is performed, are present. Examples are the visualization of the gallbladder, thyroid abnormalities, accessory spleens, recent CVAs, activity at the site of a recent surgical incision, etc. Many of these have been reviewed by Gibril et al (7). Also, chest uptake after irradiation and diffuse breast uptake in female patients can be mistaken for other pathology. In some patients, the coexistence of two different SSTR positive diseases should be considered. Diminished uptake in the spleen due to ongoing treatment with octreotide may occur, which may accompanied by a lower liver uptake. In case of hepatic metastases, this phenomenon may be misinterpreted as a better uptake in the liver metastases.

## IMAGING RESULTS IN NEUROENDOCRINE AND OTHER TUMORS

### *Pituitary Tumors*
SSTRs are present on virtually all Growth Hormone-producing pituitary adenomas. Also, in vivo SRI is positive in most cases (8-10), but other pituitary

tumors as well as pituitary metastases from SSTR positive neoplasms, parasellar meningiomas, lymphomas, or granulomatous diseases of the pituitary may be positive. Therefore the diagnostic value of SRI in pituitary tumors is limited (11).

Controversy exists as to the relationship between the calculated tumor to background ratios in pituitary acromegaly using SRI with $[^{111}In\text{-}DTPA^0]$-octreotide and the extent of suppressibility of GH levels during octreotide treatment. Most investigators find no or only a weak correlation, however (8-10). SRI, therefore, does not seem to have a role in deciding whether or not to treat an acromegalic patient with octreotide. Because of the limited effect of octreotide on hormone secretion by clinically nonfunctioning pituitary tumors, both in vivo and in vitro, and because of the absence of tumor shrinkage in the majority of patients studied, octreotide treatment in patients with a clinically nonfunctioning pituitary tumor does not seem promising (12,13). Therefore there is no role for SRI in treatment selection (11).

### Endocrine Pancreatic Tumors
The majority of the endocrine pancreatic tumors can be visualized using SRI (Figure1, panels e,f). Therefore, SRI can be of great value in localizing tumor sites in this type of patients, also in those cases where surgery is indicated but no tumor localization can be found with conventional imaging modalities, which is frequently the case. Reported data on the sensitivity of SRI in patients with gastrinomas vary from about 60 to 90% (14-19), and part of the discrepancy in results is likely due to insufficient scanning technique (especially short acquisition time), not performing SPECT studies, or injection of relatively low doses of $[^{111}In\text{-}DTPA^0]$-octreotide, all of which lead to a poorer performance of SRI.

Using ultrasound, CT, MRI, and/or angiography, endocrine pancreatic tumors can be localized in about 50% of cases (20,21). Endoscopic ultrasound has been reported to be very sensitive in detecting endocrine pancreatic tumors, also when CT or transabdominal ultrasound fail to demonstrate the tumor (22). Studies comparing the value of endoscopic ultrasonography with SRI in the same patients point to more favourable results for SRI (16,18). Gibril et al. (17) published the results of a prospective study in 80 patients with Zollinger-Ellison syndrome, comparing SRI with a variety of other imaging techniques as well as angiography. They found that SRI was as sensitive as all of the other imaging studies combined and advocated its use as the first imaging method to be used in these patients because of its sensitivity, simplicity, and cost-effectiveness. Lebtahi et al. (23) reported that the results of SRI in 160 patients with GEP tumors modified patient classification and surgical therapeutic strategy in 25% of patients. Also, Termanini et al. (24) reported that SRI altered patient management in 47% of 122 patients with gastrinomas. However, results in patients with insulinomas are disappointing, possibly because part of these tumors may either be SSTR negative or contain SSTRs that do not bind octreotide (25).

## Carcinoids

Reported values for the detection of known carcinoid tumor localizations vary from 80 to nearly 100% (26-28) (Figure 1, panel i). Also, the detection of unexpected tumor sites, not suspected with conventional imaging, is reported by several investigators (26-29). Treatment with octreotide may cause a relief of symptoms and a decrease of urinary 5-HIAA levels in patients with the carcinoid syndrome (30). In patients with the carcinoid syndrome, SRI, in consequence of its ability to demonstrate SSTR positive tumors, could therefore be used to select those patients who are likely to respond favourably to octreotide treatment. On the other hand, only for those patients who have SSTR negative tumors, chemotherapy is effective (31).

The impact on patient management is fourfold: SRI may detect resectable tumors that would be unrecognized with conventional imaging techniques; it may prevent surgery in patients whose tumors have metastasized to a greater extent than can be detected with conventional imaging; it may direct the choice of therapy in patients with inoperable tumors; and it may be used to select patients for peptide receptor radionuclide therapy (PRRT, see below).

## Paragangliomas

In virtually all patients with paragangliomas, tumors are readily visualized (32) (Figure1, panel d). Unexpected additional paraganglioma sites are frequently found. One of the major advantages of SRI is that it provides information on potential tumor sites in the whole body in patients with paraganglioma. It could thus be used as a screening test, to be followed by CT scanning, MRI, or ultrasound of the sites at which abnormalities are found.

## Medullary Thyroid Carcinoma and Other Thyroid Cancers

In patients with medullary thyroid carcinoma (MTC) the sensitivity of SRI to detect tumor localizations is 50-70% (33,34). In a series of 17 patients with MTC whom we studied (33), the ratio of CT over CEA levels was significantly higher in patients in whom SRI was successfully applied. This may imply that SSTRs can be detected in vivo on the more differentiated forms of MTC. Also, SRI is more frequently positive in patients with high serum tumor markers and large tumors (34,35), and seems therefore less suitable to demonstrate microscopic disease (34,36).

Although papillary, follicular and anaplastic thyroid cancers, and also Hürthle-cell carcinomas, do not belong to the group of classical neuroendocrine tumors, the majority of patients with these cancers show uptake of radiolabeled octreotide during SRI (37-39). Interestingly, also differentiated thyroid cancers that do not take up radioactive iodine, may show radiolabeled octreotide accumulation (37,38). In some patients, this could open new therapeutic options:

operation, if the number of observed lesions is limited, or, alternatively, PRRT if the uptake is sufficient.

**Small Cell Lung Cancer (SCLC)**
With SRI the primary tumors can be demonstrated in virtually all patients with SCLC (40-44). Part of the known metastases may be missed, however. The unexpected finding of especially brain metastases is reported by several authors (40-42). Others however, have reported the lack of any additional information with SRI (45).

Of special interest are two groups of patients in whom the additional information provided by SRI may have therapeutical consequences: those in whom unexpected cerebral metastases are found, and those in whom the additional information leads to upstaging from Limited Disease (LD) to Extensive Disease (ED). Adding SRI to the staging protocol in patients with SCLC would lead to an upstaging in 5 of 14 patients (36%) out of a group of 26 untreated patients whom we studied (40) who seemingly had LD with conventional imaging only. With conventional imaging, cerebral metastases were detected in 2 patients. SRI suggested cerebral metastases in these 2, and in another 5 patients as well. In four of them, cerebral metastases became manifest within one year.

Inclusion of SRI in the staging protocol of patients with SCLC could therefore lead to upstaging in part of the patients with LD. The cost increase as compared to a conventional work-up only must be weighed against an unnecessary treatment in part of patients with LD (i.e. local chest radiotherapy if a complete remission is achieved) (46). Applying SRI in the work-up of patients with SCLC would demonstrate otherwise undetected brain metastases. From a radiotherapeutical point of view, it would be preferable to irradiate brain metastases when they are small. Therefore, the cost increase compared to the conventional work-up is justified by the therapeutical consequences: irradiation of the brain at an early stage which may lead to a postponement of neurological symptoms and a better quality of life.

**Breast Cancer**
SRI localized 39 of 52 primary breast cancers (75%) (47). Imaging of the axillae showed non-palpable cancer-containing lymph nodes in 4 of 13 patients with subsequently histologically proven metastases. A special remark has to be made with respect to the observation of bilateral and diffuse, physiological breast uptake in normal females. This faint uptake is present in about 15 percent of patients 24 h p.i. and is clearly different from the more localized accumulation at the site of breast cancer. At the moment the basis for this finding is unkown. In the follow-up after a mean of 2.5 yr, SRI in 28 of the 37 patients with an originally SSTR positive cancer, was positive in 2 patients with clinically-recognized metastases, as well as in 6 of the remaining 26 patients who were symptom-free. SRI may be of value in selecting patients for clinical trials

with SST analogues or other medical treatments. Furthermore, SRI is sensitive for detecting recurrences of SSTR positive breast cancer.

**Malignant Lymphomas**

Although in many patients with non-Hodgkin Lymphoma (NHL) one or more lesions may be SSTR positive, receptor negative lesions also occur in a substantial number of patients (48). Therefore, the role for SRI in patients with NHL is limited. In 56 consecutive untreated patients with histologically proven Hodgkin's disease the results of SRI were compared with physical and radiological examinations (49). SRI was positive in 55/56 (98%) patients at sites of documented disease. In 20 patients SRI disclosed lymphoma localizations not revealed following procedures of conventional staging. As a result in 12 patients (21%) SRI produced a change of stage and in seven patients (13%) the additional information led to a change of treatment. Therefore, SRI seems to be promising in the clinical staging and management of patients with Hodgkin's disease.

**Melanoma**

Positive octreotide scintigrams have been reported in 16 out of 19 patients with melanoma (50). The impact of SRI on staging and patient management remains to be determined.

**Neuroblastomas and Pheochromocytomas**

In about 90 % of patients with neuroblastoma SRI visualized tumor deposits. Also, about 85 % of pheochromocytomas were SSTR positive in vivo (51). A drawback of the use of SRI for localization of this tumor in the adrenal gland is the relatively high radioligand accumulation in the kidneys. MIBG scintigraphy is preferred for its localization in this region. Discrepancies between SRI and MIBG in staging of malignant pheochromocytomas have been observed (52). The importance of this complementary radioligand accumulation, both diagnostic and therapeutic, will have to be investigated in future studies.

**Cushing's Syndrome**

In a study of 19 patients with Cushing's syndrome, none of the pituitary adenomas of 8 patients with Cushing's disease or the adrenal adenoma of another patient could be visualized with SRI (53). In 8 of the other 10 patients the primary ectopic corticotropin or CRH secreting tumors were successfully identified with SRI. Tumors smaller than 1 cm can be visualized (54,55). SRI can therefore be included as a diagnostic step in the work-up of Cushing's syndrome with a suspected ectopic corticotropin or CRH-secreting tumor. Others, however, conclude that although SRI may be helpful in selected cases, it is not a significant advance over conventional imaging (56).

**Brain Tumors**

SRI localizes meningiomas in virtually all patients (57,58). The majority of well-differentiated astrocytomas (grade I and II) are SSTR positive; whereas the undifferentiated glioblastomas (grade IV) are receptor negative. An inverse relationship between the presence of SST and epidermal growth factor (EGF) receptors has been observed. In grade III astrocytomas both receptors can be found (59). Astrocytomas have been visualized with SRI (58). A pre-requisite for the localization with this radioligand is a locally open blood-brain barrier. Especially in the lower graded astrocytomas this barrier may be unperturbed. Therefore, the grading of glia-derived brain tumors with SRI is at this moment not promising.

**Other Diseases**

In vivo SRI is also positive in a number of granulomatous and autoimmune diseases, like sarcoidosis, tuberculosis, Wegener's granulomatosis, DeQuervain's thyroiditis, aspergillosis, Graves' hyperthyroidism and Graves' ophthalmopathy (60-64). It is expected that SRI may contribute to a more precise staging and a better evaluation of several of these diseases.

In a cross-sectional study in 46 patients with sarcoidosis, known mediastinal, hilar, and interstitial disease were recognized in 36 of 37 patients (64). Also, such pathology was found in 7 other patients who had normal chest X-rays. In 5 of these, SRI pointed to interstitial disease. SRI was repeated in 13 patients. In 5 of 6 patients who had a chest X-ray monitored improvement of disease activity, SRI also showed a decrease of pathologic uptake. In 2 of 5 patients in whom the chest X-ray was unchanged, but serum ACE concentrations decreased and lung function improved, a normalisation on octreotide scintigrams was found. To determine the value of SRI in the follow-up of patients with sarcoidosis a prospective longitudinal study will have to be performed. In another recent study in 18 patients with sarcoidosis it is concluded that SRI is more accurate than gallium scintigraphy for evaluating the extent of the disease, and also for evaluating the response to therapy (65).

Summarizing, in virtually all patients with sarcoidosis, granulomas can be visualized. Uveitis may be the presenting symptom of sarcoidosis. In our experience, in patients with uveitis, octreotide scintigraphy can not infrequently be of help because of the typical pattern of uptake in the mediastinum, lung hili, and parotid glands that can be seen in patients that eventually appear to have sarcoidosis, even when the chest X-ray or CT are normal (Figure 1, panels g,h). Thus, octreotide scintigraphy can be used to reach a diagnosis, and also influence the decision how to treat this type of patient.

In Graves' hyperthyroidism accumulation of radiolabeled octreotide in the thyroid gland is markedly increased and correlates with serum levels of free thyroxine and thyrotropin binding inhibiting immunoglobulins. In vitro studies showed that the follicular cells express SSTRs in Graves' disease (66). In clinically active Graves' ophthalmopathy the orbits show accumulation of radioactivity 4 h and 24 h after injection of $[^{111}In-DTPA^0]$-octreotide (61,63). SPECT is required for a proper interpretation of orbital scintigraphy. There is also a correlation between orbital [$^{111}$In-DTPA$^0$]-octreotide uptake and Clinical Activity Score and Total Eye Score (61,67). Also, uptake is high in clinically active disease, but low if ophthalmopathy is inactive (62,63). The clinical value of SRI in Graves' disease has yet to be established. Possibly this technique could select those patients with Graves' ophthalmopathy who might benefit from treatment with octreotide (62,67).

**Surgical Probe**
Apart from its use for scintigraphic imaging, the application of radiolabeled SST analogues may be clinically useful in another way: after the injection of [$^{111}$In-DTPA$^0$]-octreotide, tumor localizations can be detected by the surgeon by means of a probe which is used during the operation (68,69). This may especially be of value if small tumors with a high receptor density are present, like for instance gastrinomas.

**PEPTIDE RECEPTOR RADIONUCLIDE THERAPY (PRRT)**
In patients with SSTR positive pathology, SRI is useful if it can localize the otherwise undetectable disease, or if it can be used for treatment selection (usually the choice between symptomatic treatment with SST analogues or other medical treatment). In patients with known metastatic disease, however, in whom little or no treatment alternatives are available, SRI has a limited role. This situation changes, however, if imaging has a sequel in treatment. Therefore, the option of treatment with radiolabeled SST analogues may become the impetus to SRI (see Chapter 15).

## Conclusions and Perspective

[$^{111}$In-DTPA$^0$]-octreotide is a radiopharmaceutical with a great potential for the visualization of SSTR positive tumors. The overall sensitivity of SRI to localize neuroendocrine tumors is high. In a number of neuroendocrine tumor types, inclusion of SRI in the localization or staging procedure may be very rewarding, either in terms of cost-effectiveness, patient management, or quality of life. The value of SRI in patients with other tumors, like breast cancer, malignant lymphomas, or in patients with granulomatous diseases, has to be established. The development of PRRT is expected to stimulate peptide receptor imaging.

## REFERENCES

1.  Patel Y.C., Amherdt M., Orci L. Quantitative electron microscopic autoradiography of insulin, glucagon and somatostatin binding sites on islets. Science 1982; 217:1155-6.
2.  Reubi J.C., Kvols L.K., Waser B., et al. Detection of somatostatin receptors in surgical and percutaneous needle biopsy samples of carcinoids and islet cell carcinomas. Cancer Res 1990; 50:5969-77.
3.  Sreedharan S.P., Kodama K.T., Peterson K.E., Goetzl E.J. Distinct subsets of somatostatin receptors on cultured human lymphocytes. J Biol Chem 1989; 264:949-53.
4.  Reubi J.C., Waser B., Vanhagen M., et al. In vitro and in vivo detection of somatostatin receptors in human malignant lymphomas. Int J Canc 1992; 50:895-900.

5.  Van Uden A., Steinmeijer M.V.J., De Swart J., et al. Imaging with octreoscan: haste makes waste. Eur J Nucl Med 1999; 26:1022.

6.  Reubi J.C., Horisberger U., Studer U.E., Waser B., Laissue JA. Human kidney as target for somatostatin: high affinity receptors in tubules and vasa recta. J Clin Endocrinol Metab 1993; 77:1323-28.

7.  Gibril F., Reynolds J.C., Chen C.C., et al. Specificity of somatostatin receptor scintigraphy: a prospective study and effects of false-positive localizations on management in patients with gastrinomas. J Nucl Med 1999; 40:539-53.

8.  Plockinger U., Reichel M., Fett U., Saeger W., Quabbe H.J. Preoperative octreotide treatment of growth hormone-secreting and clinically nonfunctioning pituitary macroadenomas: effect on tumor volume and lack of correlation with immunohistochemistry and somatostatin receptor scintigraphy. J Clin Endocrinol Metab 1994; 79:1416-23.

9.  Legovini P., De Menis E., Billeci D., et al. 111Indium-pentetreotide pituitary scintigraphy and hormonal responses to octreotide in acromegalic patients. J Endocrinol Invest 1997; 20:424-8.

10. Oppizzi G., Cozzi R., Dallabonzana D., et al. Scintigraphic imaging of pituitary adenomas: an in vivo evaluation of somatostatin receptors. J Endocrinol Invest 1998; 21:512-9.

11. Kwekkeboom D.J., de Herder W.W., Krenning E.P. Receptor imaging in the diagnosis and treatment of pituitary tumors. J Endocrinol Invest 1999; 22:80-8.

12. De Bruin T.W.A., Kwekkeboom D.J., Van 't Verlaat J.W., et al. Clinically nonfunctioning pituitary adenoma and octreotide response to long term high dose treatment, and studies in vitro. J Clin Endocrinol Metab 1992; 75:1310-7.

13. Merola B., Colao A., Ferone D., et al. Effects of a chronic treatment with octreotide in patients with functionless pituitary adenomas. Horm Res 1993; 40:149-55.

14. Kwekkeboom D.J., Krenning E.P., Oei H.Y., van Eyck C.H.J., Lamberts S.W.J. Use of radiolabeled somatostatin to localize islet cell tumors. In: Mignon M., Jensen R.T., eds. Frontiers of gastro-intestinal research, Vol 23. Endocrine tumors of the pancreas. New York: Karger, 1995: pp 298-308.

15. Cadiot G., Lebtahi R., Sarda L., et al. Preoperative detection of duodenal gastrinomas and peripancreatic lymph nodes by somatostatin receptor scintigraphy. Gastroenterology 1996; 111:845-54.

16. De Kerviler E., Cadiot G., Lebtahi R., Faraggi M., Le Guludec D., Mignon M. Somatostatin receptor scintigraphy in forty-eight patients with the Zollinger-Ellison syndrome. Eur J Nucl Med 1994; 21:1191-7.

17. Gibril F., Reynolds J.C., Doppman J.L., et al. Somatostatin receptor scintigraphy: its sensitivity compared with that of other imaging methods in detecting primary and metastatic gastrinomas. A prospective study. Ann Intern Med 1996; 125:26-34.

18. Zimmer T., Stolzel U., Bader M., et al. Endoscopic ultrasonography and somatostatin receptor scintigraphy in the preoperative localisation of insulinomas and gastrinomas. Gut 1996; 39:562-8.

19. Alexander H.R., Fraker D.L., Norton J.A., et al. Prospective study of somatostatin receptor scintigraphy and its effect on operative outcome in patients with Zollinger-Ellison syndrome. Ann Surg 1998; 228:228-38.

20. Lunderquist A. Radiologic diagnosis of neuroendocrine tumors. Acta Oncol 1989; 28:371-2.

21. Doherty G.M., Doppman J.L., Shawker T.H., et al. Results of a prospective strategy to diagnose, localize, and resect insulinomas. Surgery 1991; 110:989-97.

22. Rösch T., Lightdale C.J., Botet J.F., et al. Localization of pancreatic endocrine tumors by endoscopic ultrasonography. N Engl J Med 1992; 326:1721-6.

23. Lebtahi R., Cadiot G., Sarda L., et al. Clinical impact of somatostatin receptor scintigraphy in the management of patients with neuroendocrine gastroenteropancreatic tumors. J Nucl Med 1997; 38:853-8.

24. Termanini B., Gibril F., Reynolds J.C., et al. Value of somatostatin receptor scintigraphy: a prospective study in gastrinoma of its effect on clinical management. Gastroenterology 1997; 112:335-47.

25. Lamberts S.W.J., Hofland L.J., Van Koetsveld P.M., et al. Parallel in vivo and in vitro detection of functional somatostatin receptors in human endocrine pancreatic tumors.

Consequences with regard to diagnosis, localisation and therapy. J Clin Endocrinol Metab 1990; 71:566-74.

26. Kwekkeboom D.J., Krenning E.P., Bakker W.H., et al. Somatostatin analogue scintigraphy in carcinoid tumors. Eur J Nucl Med 1993; 20:283-92.

27. Westlin J.E., Janson E.T., Arnberg H., Ahlstrom H., Oberg K., Nilsson S. Somatostatin receptor scintigraphy of carcinoid tumours using the [111In-DTPA-D-Phe1]-octreotide. Acta Oncol 1993; 32:783-6.

28. Schillaci O., Scopinaro F., Angeletti S., et al. SPECT improves accuracy of somatostatin receptor scintigraphy in abdominal carcinoid tumors. J Nucl Med 1996; 37:1452-6.

29. Ahlman H., Wängberg B., Tisell L.E., Nilsson O., Fjälling M., Forssell-Aronsson E. Clinical efficacy of octreotide scintigraphy in patients with midgut carcinoid tumours and evaluation of intraoperative scintillation detection. Br J Surg 1994; 81:1144-9.

30. Kvols L.K., Moertel C.G., O'Connell M.J., et al. Treatment of the malignant carcinoid syndrome. N Eng J Med 1986; 315:663-6.

31. Kvols L.K. Medical oncology considerations in patients with metastatic neuroendocrine carcinomas. Semin Oncol 1994; 21(suppl 13):56-60.

32. Kwekkeboom D.J., Van Urk H., Pauw K.H., et al. Octreotide scintigraphy for the detection of paragangliomas. J Nucl Med 1993; 34:873-8.

33. Kwekkeboom D.J., Reubi J.C., Lamberts S.W.J., et al. In vivo somatostatin receptor imaging in medullary thyroid carcinoma. J Clin Endocrinol Metab 1993; 76:1413-7.

34. Tisell L.E., Ahlman H., Wängberg B., et al. Somatostatin receptor scintigraphy in medullary thyroid carcinoma. Br J Surg 1997; 84:543-7.

35. Berna L., Chico A., Matias-Guiu X., et al. Use of somatostatin analogue scintigraphy in the localization of recurrent medullary thyroid carcinoma. Eur J Nucl Med 1998; 25:1482-8.

36. Adams S., Baum R.P., Hertel A., Schumm-Draeger P.M., Usadel K.H., Hör G. Comparison of metabolic and receptor imaging in recurrent medullary thyroid carcinoma with histopathological findings. Eur J Nucl Med 1998; 25:1277-83.

37. Postema P.T.E., De Herder W.W., Reubi J.C., et al. Somatostatin receptor scintigraphy in non-medullary thyroid cancer. Digestion 1996; 1(suppl):36-7.

38. Baudin E., Schlumberger M., Lumbroso J,, Travagli I P., Caillou B., Parmentier C. Octreotide scintigraphy in patients with differentiated thyroid carcinoma: contribution for patients with negative radioiodine scan. J Clin Endocrinol Metab 1996; 81:2541-4.

39. Gulec S.A., Serafini A.N., Sridhar K.S., et al. Somatostatin receptor expression in Hurthle cell cancer of the thyroid. J Nucl Med 1998; 39:243-5.

40. Kwekkeboom D.J., Kho G.S., Lamberts S.W., Reubi J.C., Laissue J.A., Krenning E.P. The value of octreotide scintigraphy in patients with lung cancer. Eur J Nucl Med 1994; 21:1106-13.

41. O'Byrne K.J., Ennis J.T., Freyne P.J., Clancy L.J., Prichard J.S., Carney D.N. Scintigraphic imaging of small-cell lung cancer with [111In]pentetreotide, a radiolabelled somatostatin analogue. Br J Cancer 1994; 69:762-6.

42. Bombardieri E., Crippa F., Cataldo I., et al. Somatostatin receptor imaging of small cell lung cancer (SCLC) by means of 111In-DTPA octreotide scintigraphy. Eur J Cancer 1995; 31A:184-8.

43. Berenger N., Moretti J.L., Boaziz C., Vigneron N., Morere J.F., Breau J.L. Somatostatin receptor imaging in small cell lung cancer. Eur J Cancer 1996; 32A:1429-31.

44. Reisinger I., Bohuslavitzki K.H., Brenner W., et al. Somatostatin receptor scintigraphy in small-cell lung cancer: results of a multicenter study. J Nucl Med 1998; 39:224-7.

45. Kirsch C.M., von Pawel J., Grau I., Tatsch K. Indium-111 pentetreotide in the diagnostic work-up of patients with bronchogenic carcinoma. Eur J Nucl Med 1994; 21:1318-25.

46. Kwekkeboom D.J., Lamberts S.W.J., Habbema J.D., Krenning E.P. Cost-effectiveness analysis of somatostatin receptor scintigraphy. J Nucl Med 1996; 37:886-92.

47. Van Eijck C.H., Krenning E.P., Bootsma A., et al. Somatostatin-receptor scintigraphy in primary breast cancer. Lancet 1994; 343:640-3.

48. Van Hagen P.M., Krenning E.P., Reubi J.C., et al. Somatostatin analogue scintigraphy of malignant lymphomas. Br J Haemat 1993; 83:75-9.

49. Van den Anker-Lugtenburg P.J., Krenning E.P., Oei H.Y., et al. Somatostatin receptor scintigraphy in the initial staging of Hodgkin's disease. Br J Haematol 1996; 93:96-103.

214

50. Hoefnagel C.A., Rankin E.M., Valdés Olmos R.A., Israëls S.P., Pavel S., Janssen A.G.M. Sensitivity versus specificity in melanoma imaging using iodine-123 iodobenzamide and indium-111 pentetreotide. Eur J Nucl Med 1994; 21:587-8.

51. Krenning E.P., Kwekkeboom D.J., Bakker W.H., et al. Somatostatin receptor scintigraphy with [$^{111}$In-DTPA-D-Phe$^1$]- and [$^{123}$I-Tyr-3-]-octreotide: the Rotterdam experience with more than 1000 patients. Eur J Nucl Med 1993; 20:716-31.

52. Tenenbaum F., Lumbroso J., Schlumberger M., et al. Comparison of radiolabeled octreotide and meta-iodobenzylguanidine (MIBG) scintigraphy in malignant pheochromocytoma. J Nucl Med 1995; 36:1-6.

53. De Herder W.W., Krenning E.P., Malchoff C.D., et al. Somatostatin receptor scintigraphy: its value in tumor localization in patients with the Cushing syndrome caused by ectopic cortictropin and/or CRH secretion. Am J Med 1994; 96:305-12.

54. Philipponneau M., Nocaudie M., Epelbaum J., et al. Somatostatin analogs for the localization and preoperative treatment of an ACTH-secreting bronchial carcinoid tumor. J Clin Endocrinol Metab 1994; 78:20-4.

55. Weiss M., Yellin A., Husza'r M., Eisenstein Z., Bar-Zif J., Krausz Y. Localization of adrenocorticotropic hormone-secreting bronchial carcinoid tumor by somatostatin-receptor scintigraphy. Ann Intern Med 1994; 121:198-9.

56. Torpy D.J., Chen C.C., Mullen N., et al. Lack of utility of (111)In-pentetreotide scintigraphy in localizing ectopic ACTH producing tumors: follow-up of 18 patients. J Clin Endocrinol Metab 1999; 84:1186-92.

57. Haldemann A.R., Rosler H., Barth A., et al. Somatostatin receptor scintigraphy in central nervous system tumors: role of blood-brain barrier permeability. J Nucl Med 1995; 36:403-10.

58. Schmidt M., Scheidhauer K., Luyken C., et al. Somatostatin receptor imaging in intracranial tumours. Eur J Nucl Med 1998; 25:675-86.

59. Reubi J.C., Horisberger U., Lang W., Koper J.W., Braakman R., Lamberts S.W.J. Coincidence of EGF receptors and somatostatin receptors in meningiomas but inverse, differentiation-dependent relationship in glial tumors. Am J Pathol 1989; 134;337-44.

60. Vanhagen P.M., Krenning E.P., Reubi J.C., et al. Somatostatin analogue scintigraphy in granulomatous diseases. Eur J Nucl Med 1994; 21:497-502.

61. Postema P.T.E., Krenning E.P., Wijngaarde R., et al. [$^{111}$In-DTPA-D-Phe$^1$]-octreotide scintigraphy in thyroidaal and orbital Graves' disease: a parameter for disease activity? J Clin Endocrinol Metab 1994; 79:1845-51.

62. Krassas G.E., Dumas A., Pontikides N., Kaltsas T. Somatostatin receptor scintigraphy and octreotide treatment in patients with thyroid eye disease. Clin Endocrinol (Oxf) 1995; 42:571-80.

63. Kahaly G., Görges R., Diaz M., Hommel G., Bockisch A. Indium-111-pentetreotide in Graves' disease. J Nucl Med 1998; 39:533-6.

64. Kwekkeboom D.J., Krenning E.P., Kho G.S., Breeman W.A.P., Van Hagen P.M. Octreotide scintigraphy in patients with sarcoidosis. Eur J Nucl Med 1998; 25:1284-92.

65. Lebtahi R., Crestani B., Belmatoug N., et al. Somatostatin receptor scintigraphy and gallium scintigraphy in patients with sarcoidosis. J Nucl Med 2001; 42:21-6.

66. Reubi J.C., Waser B., Friess H., Krenning E.P., Büchler M., Laissue J. Regulatory peptide receptors in goiters of the human thyroid. J Nucl Med 1997; 38(suppl):266P.

67. Gerding M.N., van der Zant F.M., van Royen E.A., et al. Octreotide-scintigraphy is a disease-activity parameter in Graves' ophthalmopathy. Clin Endocrinol (Oxf) 1999; 50:373-9.

68. Modlin I.M., Cornelius E., Lawton G.P. Use of an isotopic somatostatin receptor probe to image gut endocrine tumors. Arch Surg 1995; 130:367-73.

69. Wangberg B., Forssell-Aronsson E., Tisell L.E., Nilsson O., Fjalling M., Ahlman H. Intraoperative detection of somatostatin-receptor-positive neuroendocrine tumours using indium-111-labelled DTPA-D-Phe1-octreotide. Br J Cancer 1996; 73:770-5.

# 14
# SOMATOSTATIN ANALOGS IN THE TREATMENT OF PITUITARY TUMORS

Annamaria Colao

*Department of Molecular and Clinical Endocrinology and Oncology, "Federico II" University of Naples, Naples, Italy*

## ABSTRACT

Recently the medical approach to patients with secreting and nonsecreting pituitary adenomas has received great impulse thanks to the availability of new somatostatin analogs provided with slow release, such as lanreotide (LAN) and octreotide-LAR (OCT-LAR). In acromegaly, disease control (GH $\leq 2.5$ $\mu g/l$ as fasting value or $\leq 1$ $\mu g/l$ after glucose load, together with age-normalized IGF-I) is achievable in more than half patients under treatment with LAN and OCT-LAR. Improvement of cardiomyopathy, sleep apnea and arthropathy has been reported during treatment with these somatostatin analogs. LAN displayed beneficial effects also in TSH-secreting adenomas, while the results of treatment with these peptides in patients with clinically nonfunctioning adenomas are still rather limited and controversial.

## INTRODUCTION

In recent years, new somatostatin (SST) analogs (SSTA) provided in slow release formulations, such as lanreotide (LAN) and octreotide (OCT)-LAR (OCT-LAR) have greatly improved the pharmacotherapy to pituitary adenomas. In particular, SSTA have been successfully employed in patients with GH- and TSH-secreting tumors. Whether these drugs should be employed as first line therapy or after unsuccessful surgery in patients with acromegaly in currently under debate. Conversely, SSTA are generally considered beneficial in patients affected with either GH- or TSH-secreting tumors with persistent disease after surgery. Results in patients with clinically nonfunctioning adenomas are, conversely, rather disappointing and no effect was proven in PRL-secreting and ACTH-secreting adenomas.

In this chapter, the most recent studies pointing out the effects of LAN and OCT-LAR in patients with pituitary adenomas are reviewed with particular emphasis on debated issues such as the pre-surgical treatment and their potential use as first line therapy in acromegaly.

## SST ANALOGS IN GH-SECRETING PITUITARY ADENOMAS

The objectives of treating acromegaly are the tumor removal with resolution of its mass effects, the restoration of physiological GH secretion (both basal and stimulated) and the relief of symptoms directly caused by GH excess. The treatment is performed also to possibly prevent the progressive disfigurement, bone expansion, osteoarthritis and cardiomyopathy, which are disabling long-term consequences, hypertension, insulin resistance, diabetes mellitus and lipid abnormalities, that are risk factors for vascular damage. The prognosis of acromegaly is poor since the mortality rate of untreated patients is double that of healthy subjects after the age of 45 yrs (1-6). Whether the suppression of GH and IGF-I is able to reverse the poor long-term outcome is still questioned. To date the disease is considered controlled when fasting GH levels are ≤2.5 $\mu$g/l, glucose-load suppressed GH levels are ≤1 $\mu$g/l and IGF-I levels are normalized for age (7). When discussing the successful outcome of LAN and OCT-LAR treatment, these criteria are considered.

The currently available treatment options for acromegaly include surgery, irradiation and pharmacological suppression of GH levels by means of dopamine-agonists (DA) or SSTA, either alone or in combination. Surgical removal of the pituitary adenoma still remains the first therapeutic option in most cases (3,5,8,9), although primary pharmacotherapy has been recently proposed (10,11). Irradiation or pharmacological suppression of GH excess can be used following unsuccessful surgery or as individualized primary therapy in elderly patients (5,8,9). Selective $D_2$ DA, like cabergoline, have been recently shown to induce a higher GH and IGF-I suppression than bromocriptine (12-14). However, DA were universally recognized less potent than SSTA in controlling acromegaly.

OCT, the first long-acting synthetic SST analog introduced in the clinical practice, has a half-life of 80-100 min (15). Its use to treat acromegaly was started in the middle of the 80's (16). Many subsequent studies demonstrated that s.c. OCT reduced GH concentrations in over 90% of patients suppressing GH levels <5 $\mu$g/l in half of them (5, 8, 9, 15). A combined analysis of 557 patients treated worldwide showed that OCT administration normalized IGF-I levels in 48.5% and reduced tumor size of at least 20% of baseline size, in 40.3% of patients (5). Soon after the administration of the first doses of s.c. OCT clinical signs and symptoms, in particular headache, hyperhidrosis and joint pain, are significantly reduced (3-5, 8, 9, 15). The long-term treatment with s.c. OCT was also shown to improve obstructive and central sleep apnea, one of the most severe complications of acromegaly: both the number of apneic episodes and the degree of blood oxygen desaturation were reduced after treatment (17). Beneficial effects of both short-term and long-term s.c. OCT treatment were also demonstrated in decreasing cardiac size (18-21), and improving the diastolic function (18,21). A remarkable improvement of systolic function was obtained in patients with overt heart failure refractory to cardiac therapies (22). Moreover, in patients without symptoms of cardiac disease, a significant increase of the left ventricular ejection fraction on effort was observed after 1 year of treatment with s.c. OCT only if they achieved disease control (23). These data have been confirmed by a recent 5-year prospective study: only patients with controlled disease improved their systolic function while no change was found in those showing mild GH and IGF-I hypersecretion during the follow-up (24). Arthropathy, which is one of the major causes of morbidity for the disease (25-27), was considered not to be modified by the treatment of acromegaly. Indeed, ultrasonography demonstrated that even in patients cured from the disease for at least 5 years, joint thickness was higher than in controls (28). However, in contrast with previous information, after 6 months of OCT treatment, a significant decrease of shoulder and wrist cartilage thickness was found (28). Interestingly, the decrease in thickness of shoulder and wrist cartilage, two examples of not weight-bearing joints, was more pronounced than that measured at the level of knees (28). Acromegalic patients were also found to have an age-independent increase in prostate volume (29) that was significantly reduced after s.c. OCT treatment (30).

The most frequent side effects of s.c. OCT treatment were related to a poor gastro-intestinal tolerance: diarrhea, steatorrhea were frequently reported by patients at treatment beginning, but generally subsided with treatment continuation. As delayed side effects was reported an increased frequency of gallstones and gallbladder abnormalities, due to suppressed cholecystokinin release and impaired gall bladder emptying. However, these abnormalities (sediment, sludge, microlithiasis and gallstones) seem to be associated with low morbidity, and can be managed conservatively as in other patients with asymptomatic cholelitiasis (5).

More recently, SSTA were provided in slow-release formulations to improve patients' compliance to long-term treatment. LAN was first shown to be effective in patients previously responsive to s.c. OCT and, as expected, it was demonstrated to be better tolerated (31-36). In a 3-year study, LAN treatment induced GH and IGF-I normalization in 14 of 22 patients but 9 of them had to increase the injection frequency to one every 10 days for an optimal control of the disease (37). Similarly to side effects encountered with s.c. OCT treatment, loose stools, nausea and/or abdominal pain were reported by half of the patients treated with LAN: side effects appeared within 1-2 days after the injection and ~20% of patients developed new gallstones (31-36). In a recent survey study performed in 118 patients with acromegaly treated for 24 months with LAN, 25 of whom newly diagnosed, we demonstrated that disease control was achieved in 77% of patients after 24 months of therapy (38). Reduction in tumor size was obtained in 5 out of 23 naive patients (22%), and in 5 (5.9%) of 84 OCT-pretreated patients with evident tumor remnant (38). In our cohort of patients LAN treatment was well tolerated: only 2 patients (1.7%) withdrew from treatment due to severe side-effects (38). A decrease of the left ventricular mass (39,40), together with an improvement of the diastolic function, were also demonstrated during LAN treatment. Recently, systolic blood pressure and heart rate were shown to be significantly decreased already after 1 week, while ejection fraction and shortening fraction, exercise capacity and $VO_2$ at anaerobic threshold were shown to be increased after 2 weeks of LAN administration (41). In a small cohort of patients prospectively followed after LAN treatment given as primary therapy for 12 months, a significant decrease in shoulder, wrist and knee joint thickness was achieved (42). Prostate volume significantly decreased after 2 years of LAN treatment in acromegalic men aged less than 50 yrs (Fig.1) (43).

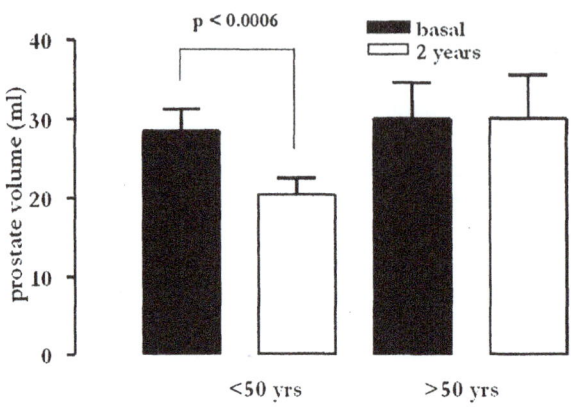

*Figure 1*: *Prostate volume measured by transrectal ultrasonography before and after 2 years of lanreotide treatment in 23 men with acromegaly (43).*

Finally, after 2 weeks from LAN injection a decrease of urinary albumin excretion (UAE) was also demonstrated (44). Lanreotide Autogel (LAN-ATG) is a new long-acting aqueous preparation of lanreotide for the treatment of acromegaly and is administered by deep sc injection from a small volume, pre-filled syringe. LAN-ATG, administered every 28 days to 107 patients previously responsive to lanreotide 30 mg, im controlled acromegalic symptoms as effectively as the latter (45). After 3 injections of LAN-ATG, GH and IGF-I levels were comparable with those recorded at the end of lanreotide 30 mg treatment (45). GH levels below 2.5 ng/ml and age-/sex-normalized IGF-I were achieved in 33% and 39% of patients during lanreotide 30 mg and LAN-ATG treatment, respectively. Diarrhea, abdominal pain, and nausea were reported by 38%, 22%, and 18% of patients during lanreotide 30 mg treatment and by 29%, 17%, and 9% of patients, respectively, during LAN-ATG treatment (45).

OCT-LAR is the other slow-release formulation applied in the treatment of acromegaly. GH levels were suppressed <5 $\mu g/l$ in 86-100%, <2 $\mu g/l$ in 39-75% and <1 $\mu g/l$ in 24-40% of patients (46-49). Notable tumor shrinkage was also reported, mostly in patients who had not undergone surgery prior to OCT-

LAR treatment (46). In order to better clarify the effects of long-term treatment of OCT-LAR on hormone profile and tumor mass we followed 36 patients with active acromegaly for 2 years (50). Serum GH and IGF-I levels significantly decreased as early as after the first injection of OCT-LAR and progressively declined during the 12-24 months of treatment both in the 15 *de novo* and in the 21 operated patients. After 24 months, disease control was achieved in 70% of the entire series of patients (50), without any difference between *de novo* and operated patients (Table 1).

**Table 1**: *Effect of 24 month OCT-LAR as primary treatment in 15 de novo compared to 21 previously operated patients with acromegaly (50).*

|  | de novo | operated | p |
|---|---|---|---|
| Basal GH levels ($\mu$g/l) | 55.1±10.8 | 19.2±2.9 | < 0.001 |
| Nadir GH levels ($\mu$g/l) | 2.6±0.8 | 2.3±0.6 | 0.5 |
| Basal IGF-I levels ($\mu$g/l) | 861.3±75.2 | 729.5±41.4 | 0.8 |
| Nadir GH levels ($\mu$g/l) | 335.1±41.1 | 277.4±36. | 0.6 |
| GH control (%) | 73.3% | 76.2% | 0.9 |
| IGF-I control (%) | 53.3% | 71.5% | 0.08 |

Interestingly, a decrease in tumor volume was observed in 12 of 15 (80%) *de novo* patients (Fig.2), while no shrinkage was detected in 7 of 9 operated patients; no patient had tumor re-expansion during OCT-LAR treatment.

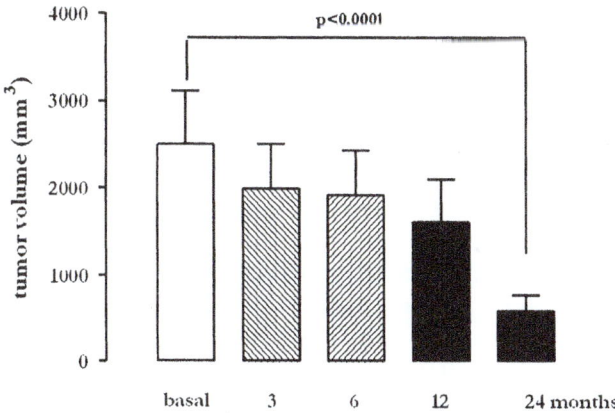

**Figure.2**: *Tumor volume measured by MRI in de novo patients before (n=15) and during 24 months of OCT-LAR treatment. Note that only 7 patients were included in the 24-month follow-up (49).*

Both in our studies (50) and in others (46-49), OCT-LAR was well tolerated and the mild to moderate side effects experienced by up to 50% of the patients were of short duration and often subsided with continued drug administration. More recently, an open prospective study of 27 patients with newly diagnosed

220

acromegaly, conducted in nine endocrine centers in the United Kingdom (51), confirmed our previous results. All 27 tumors shrank during sc OCT treatment: in microadenomas median tumor volume decreased by 49% (range, 12-73), and in macroadenomas by 43% (range, 6-92) (51). In the 15 patients given OCT-LAR there was a further tumor volume reduction of 24% (51). The treatment with OCT-LAR induced a significant decrease of left ventricular mass index, interventricular septum thickness and left ventricular posterior wall thickness in 15 *de novo* patients already after 3 months (52): LVMi decreased similarly in patients with or without disease control (Fig.3).

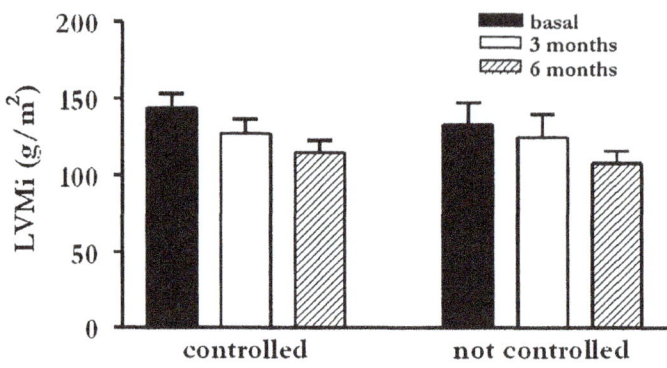

**Figure 3**: *Left ventricular mass index (LVMi) in 9 patients with controlled disease and in 6 patients with persistently high GH and IGF-I levels (50).*

Among the 11 patients with left ventricular hypertrophy, 6 normalized their LVMi after treatment. However, the effect of OCT-LAR on recovering left ventricular hypertrophy (LVH) and systolic performance depend also on patients' age. We have recently reported (53) that the LVM index decreased significantly in both in young patients (aged <40 yrs) (124.4 ± 5.8 vs. 103.4 ± 3.9 g/m2; P = 0.01) and middle-aged patients (aged 40-60 yrs) (140.9 ± 7.9 vs. 117.8 ± 6.6 g/m2; P = 0.03) but LVH disappeared in all the young and 50% of the middle-aged patients. Similarly, LVEF at rest and at peak exercise increased significantly in both groups but the response on effort normalized in 80% of the young and 50% of the middle-aged patients (53). As far as side effects during long-term treatment are concerned, the incidence of gallbladder abnormalities seems to be higher in patients who had previously undergone long-term treatment with s.c. OCT (46). In a recent study, a 3 times increase of fasting gall-bladder volume was found during OCT-LAR but post-prandial cholecystokinin release and gall-bladder emptying were severely suppressed, suggesting that these two latter findings were major cause of gallstone formation (54).

When the efficacy of OCT-LAR and LAN treatments in acromegaly were compared, OCT-LAR was suggested to have some advantages, although the differences were not great (55). In another study, including 125 patients with acromegaly OCT-LAR at a dose of 20 mg administered once monthly was more effective than LAN 30 mg administered 2 or 3 times monthly in reducing GH and IGF-I (56).

The combined administration of SSTA with DA was reported to be more effective than the two drugs administered separately. In one study including 10 patients poorly responsive to LAN, the combined administration of LAN+cabergoline induced the control of the disease in half of them (57). One explanation could be that receptors from different G protein-coupled receptor families interact through oligomerization: $D_2$ receptor and SST subtype 5 receptor interact physically through hetero-oligomerization to create a novel receptor with enhanced functional activity (58).

Since the treatment with s.c. OCT was shown to reduce tumor size by 20% in 40.3% of patients (5,8), this latter effect of OCT treatment was claimed to favor the complete removal of macroadenomas at surgery reducing the intra-operative manipulation of adjacent structures. In particular, Spinas et al. (59) and Barkan et al. (60) first noted the suitability of s.c. OCT in the pre-surgical treatment of acromegalic patients. Subsequently, some studies confirmed their data (61-63) while other studies failed to demonstrate any beneficial effect of a pre-surgical treatment with s.c. OCT on the surgical outcome of GH-secreting adenomas (64-66). To this extent, we have proposed that a short-lasting pre-surgical treatment with OCT could improve the clinical status of acromegalic patients and reduce post-operative complications (67).

It is still unknown whether the poor perspectives of acromegalic patients in terms of mortality from cardio-respiratory or neoplastic diseases can be reversed by suppressing GH and IGF-I levels. One preliminary study suggested that normal survival occurred in patients with post-treatment GH levels below 2.5 $\mu$g/l, but not in those with higher levels (68). To date the best medical approach seems to be SSTA, even if marked variability in individual patient responses to depot SSTA existed. It has been suggested that for OCT-LAR many patients may be as adequately controlled with 6 weekly injections as with 4 weekly injections (69). However, since a small but not negligible proportion (about 15-20%) of acromegalic patients do not respond well to SSTA, indicating that the receptor subtypes can be abnormally expressed or regulated, the future research should provide selective analogs able to bind the other SSTR subtypes. A novel, high affinity, enzymatically stable, and long-acting SSTA, named PTR-3173, which binds with nanomolar affinity to human SST receptors subtypes 2, 4, and 5 was recently identified (70). In the human carcinoid-derived cell line BON-1, PTR-3173 inhibited forskolin-stimulated cAMP accumulation as efficiently as OCT (70). In hormone secretion studies with rats, PTR-3173 was shown to be 1000-fold and more than 10,000-fold more potent in inhibiting GH release than glucagon and insulin release, respectively, suggesting that it could be a highly selective SSTA for the *in vivo* inhibition of GH secretion (70). Finally were recently another approach to acromegaly was proposed. Pegvisomant is a genetically engineered GH-receptor antagonist that blocks the action of GH. In a 12-week, randomized, double-blind study of three daily doses of pegvisomant (10 mg, 15 mg, and 20 mg) and placebo, Pegvisomant given s.c. in 112 patients with acromegaly normalized IGF-I levels in, 54, 81, and 89% vs.10% of patients (71).Pegvisomant was also shown to be a useful treatment for patients with GH-secreting tumors resistant to octreotide (72). Lastly, antibody formation against OCT has been reported more frequent than previously believed (73). It depends primarily on drug exposure time and route of administration. However, it does not seem to alter the GH/IGF-I status in treated acromegalic patients and induces only mild local reactions in some patients.

## TSH-SECRETING PITUITARY ADENOMAS

This adenoma histotype is rare. Frequently they are macroadenomas at diagnosis presenting with mass effect symptoms such as headache, visual disturbance, together with variable symptoms and signs of hyperthyroidism (74,75). Total surgical removal of the pituitary tumor remains the treatment of choice for TSH-secreting adenomas (74,76,77). However, surgery often is not followed by cure since large tumors are frequently found (74,76-78). In fact, serum $\alpha$-subunit ($\alpha$ SU) and/or TSH levels (either as basal values or under $T_3$ suppression), as markers of complete tumor removal, do not decrease immediately after surgery in many cases (78). The success of additional pituitary radiotherapy is achieved in less than 40% of cases (73, 76, 77, 79). Since SST inhibits TSH secretion both in physiological conditions (80,81) and in TSH-secreting pituitary adenomas (82), and since TSH-secreting adenomas express SST receptors (83), SSTA have been used to normalize or at least decrease TSH and thyroid hormone levels in patients with TSH-secreting pituitary tumors to improve clinical signs. OCT given s.c. induced an acute TSH suppression also reducing $T_4$ deiodination in acromegalic patients (84). In more than 90% of TSH-secreting adenomas, OCT s.c. suppressed TSH secretion and in about 50% of cases adenoma size was reduced (74,79,85-89). OCT treatment was also considered useful preoperatively as it allowed an easier tumor removal (90). The dose of OCT in patients with TSH-secreting adenomas to achieve TSH normalization was reported to be lower than that needed to suppress GH in GH-secreting adenomas (74,86,91). In four patients with thyrotropinoma a single im injection of 30 mg LAN normalized TSH and thyroid hormone levels for 10–15 days (92). LAN at the dose of 0.5 mg was demonstrated to inhibit acutely TSH secretion in TSH-secreting adenomas in the same amount of 0.15 mg of OCT (92). Recently, the chronic effects of 30 mg LAN were investigated in a larger cohort of 18 patients with hyperthyroidism related to a TSH-secreting pituitary adenoma (93). Clinical signs of hyperthyroidism improved within 1 month in all patients. During LAN therapy, serum TSH levels decreased from 2.7±0.3 to 1.9±0.3 mU/l with a parallel significant decrease of α SU, fT₄ and fT₃ levels (93). Conversely, the adenoma size did not significantly change. As in patients with GH-secreting adenoma the most common side-effects were pain at the injection point, abdominal cramps, and diarrhea, but they were mild and transient and did not induce treatment withdrawal (93).

## CLINICALLY NONFUNCTIONING PITUITARY ADENOMAS

Clinically non-functioning pituitary adenomas (NFA) represent a very heterogeneous group of tumors. A consistent proportion of them (up to 90%) indeed secrete low amounts of intact FSH and LH and/or their α- and β-subunits either *in vitro* or *in vivo* (94,95). The first approach in these adenomas is trans-sphenoidal or trans-cranial surgery to remove tumor mass and decompress parasellar structures. Medical therapy has its rationale after surgery to delay radiotherapy and the potential occurrence of hypopituitarism. The treatment with SSTA, mostly with s.c. OCT, was based on the evidence that these tumors express SST-receptors both *in vitro* (96-98) and *in vivo* (99-101), as exemplary shown in Fig.4.

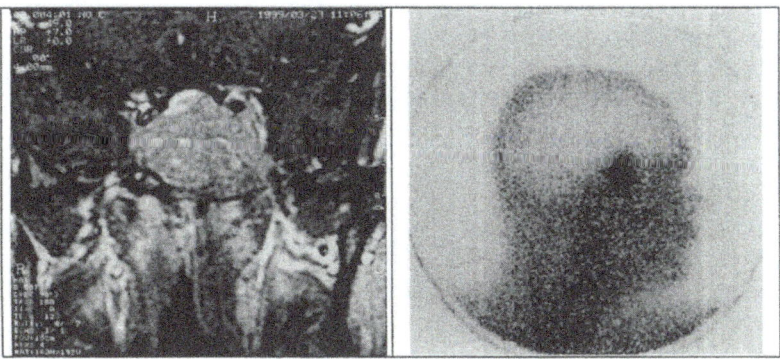

*Figure 4: Intrasellar, suprasellar and bilateral parasellar clinically nonfunctioning macroadenoma at MRI (left) and ¹¹¹In-DTPA-D-phe¹-octreotide (right). (Octreoscan was kindly provided by S. Lastoria, National Cancer Institute, Naples)*

Several clinical trials reported, however, that tumor reduction could be observed only in 11-13% of cases, indicating a weak correlation between SST-receptor expression and treatment efficacy with OCT in these patients (99-101). In 9 patients with NFA we correlated the uptake of ¹¹¹In-DTPA-D-phe¹-octreotide with the percent suppression of α-subunit levels after 6-12 month of s.c. OCT treatment (0.3-0.6 mg/day) (102). Normalization of circulating αSU levels was achieved in 5 of 6 patients, respectively, with moderate-to-intense uptake of the radiotracer (Table 2). Two patients (22.2%) had a significant tumor shrinkage (≥30% of baseline size) during long-term OCT therapy (103).

The percent suppression of α-subunit levels was significantly correlated with the scintigraphic results (102).

*Table 2: Effect of 6-12 months of s.c. OCT treatment on α-subunit levels and tumor mass in 9 patients with NFA undergoing pituitary scintigraphy with $^{111}$In-DTPA-D-phe$^1$-octreotide (99).*

| | α-subunit levels (mU/l) | | Tumor shrinkage |
|---|---|---|---|
| | *basal* | *nadir* | |
| Faint uptake (n=3) | 1.3±0.1 | 1.1±0.1 | 1 |
| Moderate uptake (n=2) | 2.6±1.9 | 0.2±0.1 | 0 |
| Intense uptake (n=4) | 4.8±1.5 | 0.6±0.3 | 2 |

A negative predictive value of $^{111}$In-DTPA-D-phe$^1$-octreotide scintigraphy was also reported in 29 NFA treated for 1 month with s.c. OCT (103). Although the results of chronic OCT treatment were controversial, it was reported to induce a rapid improvement of headache and visual disturbances, without any change in tumor volume (104,105), together with variable inhibition of gonadotropin and αSU levels both in vitro and in vivo (106-108). The analgesic effect of s.c. OCT was likely due to the *in vitro* evidence that OCT display an inhibition of the growth of new vessels and of endothelial cells, leading to hypothesize that the visual improvement was more likely to be due to a direct effect of OCT on the retina and the optic nerve than to an effect mediated by the SST-receptors expressed on the pituitary tumor (101). Decrease of tumor volume by 30±4% were reported in some patients with NFA treated with a combined OCT+cabergoline treatment (109). There are no data on the effects of LAN or OCT-LAR in NFA. A very recent in vitro study demonstrated that both SST and LAN inhibited human NFA cell proliferation (110). The possible mechanisms underlying this effect were likely the activation of tyrosine phosphatases and the inhibition of the activity of voltage-dependent calcium channels (109).

# CONCLUSIONS

The medical approach to pituitary adenomas has improved in recent years due to the availability of effective, well tolerated and safe drugs able to suppress endocrine hypersecretion by tumor cells and, in some instances, to reduce the tumor mass. A new selective SST-receptor subtype 2 and 5 analog (BIM 23244) was recently shown to suppress GH secretion from 5 OCT-resistant GH-secreting adenomas more efficaciously than subtypes 2 or 5 specific analogs (111). The use of more selective compounds will open new perspectives in the medical management of patients with pituitary adenomas, possibly making the surgical removal of the tumor an option for patients intolerant or resistant to pharmacotherapy, as it currently occurs in prolactinomas.

# REFERENCES

1.  Nabarro J.D.N. Acromegaly. Clin Endocrinol 1987; 26: 481-512.
2.  Rajasoorya C., Holdaway I.M., Wrightson P., Scott D.J., Ibbertson H.K. Determinants of clinical outcome and survival in acromegaly. Clin Endocrinol 1994; 41: 95-102.
3.  Melmed S., Ho K., Klibanski A., Reichlin S., Thorner M.O. Recent advances in pathogenesis, diagnosis and management of acromegaly. J Clin Endocrinol Metab 1995; 80:3395-3402.
4.  Colao A., Merola B., Ferone D., Lombardi G. Acromegaly. J Clin Endocrinol Metab 1997; 82:2777-2781.
5.  Colao A., Lombardi G. GH and PRL excess. Lancet 1998; 352:1455-1461.
6.  Orme S.M., McNally R.J., Cartwright R.A., Belchetz P.E. Mortality and cancer incidence in acromegaly: a retrospective cohort study. United Kingdom Acromegaly Study Group. J Clin Endocrinol Metab 1998; 83: 2730-2734.
7.  Giustina A., Barkan A., Casanueva F.F., Cavagnini F., Frohman L., Ho K., Veldhuis J., Wass J., Von Werder K., Melmed S. Criteria for cure of acromegaly: a consensus statement. J Clin Endocrinol Metab 2000; 85:526-529.
8.  Melmed S., Jackson I., Kleinberg D., Klibanski A. Current treatment guidelines for acromegaly. J Clin Endocrinol Metab1998; 83:2646-2652.
9.  Ferone D., Colao A., van der Lely A-J., Lamberts S.W.J. Pharmacotherapy or Surgery as Primary Treatment for Acromegaly? Drugs 2000; 17: 81-92.
10. Newman C.B., Melmed S., George A., Torigian D., Duhaney M., Snyder P., Young W., Klibanski A., Molitch M.E., Gagel R., Sheeler L., Cook D., Malarkey W., Jackson I., Vance M.L., Barkan A., Frohman L., Kleinberg D.L. Octreotide as primary therapy for acromegaly. J Clin Endocrinol Metab 1998; 83: 3034-3040.
11. Freda P.U., Wardlaw S.L. Primary therapy of acromegaly. J Clin Endocrinol Metab 1998; 83: 3031-3033
12. Colao A., Ferone D., Marzullo P., Di Sarno A., Cerbone G., Sarnacchiaro F., Cirillo S., Merola B., Lombardi G. Effect of different dopaminergic agents in the treatment of acromegaly. J Clin Endocrinol Metab 1997; 82:518-523.
13. Abs R., Versholst J., Maiter D., Van Acker K., Nobels F., Coolens J.L., Mahler C., Beckers A. Cabergoline in the treatment of acromegaly: a study in 64 patients. J Clin Endocrinol Metab 1998; 83:374-378.
14. Cozzi R., Attanasio R., Barausse M., Dallabonzana D., Orlandi P., Da Re N., Branca V., Oppizzi G., Gelli D. Cabergoline in acromegaly: a renewed role for dopamine agonist treatment? Eur J Endocrinol 1998; 139:516-521.
15. Lamberts S.W.J., van der Lely A-J., de Herder W.W., Hofland L.J. Octreotide. N Engl J Med 1996; 334:246-254.
16. Lamberts S.W.J., Uitterlinden P., Verschoor L., van Dongen K.J., del Pozo E. Long-term treatment of acromegaly with the somatostatin analogue SMS 201-995. N Engl J Med 1985; 313:1576-1580.
17. Barkan A. Acromegalic arthropathy and sleep apnea. J Endocrinol 1997; 155:S41-44.
18. Thuesen L., Christensen S.E., Weeke J., Ørskov H., Henningsen P. The cardiovascular effects of octreotide treatment in acromegaly: an echocardiographic study. Clin Endocrinol 1989; 30: 619-625.
19. Pereira J.L., Rodriguez-Puras M.J., Leal-Cerro A., Martinez A., Garcia-Luna P.P., Gavilan I., Pumar A., Astorga R. Acromegalic cardiopathy improves after treatment with increasing doses of octreotide. J Endocrinol Invest 1991; 14: 17-23
20. Lim M.J., Barkan A.L., Buda A.J. Rapid reduction of left ventricular hypertrophy in acromegaly after suppression of growth hormone hypersecretion. Ann Intern Med 1992; 117:719-726.
21. Merola B., Cittadini A., Colao A., Ferone D., Fazio S., Sabatini D., Biondi B., Saccà L., Lombardi G. Chronic treatment with the somatostatin analog octreotide improves cardiac abnormalities in acromegaly. J Clin Endocrinol Metab 1993; 77: 790-3.

226

22. Chanson P., Timsit J., Masquet C., Warnet A., Guillausseau P.J., Birman P., Harris A.G., Lubetzki J. Cardiovascular effects of the somatostatin analog octreotide in acromegaly. Ann Intern Med 1990; 113: 921-925.

23. Colao A., Cuocolo A., Marzullo P., Nicolai E., Ferone D., Florimonte L., Salvatore M., Lombardi G. Effects of one-year treatment with octreotide on cardiac performance in patients with acromegaly. J Clin Endocrinol Metab 1999; 84: 17-23.

24. Colao A., Cuocolo A., Marzullo P., Nicolai E., Ferone D., Della Morte A.M., Pivonello R., Salvatore M., Lombardi G. Is the acromegalic cardiomyopathy reversible? Effect of 5 year normalization of growth hormone and insulin-like growth factor-I levels on cardiac performance. J Clin Endocrinol Metab 2001; 86: April

25. Waine H., Bennet G.A., Bauer, W. Joint disease associated with acromegaly. Am J Med Sci 1945; 209: 671-678.

26. Kellgren J.H., Ball, J., Tutton, G.K. The articular and other limb changes in acromegaly. Q J Med 1952; 21: 405-423.

27. Bluestone R., Bywaters E., Hartog M., Holt P.J.L., Hyde S. Acromegalic arthropathy. Ann Rheum Dis 1971; 30: 43-258.

28. Colao A., Marzullo P., Vallone G., Marinò V., Annecchino M., Ferone D., De Brasi D., Scarpa R., Oriente P., Lombardi G. Reversibility of joint thickening in acromegalic patients: an ultrasonography study. J Clin Endocrinol Metab 1998; 83:2121-2125

29. Colao A., Marzullo P., Spiezia S., Ferone D., Giaccio A., Cerbone G., Pivonello R., Di Somma C., Lombardi G. Effect of growth hormone (GH) and insulin-like growth factor-1 on prostate diseases: an ultrasonographic and endocrine study in acromegaly, GH-deficiency and healthy subjects. J Clin Endocrinol Metab 1999; 84: 1986-1991.

30. Colao A., Marzullo P., Ferone D., Spiezia S., Cerbone G., Marinò V., Di Sarno A., Lombardi G. Prostatic hyperplasia: an unknown feature of acromegaly. J Clin Endocrinol Metab 1998; 83:2606-2607.

31. Morange I., de Boisvilliers F., Chanson P., Lucas B., DeWailly D., Catus F., Thomas F., Jaquet P.: Slow release lanreotide treatment in acromegalic patients previously normalized by octreotide. J Clin Endocrinol Metab 1994; 79:145-151.

32. Giusti M., Gussoni G., Cuttica C.M., Giordano G., and the Italian Multicenter Slow Release Lanreotide Study Group: Effectiveness and tolerability of slow release lanreotide treatment in active acromegaly; six month report on an Italian multicenter study. J Clin Endocrinol Metab 1996; 81:2089-2097.

33. Colao A., Marzullo P., Ferone D., Marinò V., Pivonello R., Di Somma C., Di Sarno A., Giaccio A., Lombardi G.: Effectiveness and tolerability of slow release lanreotide treatment in active acromegaly. J Endocrinol Invest 1999; 22: 40-47.

34. Suliman M., Jenkins R., Ross R., Powell T., Battersby R., Cullen D.R. Long-term treatment of acromegaly with the somatostatin analogue SR-lanreotide. J Endocrinol Invest 1999 22: 409-418.

35. Cannavò S., Squadrito S., Curto L., Almoto B., Vieni A., Trimarchi F. Results of a two year treatment with slow-release lanreotide. Horm Metab Res 2000; 32: 224-9.

36. Verhelst J.A., Pedroncelli A.M., Abs R., Montini M., Vanderweghe M.V., Albani G., Maiter D., Pagani M.D., Legros J.J., Gianola D., Bex M., Pop Mockel J., Pagani G. Slow-release lanreotide in the treatment of acromegaly: a study in 66 patients. Eur J Endocrinol 2000; 143: 577-584.

37. Caron P., Morange-Ramos I., Cogne M., Jaquet P. Three year follow up of acromegalic patients treated with intramuscular slow-release lanreotide. J Clin Endocrinol Metab 1997; 82:18-22

38. Baldelli R,, Colao A,, Razzore P,, Jaffrain-Rea M-L,, Marzullo P,, Ciccarelli E,, Ferretti E,, Ferone D,, Gaia D,, Camanni F,, Lombardi G,, Tamburrano G. Two-year follow-up of acromegalic patients treated with SR-lanreotide 30 mg. J Clin Endocrinol Metab 2000; 85: 4099-4103.

39. Baldelli R., Ferretti E., Jaffrain-Rea M.L., Iacobellis G., Minniti G., Caracciolo B., Moroni C., Cassone R., Gulino A., Tamburrano G. Cardiac effects of slow-release lanreotide, a slow-release somatostatin analog, in acromegalic patients. J Clin Endocrinol Metab 1999; 84: 575-532.

40. Hradec J., Kral J., Janota T., Krsek M., Hana V., Marek J., Malik M. Regression of acromegalic left ventricular hypertrophy after lanreotide (a slow release somatostatin analog). Am J Cardiol 1999; 83: 1506-1509.
41. Colao A., Marzullo P., Vallone G., Giaccio A., Ferone D., Scarpa R., Rossi E., Smaltino F., Lombardi G. Ultrasonographic evidence of joint thickening reversibility in acromegalic patients treated with lanreotide for 12 months. Clin Endocrinol 1999; 51: 611-618.
42. Colao A., Marzullo P., Spiezia S., Giaccio A., Ferone D., Cerbone G., Di Sarno A., Lombardi G. Effect of two years of growth hormone-insulin-like growth factor-I suppression on prostate diseases in acromegalic patients. J Clin Endocrinol Metab 2000; 85: 3754-3761.
43. Manelli F., Desenzani P., Boni E., Bugari G., Negrini F., Romanelli G., Grassi V., Giustina A. Cardiovascular effects of a single slow release lanreotide injection in patients with acromegaly and left ventricular hypertrophy. Pituitary 1999; 2:205-10.
44. Manelli F., Bossoni S., Burattin A., Doga M., Solerte S.B., Romanelli G., Giustina A. Exercise-induced microalbuminuria in patients with active acromegaly: acute effects of slow-release lanreotide, a long-acting somatostatin analog. Metabolism 2000; 49:634-9.
45. Caron P., Beckers A., Cullen D.R., Goth M.I., Gutt B., Laurberg P., Pico A.M., Valimaki M., Zgliczynski W. Efficacy of the new long-acting formulation of lanreotide (lanreotide Autogel) in the management of acromegaly. J Clin Endocrinol Metab. 2002; 87:99-104.
46. Gillis J.C., Noble S., Goa K.L. Octreotide long-acting release (LAR). A review of its phamacological properties and therapeutic use in the management of acromegaly. Drugs 1997; 53:681-699.
47. Stewart P.M., Kane K.F., Stewart S.E., Lancranjan I., Sheppard M.C. Depot long-acting somatostatin analog (Sandostatin-LAR) is an effective treatment for acromegaly. J Clin Endocrinol Metab 1995; 80:3267-3272.
48. Flogstad A., Halse J., Bakke S., Lancranjan I., Marbach P., Bruns C., Jervell J. Sandostatin LAR in acromegalic patients: long term treatment. J Clin Endocrinol Metab 1997; 82:23-28
49. Lancranjan I., Atkinson A.B. and the Sandostatin® LAR® group. Results of a European multicenter study with Sandostatin® LAR® in acromegalic patients. Pituitary 1999; 1: 105-114.
50. Colao A., Ferone D., Marzullo P., Cappabianca P., Cirillo S., Boerlin V., Lancranjan I., Lombardi G. Long-term effects of depot long-acting somatostatin analog octreotide on hormone levels and tumor mass in acromegaly. J Clin Endocrinol Metab 2001; 86: in press
51. Bevan J.S., Atkin S.L., Atkinson A.B., Bouloux P.M., Hanna F., Harris P.E., James R.A., McConnell M., Roberts G.A., Scanlon M.F., Stewart P.M., Teasdale E., Turner H.E., Wass J.A., Wardlaw J.M. Primary medical therapy for acromegaly: an open, prospective, multicenter study of the effects of subcutaneous and intramuscular slow-release octreotide on growth hormone, insulin-like growth factor-I, and tumor size. J Clin Endocrinol Metab. 2002; 87:4554-63.
52. Colao A., Marzullo P., Ferone D., Spinelli L., Cuocolo A., Bonaduce D., Salvatore M., Boerlin V., Lancranjan I., Lombardi G. Cardiovascular effects of depot long-acting somatostatin analog Sandostatin LAR® in acromegaly. J Clin Endocrinol Metab 2000; 86: 3132-3140.
53. Colao A., Marzullo P., Cuocolo A., Spinelli L., Pivonello R., Bonaduce D., Salvatore M., Lombardi G. Reversal of acromegalic cardiomyopathy in young but not in middle-aged patients after 12 months of treatment with the depot long-acting somatostatin analogue octreotide. Clin Endocrinol (Oxf). 2003; 58:169-76.
54. Moschetta A., Stolk M.F., Rehfeld J.F., Portincasa P., Slee P.H., Koppeschaar H.P., Van Erpecum K.J., Vanberge-Henegouwen G.P. Severe impairment of postprandial cholecystokinin release and gall-bladder emptying and high risk of gallstone formation in acromegalic patients during Sandostatin LAR. Aliment Pharmacol Ther 2001; 15:181-185.
55. Kendall-Taylor P., Miller M., Gebbie J., Turner S., al-Maskari M. Long-acting octreotide LAR compared with lanreotide SR in the treatment of acromegaly. Pituitary 2000; 3:61-5.
56. Chanson P., Boerlin V., Ajzenberg C., Bachelot Y., Benito P., Bringer J., Caron P., Charbonnel B., Cortet C., Delemer B., Escobar-Jimenez F., Foubert L., Gaztambide S., Jockenhoevel F., Kuhn J.M., Leclere J., Lorcy Y., Perlemuter L., Prestele H., Roger P., Rohmer V., Santen R., Sassolas G., Scherbaum W.A., Schopohl J., Torres E. Comparison

of octreotide acetate LAR and lanreotide SR in patients with acromegaly. Clin Endocrinol 2000; 53:577-86

57. Marzullo P., Ferone D., Di Somma C., Pivonello R., Filippella M., Lombardi G., Colao A Efficacy of combined treatment with lanreotide and cabergoline in selected therapy-resistant acromegalic patients. Pituitary 1999; 1:115-20

58. Rocheville M., Lange D.C., Kumar U., Patel S.C., Patel R.C., Patel Y.C. Receptors for dopamine and somatostatin: formation of hetero-oligomers with enhanced functional activity. Science 2000; 288:154-157.

59. Spinas G.A., Zapf J., Landolt A.M., Stuckmann G., Froesch E.R. Pre-operative treatment of 5 acromegalics with a somatostatin analogue: endocrine and clinical observations. Acta Endocrinol 1987; 114:249-256.

60. Barkan A., Lloyd R.V., Chandler W.F. Preoperative treatment of acromegaly with long-acting somatostatin analog SMS 201-995: shrinkage of invasive pituitary macroadenomas and improved surgical remission rate. J Clin Endocrinol Metab 1988; 67:1040-1048.

61. Stevenaert A., Harris A.G., Kovacs K., Beckers A. Presurgical octreotide treatment in acromegaly. Metabolism 1992; 41:51-58.

62. Lucas Morante T., Garcia Uria L., Estrada J., Saucedo G., Cabello A., Alcaniz J., Barcelo B. Treatment of invasive growth hormone pituitary adenomas with long-acting somatostatin analog SMS 210-995 before transsphenoidal surgery. J Neurosurg 1994; 81: 10-14.

63. Waśko R., Ruchala M., Sawicka J., Kotwocka M., Liebert W., Sowiński J. Short-term pre-surgical treatment with somatostatin analogues, octreotide and lanreotide, in acromegaly. J Endocrinol Invest 2000; 23: 12-18.

64. Plockinger U., Reichel M., Fett U., Saeger W., Quabbe H-J. Preoperative octreotide treatment of growth hormone-secreting and clinically nonfunctioning pituitary macroadenomas: Effects on tumor volume and lack of correlation with immunohistochemistry and somatostatin receptor scintigraphy. J Clin Endocrinol Metab 1994; 79:1416-1423.

65. Ezzat S., Horvath E., Harris A.G., Kovacs K. Morphological effects of octreotide on growth hormone-producing pituitary adenomas. J Clin Endocrinol Metab 1994; 79:113-118.

66. Biermasz N.R., van Dulken H., Roelfsema F. Direct postoperative and follow-up results of transsphenoidal surgery in 19 acromegalic patients pretreated with octreotide compared to those in untreated matched controls. J Clin Endocrinol Metab 1999; 84:3551-5

67. Colao A., Ferone D., Cappabianca P., de Caro M.L., Marzullo P., Monticelli A., Alfieri A., Merola B., Calì A., de Divitiis E., Lombardi G. Effect of octreotide pretreatment on surgical outcome in acromegaly. J Clin Endocrinol Metab 1997;82:3308-3314.

68. Bates A., vant'Hoff W., Jones J., Clayton R. An audit of outcome of treatment in acromegaly. Q J Med 1993; 86;293-299.

69. Jenkins P.J., Akker S., Chew S.L., Besser G.M., Monson J.P., Grossman A.B. Optimal dosage interval for depot somatostatin analogue therapy in acromegaly requires individual titration. Clin Endocrinol 2000 53:719-724

70. Afargan M., Janson E.T., Gelerman G., Rosenfeld R., Ziv O., Karpov O., Wolf A., Bracha M., Shohat D., Liapakis G., Gilon C., Hoffman A., Stephensky D., Oberg K. Novel Long-Acting Somatostatin Analog with Endocrine Selectivity: Potent Suppression of Growth Hormone But Not of Insulin. Endocrinology 2001; 142:477-486

71. Trainer P.J., Drake W.M., Katznelson L., Freda P.U., Herman-Bonert V., van der Lely A-J., Dimaraki E.V., Stewart P.M., Friend K.E., Vance M.L., Besser G.M., Scarlett J.A., Thorner M.O., Parkinson C., Klibanski A., Powell J.S., Barkan A.L., Sheppard M.C., Malsonado M., Rose D.R., Clemmons D.R., Johannsson G., Bengtsson B.A., Stavrou S., Kleinberg D.L., Cook D.M., Phillips L.S., Bidlingmaier M., Strasburger C.J., Hackett S., Zib K., Bennett W.F., Davis R.J. Treatment of acromegaly with the growth hormone-receptor antagonist pegvisomant. N Engl J Med 2000; 342:1171-7

72. Herman-Bonert V.S., Zib K., Scarlett J..A, Melmed S. Growth hormone receptor antagonist therapy in acromegalic patients resistant to somatostatin analogs. J Clin Endocrinol Metab 2000; 85:2958-61

73. Kaal A., Orskov H., Nielsen S., Pedroncelli A.M., Lancranjan I., Marbach P., Weeke J. Occurrence and effects of octreotide antibodies during nasal, subcutaneous and slow release intramuscular treatment. Eur J Endocrinol 2000; 143:353-61

74. Beck-Peccoz P., Brucker-Davis F., Persani L., Smallridge R.C., Weintraub B.D. Thyrotropin-secreting pituitary tumors. Endocr Rev 1996;17:610-638.
75. Brucker-Davis F., Oldfield E.H., Skarulis M.C., Doppman J.L., Weintraub B.D. Thyrotropin-secreting pituitary tumors: diagnostic criteria, thyroid hormone sensitivity, and treatment outcome in 25 patients followed at the National Institute of Health. J Clin Endocrinol Metab 1999; 84:476-486.
76. Smallridge R.C. Thyrotropin-secreting pituitary tumors. Endocrinol Metab. 1987; 16:765-791.
77. Mc Cutcheon I.E., Weintraub B.D., Oldfield E.H. Surgical treatment of thyrotropin secreting adenomas. J Neurosurg. 1990; 73:674-683.
78. Losa M., Giovanelli M., Persani L., Mortini P., Faglia G., Beck-Peccoz P. Criteria of cure and follow-up of central hyperthyroidism due to thyrotropin-secreting pituitary adenomas. J Clin Endocrinol Metab 1996; 81:3084-3090.
79. Chanson P., Weintraub B.D., Harris A.G. Treatment of TSH-secreting pituitary adenomas with octreotide: a follow-up of 52 patients. Ann Intern Med 1993; 119:236-240.
80. Siler T.M., Yen S.S.C., Vale W., Guillemin R. Inhibition by somatostatin of the release of TSH induced in man by thyrotropin-releasing factor. J Clin Endocrinol Metab. 1974; 38:742-745.
81. Weeke J., Hansen A.P., Lundaek K. Inhibition by somatostatin of basal levels of serum thyrotropin (TSH) in normal men. J Clin Endocrinol Metab. 1975; 41:168-171
82. Reschini E., Giustina G., Cantalemessa L., Peracchi M. Hyperthyroidism with elevated plasma TSH levels and pituitary tumor: study with somatostatin. J Clin Endocrinol Metab. 1976; 46:924-927.
83. Losa M., Magnani P., Mortini P., Persani L., Acerno S., Giugni E., Songini C., Fazio F., Beck-Peccoz P., Giovanelli M. Indium-111 pentetreotide single-photon emission tomography in patients with TSH-secreting pituitary adenomas: correlation with the effect of a single administration of octreotide on serum TSH levels. Eur J Nucl Med 1997; 24:728-731.
84. Christensen S.E., Weeke J., Kaal A., Harris A.G., Orskov H. SMS 201–995 and thyroid function in acromegaly: acute, intermediate and long-term effects. Horm Metab Res. 1992; 24:237-239.
85. Beck-Peccoz P., Mariotti S., Guillausseau P.J. et al. Treatment of thyrotropin with the somatostatin analog SMS 201–295 J Clin Endocrinol Metab. 1989; 68:208-214.
86. Chanson P., Warnet A. Treatment of thyroid-stimulating hormone- secreting adenomas with octreotide. Metabolism. 1992; 4:62-65.
87. Comi R.J., Gesundheit N., Murray L., Gorden P., Weintraub B.D. Response of thyrotropin-secreting pituitary adenomas to a long-acting somatostatin analogue. N Engl J Med. 1992; 317:12-17.
88. Gesundheit N, Petrik PA, Nissim M, et al. Thyrotropin-secreting pituitary adenomas: clinical and biochemical heterogeneity. Ann Intern Med. 1989; 111:827-835
89. Guillausseau PJ, Chanson P, Timsit J, et al. Visual improvement with SMS 201–995 in a patient with a thyrotropin-secreting adenoma. N Engl J Med. 1987; 317:53-54.
90. Warnet A, Lajeunie E, Gelbert F, et al. Shrinkage of a primary thyrotropin-secreting pituitary adenoma treated with the long-acting somatostatin analogue octreotide (SMS 201–995). Acta Endocrinol (Copenh). 1991; 124:487-491
91. Wemeau L, Dewailly D, Leroy R, et al. Long term treatment with a somatostatin analog SMS 201–995 in a patient with a thyrotropin and growth hormone-secreting pituitary adenoma. J Clin Endocrinol Metab. 1988; 66:636-639.
92. Gancel A., Vuillermet P., Legrand A., Catus F., Thomas F., Kuhn J.M. Effects of a slow-release formulation of the new somatostatin analogue lanreotide in TSH-secreting pituitary adenomas. Clin Endocrinol 1994; 40:421-428.
93. Kuhn J.M., Arlot S., Lefebvre H., Caron P., Cortet-Rudelli C., Archambaud F., Chanson P., Tabarin A., Goth M.I., Blumberg J., Catus F., Ispas S., Beck-Peccoz P. Evaluation of the Treatment of Thyrotropin-Secreting Pituitary Adenomas with a Slow Release Formulation of the Somatostatin Analog Lanreotide J Clin Endocrinol Metab 2000; 851487-1491
94. Katnelson L., Alexander J.M., Klibanski A. Clinically nonfunctioning pituitary adenomas. J Clin Endocrinol Metab 1993; 76:1089-1094.

230

95. Freda P.U., Wardlaw S.L. Diagnosis and treatment of pituitary tumors. J Clin Endocrinol Metab 1999; 84:3859-3866.
96. Ikuyama S., Nawata H., Kato K., Karashima T., Ibayashi H., Nakagaki H. Specific somatostatin receptors on human pituitary adenoma cell membranes. J Clin Endocrinol Metab 1985; 61: 98-103
97. Reubi J.C., Heitz P.U., Landolt A.M. Visualization of somatostatin receptors and correlation with immunoreactive growth hormone and prolactin in human pituitary adenomas: evidence for different tumor subclasses. J Clin Endocrinol Metab 1987; 65: 65-73.
98. Greenman Y., Melmed S. Heterogeneous expression of two somatostatin receptor subtypes in pituitary tumors. J Clin Endocrinol Metab 1994; 78: 398-403.
99. Faglia G., Bazzoni N., Spada A., Arosio M., Ambrosi B., Spinelli F., Sara R., Bonino C., Lunghi F. In vivo detection of somatostatin receptors in patients with functionless pituitary adenomas by means of a radio-iodinated analog of somatostatin [($^{123}$I)SDZ204-090]. J Clin Endocrinol Metab 1991;73:850-856.
100. Plockinger U., Reichel M., Fett U., Saeger W., Quabbe H.J. Preoperative octreotide treatment of growth hormone-secreting and clinically nonfunctioning pituitary macroadenomas: effect on tumor volume and lack of correlation with immunohistochemistry and somatostatin receptor scintigraphy. J Clin Endocrinol Metab 1994; 79:1416-1423.
101. Lamberts S.W.J., de Herder W.W., van der Lely A.J., Hofland L.J. Imaging and medical management of clinically nonfunctioning pituitary tumors. Endocrinologist 1995; 5: 448-451.
102. Colao A., Lastoria S., Ferone D., Varrella P., Marzullo P., Pivonello R., Cerbone G., Acampa W., Salvatore M., Lombardi G. Pituitary uptake of In-111-DTPA-D-Phe$^1$-octreotide in the normal pituitary and in pituitary adenomas. J Endocrinol Invest 1999; 22: 176-183
103. Borson-Chazot F., Houzard C., Ajzenberg C., Nocaudie M., Duet M., Mundler O., Marchandise X., Epelbaum J., Gomez De Alzaga M., Schafer J., Meyerhof W., Sassolas G., Warnet A. Somatostatin receptor imaging in somatotroph and non-functioning pituitary adenomas: correlation with hormonal and visual responses to octreotide. Clin Endocrinol 1997; 47: 589-598.
104. Warnet A., Harris A.G., Renard E., Martin D., James-Deidier A., Chaumet-Riffaud P., and the French multicenter octreotide study group: A prospective multicenter trial of octreotide in 24 patients with visual defects caused by nonfunctioning and gonadotropin-secreting pituitary adenomas. Neurosurgery 1997;41:786-797.
105. Gasperi M., Petrini L., Pilosu R., Nardi M., Marcello A.A., Mastio F., Bartalena L., Martino E. Octreotide treatment does not affect the size of most non-functioning pituitary adenomas. J Endocrinol Invest 1993; 16:541-543.
106. De Bruin T.W.A., Kwekkeboom D.J., van't Verlaat J.W., Reubi J.C., Krenning E.P., Lamberts S.W., Croughs R.J. Clinically nonfunctioning pituitary adenoma and octreotide response to long term high dose treatment, and studies in vitro. J Clin Endocrinol Metab. 1992; 75.1310-1317.
107. Klibanski A., Alexander J.M., Bikkal H.A., Hsu D.W., Swearingen B., Zervas N.T. Somatostatin regulation of glycoprotein hormone and free subunit secretion in clinically nonfunctioning and somatotroph adenomas in vitro. J Clin Endocrinol Metab 1991; 73: 1248-1255.
108. Katznelson L., Oppenheim D.S., Coughlin J.F., Kliman B., Schoenfeld D.A., Klibanski A. Chronic somatostatin analog administration in patients with a-subunit-secreting pituitary tumors. J Clin Endocrinol Metab 1992; 75: 1318-1325.
109. Andersen M., Bjerre P., Schrøder H.D., Edal A., Høilund-Carlsen P.F., Pedersen P.H., Hagen C. In vivo secretory potential and the effect of combination therapy with octreotide and cabergoline in patients with clinically non-functioning pituitary adenomas: Clin Endocrinol 2001; 54: 23-30
110. Florio T., Thellung S., Arena S., Corsaro A., Spaziante R., Gussoni G., Acuto G., Giusti M., Giordano G., Schettini G. Somatostatin and its analog lanreotide inhibit the proliferation of

dispersed human non-functioning pituitary adenoma cells in vitro. Eur J Endocrinol 1999; 141:396-408

111. Saveanu A., Gunz G., Dufour H., Caron P., Fina F., Ouafik L., Culler M.D., Moreau J.P., Enjalbert A., Jaquet P. BIM-23244, a somatostatin receptor subtype 2- and 5-selective analog with enhanced efficacy in suppressing growth hormone (GH) from octreotide-resistant GH-secreting adenomas. J Clin Endocrinol Metab 2001; 86:140-145.

# 15
# SOMATOSTATIN RECEPTOR TARGETED-RADIO-ABLATION OF TUMORS

Marion de Jong, Roelf Valkema, Dik J Kwekkeboom and Eric P Krenning
*Department of Nuclear Medicine, Erasmus University Medical Center Rotterdam, Rotterdam, The Netherlands*

## INTRODUCTION

Somatostatin (SST) plays an important role in the physiological regulation of hormones and organs. The finding that SST inhibits hormone secretion of various glands led to the application of SST in the treatment of diseases based on for instance overproduction of hormones by tumors. The native peptide SST itself is unsuitable for routine treatment, as after intravenous administration it has a very short half-life due to rapid enzymatic degradation. Therefore, SST analogs that are more resistant to enzymatic degradation were synthesized; the molecule was modified in various ways with preservation of the biological activity of the original molecule. Introduction of D-amino acids, an amino-alcohol Thr-ol at the C-terminus and shortening of the molecule resulted e.g. in the 8 amino acids-containing SST analog octreotide, having a long and therapeutically useful plasma half-life. The diagnostic radiolabeled peptide [$^{111}$In-DTPA] octreotide (OctreoScan, $^{111}$In-pentetreotide) was approved by the FDA on June 2, 1994 for scintigraphy of patients with neuroendocrine tumors. Nowadays SST analogs are widely used in the treatment of symptoms due to neuroendocrine-active tumors, such as growth hormone-producing pituitary adenomas and gastroenteropancreatic tumors (1-4). Recently, improvement of quality of life has been demonstrated with long acting depot formulations.

SST effects are mediated by high-affinity G-protein coupled membrane receptors, integral membrane glycoproteins. Five different human SST receptor (SSTR) subtypes have been cloned (5-7). SST binds to all subtypes with high affinity, while the affinity of the different SST analogs for these subtypes differ considerably. Octreotide e.g. binds with high affinity to the SSTR subtype 2 ($sst_2$), and with lower affinities to $sst_5$ and $sst_3$. It shows no binding to $sst_1$ and $sst_4$ (6).

Peptide receptor scintigraphy with radioactive octreotide, first [$^{123}$I-Tyr$^3$]octreotide and nowadays mostly [$^{111}$In-DTPA]octreotide (OctreoScan®), appeared to be sensitive and specific for in vivo visualisation of the presence and abundance of SST receptors on the primary as well as metastatic tumor sites of a variety of neuroendocrine tumors, like carcinoids, islet cell tumors and paragangliomas (8-13).

As soon as the success of peptide receptor scintigraphy for tumor visualization became clear, the next logical step was to try to label these peptides with therapeutic radionuclides, emitting α- or β-particles, or Auger or conversion electrons, and to perform peptide receptor radionuclide therapy (PRRT) with these radiolabeled peptides, as no effective general anti-tumor treatment is available for metastasized neuroendocrine tumors and only symptomatic relief can be achieved using SST analogues and interferon.

## OCTREOTIDE

### Experimental Data
The molecular basis of the use of radiolabeled octreotide in scintigraphy and radionuclide therapy is receptor-mediated internalization and cellular retention of the radionuclide. Internalization of radiolabeled [DTPA]octreotide in SST receptor-positive tumors and tumor cell lines has been investigated (14-16), it appeared that this process is receptor-specific and temperature dependent. Receptor-mediated internalization of [$^{111}$In-DTPA]octreotide results in degradation to the final radiolabeled metabolite $^{111}$In-DTPA-D-Phe in the lysosomes (17). This metabolite is not capable of passing the lysosomal and/or other cell membrane(s), and will therefore stay in the lysosomes, causing the long retention time of $^{111}$In in sst$_2$-positive (tumor) cells.

Internalization of [$^{111}$In-DTPA]octreotide is especially important for radionuclide therapy of tumors when radionuclides emitting therapeutical particles with very short path lengths are used, like those emitting Auger electrons (e.g. $^{111}$In). These electrons are only effective in a short distance of only a few nm up to μm from their target, the nuclear DNA. Recently, Hornick et al. (18) and Wang et al. (19) described *in vitro* cellular internalization, nuclear translocation, and DNA binding of radiolabeled SST analogs, which significantly increased after prolonged exposure. So, $^{111}$In-labeled peptides are therefore suitable for both scintigraphy and radionuclide therapy, all the more so as the decay of Auger electron emitter has recently been shown to lead to a "bystander effect", an in vivo, dose-independent inhibition or retardation of tumor growth in nonradiotargeted cells by a signal produced in Auger electron-labeled cells (20).

In our preclinical radionuclide therapy studies, we used the rat pancreatic CA20948 tumor as a model for receptor-targeted scintigraphy and radionuclide therapy using radiolabeled SST analogs. This tumor is transplantable in Lewis rats, not only subcutaneously in the flank, but also metastasized to the liver. The latter is achieved by inoculation of tumor cells in the portal vein of the liver. The

CA20948 tumor has been shown to be SSTR-positive, being an excellent model to study receptor-targeted scintigraphy and radionuclide therapy in rats using radiolabeled SST analogs (21).

We performed radionuclide therapy using [$^{111}$In-DTPA]octreotide in the CA20948 liver metastases model (22). Radionuclide therapy with administrations of 370 MBq (coupled to 0.5 µg octreotide) [$^{111}$In-DTPA]octreotide after intraportal CA20948 tumor cell inoculation induced a significant decrease in the number of hepatic metastases at day 21. Co-injection with 1 mg unlabeled octreotide resulted in inhibition of the tumor response to radionuclide therapy, pointing to a receptor-dependent therapeutic effect. Also dose dependent effects of radionuclide therapy by injection of 370, 37 or 3.7 MBq [$^{111}$In-DTPA]octreotide one day after tumor inoculation were investigated (23). The 370 MBq dosage had significantly more therapeutic effects and inhibited the increase of liver weight due to tumor growth more than the 37 or 3.7 MBq doses. The findings hold promise for application of radionuclide therapy with $^{111}$In-labeled octreotide in an adjuvant, micrometastatic setting, e.g. after surgery to eradicate occult metastases.

**Clinical Data**
Fifty patients with SSTR-positive tumors were treated with multiple doses of [$^{111}$In-DTPA]octreotide in Rotterdam (24). Forty patients were evaluated after cumulative doses of at least 20 GBq up to 160 GBq. Therapeutic effects were seen in 21 patients: partial remission in 1 patient, minor remissions in 6 patients, and stabilization of previously progressive tumors in 14 patients, underscoring the therapeutic potential of Auger-emitting radiolabelled peptides. The toxicity was generally mild bone marrow toxicity, but 3 of the 6 patients who received more than 100 GBq developed a myelodysplastic syndrome or leukemia. Therefore, 100 GBq was considered the maximal tolerable dose. With a renal radiation dose of 0.45 mGy/MBq (based on previous studies) a cumulative dose of 100 GBq will lead to 45 Gy on the kidneys, twice the accepted limit for external beam radiation. However, no development of hypertension, proteinuria, or significant changes in serum creatinine or creatinine clearance were observed in the patients, including 2 patients who received 106 and 113 GBq [$^{111}$In-DTPA]octreotide without protection with amino acids, over a follow- up period of respectively 3 and 2 years. These findings show that the radiation of the short-range Auger electrons originating from the cells of the proximal tubules is not harmful for the renal function. The decrease in serum inhibin B and concomitant increase of serum FSH levels in men indicate that the spermatogenesis was impaired.

Fjalling et al. (25) described a patient with a midgut carcinoid syndrome due to metastatic spread of an ileal tumor to the liver, paraortic and mediastinal lymph nodes and to the skeleton, who was given systemic radionuclide therapy with [$^{111}$In-DTPA]octreotide. Treatment was given on three separate occasions (3.0, 3.5 and 3.1 GBq) 8 and 4 weeks apart. After each therapy, the patient experienced facial flush and pain over the skeletal lesions followed by symptomatic relief, even though no objective tumor regression was found radiologically after 5 months. After initiation of radiolabeled octreotide treatment, there was a 14% reduction of the main tumor marker, urinary 5-HIAA. After three subsequent radionuclide therapies, there was a further 31% reduction of 5-HIAA levels. No adverse reactions, other than a slight decrease in leukocyte counts, were seen.

At the Louisiana State University Medical Center in New Orleans, a clinical trial was performed to determine the effectiveness and tolerability of therapeutic doses of [$^{111}$In-DTPA]octreotide in patients with GEP tumors (26-28). GEP tumor patients who had failed all forms of conventional therapy, with worsening of tumor-related signs and symptoms and/or radiographically documented progressive disease, an expected survival less than 6 months, and SST receptor expression on the tumor as determined by the uptake on a 6.0 mCi [$^{111}$In-DTPA]octreotide scan, were treated with at least 2 monthly 180-mCi intravenous injections. Twenty-seven GEP (24 carcinoid neoplasms with carcinoid syndrome and 3 pancreatic islet cells) patients were accrued. Clinical benefit occurred in 16 (62%) patients. Objective partial responses on CT occurred in 2 (8%) patients, and significant tumor necrosis developed in 7 (27%) patients. The following transient grades 3/4 side effects were observed, respectively: leukocytes: 1/1; platelets: 0/2; hemoglobin: 3/0; bilirubin: 1/3; creatinine: 1/0; neurologic: 1/0. Myeloproliferative disease and/or myelodysplastic syndrome had not been observed in the 6 patients followed-up for 48+ months. They concluded that two doses (180 mCi each) of [$^{111}$In-DTPA]octreotide were safe, well tolerated, and improved symptoms in 62% of patients, with 8% partial radiographic responses and increased expected survival in GEP cancer patients with SST receptor-expressing tumors.

As $^{111}$In emits therapeutic Auger electrons with mean particle ranges of less than one cell diameter, radiation emitted from a receptor-positive tumor cell cannot kill neighbouring receptor-negative cells in tumors with receptor heterogeneity. Consequently, various research groups aimed to develop SST analogues that can be linked via a chelator to a therapeutic radionuclide which emits ß-particles with longer particle ranges, such as $^{90}$Y and $^{177}$Lu. In addition, new SST analogues were synthesized to improve receptor affinity.

## OCTREOTIDE ANALOGUES

### Experimental Data
Various chelator-peptide constructs have been synthesized and evaluated concerning their receptor affinity, internalization capacities, and biodistribution in vivo (29). The analogs tested included [DTPA,Tyr$^3$]octreotide and [DTPA,Tyr$^3$]octreotate (in octreotate the C-terminal threoninol has been replaced with the native amino acid threonine) in comparison to [DTPA]octreotide. Phe$^3$-residues were replaced with Tyr to increase the

hydrophylicity of the peptides. Octreotate was synthesized to investigate the effects of an additional negative charge on clearance and cellular uptake. We concluded that radiolabeled [DTPA,Tyr$^3$]octreotate and second best [DTPA,Tyr$^3$]octreotide, and also their DOTA-coupled counterparts, were most promising for scintigraphy and radionuclide therapy of octreotide receptor-positive tumors in humans (29).

DOTA is a universal chelator capable of formation of stable complexes with metals like $^{111}$In, $^{67}$Ga, $^{68}$Ga, $^{86}$Y and $^{64}$Cu for imaging as well as with $^{90}$Y (high energy β-particle emitter) and with radiolanthanides like e.g. $^{177}$Lu (low energy β-particle and gamma emitter) for receptor-mediated radionuclide therapy. Reubi et al. (30) evaluated the in vitro binding characteristics of labeled (indium, yttrium, gallium) and unlabeled [DOTA,Tyr$^3$]octreotide, [DOTA]octreotide, [DOTA]lanreotide, [DOTA]vapreotide (RC-160), [DTPA,Tyr$^3$]octreotate and [DOTA,Tyr$^3$]octreotate using cell lines transfected with the human SST receptor subtypes sst$_1$, sst$_2$, sst$_3$, sst$_4$ and sst$_5$. They found that small structural modifications, chelator substitution or metal replacement considerably affected the receptor binding affinity. A marked improvement of sst$_2$ affinity was found for [Ga-DOTA,Tyr$^3$]octreotide (IC$_{50}$ 2.5 nM) compared with the Y-labeled compound and [In-DTPA]octreotide. An excellent binding affinity for sst$_2$ in the same range was also found for [In-DTPA,Tyr$^3$]octreotate (IC$_{50}$ 1.3 nM) and for [Y-DOTA,Tyr$^3$]octreotate (IC$_{50}$ 1.6 nM). Of $^{111}$In-, $^{88}$Y- and $^{177}$Lu-labeled [DOTA,Tyr$^3$]octreotate, biodistribution and tumor uptake were compared in CA20948 tumor-bearing rats (31). An *in vivo* rapid clearance from the blood and very high specific uptake in sst$_2$-positive organs and tumor was found for all 3 radiolabeled analogs. For tumor and sst$_2$-positive organs it was found that uptake of $^{111}$In- ≈ $^{88}$Y- < [$^{177}$Lu-DOTA,Tyr$^3$]octreotate uptake, making the latter analog most promising for radionuclide therapy. Schmitt et al. (32) investigated [$^{177}$Lu-DOTA,Tyr$^3$]octreotate in nude mice with human small cell lung cancer and concluded that the tumor had a higher activity concentration compared to all measured normal tissues at all time points tested, pointing to the therapeutic potential of [$^{177}$Lu-DOTA,Tyr$^3$]octreotate for small cell lung cancer.

A comparison between radiolabeled octreotide and octreotate analogs was also performed in rats using $^{64}$Cu-labeled analogs, reaching the same conclusion, that because of its high tumor uptake in comparison to that of the other analogs tested, [Tyr$^3$]octreotate was selected for future PET imaging and targeted radiotherapy studies. New stable analogues of SST with high affinity for different SSTRs are currently being developed. An interesting example is [DOTA, 1-Nal$^3$]octreotide, which has high affinity for sst$_2$, sst$_3$ and sst$_5$ (33, 34). This compound may allow PRRT of tumors which do not bind octreotide and octreotate with high affinity, i.e. sst$_3$- and sst$_5$-positive tumors.

A problem during radionuclide therapy may be caused by the high uptake of radioactivity in the kidneys; small peptides in the blood plasma are filtered through the glomerular capillaries in the kidneys and subsequently reabsorbed by and retained in the proximal tubular cells, thereby reducing the scintigraphic sensitivity for detection of small tumors in the perirenal region and the possibilities for radionuclide-therapy. The renal uptake of radiolabeled octreotide in rats could be reduced by positively charged amino acids, e.g. with about 50 % by single intravenous administration of 400 mg/kg L- or D-lysine (35, 36). Therefore, during PRRT an infusion containing the positively charged amino acids L-lysine and L-arginine (in patients e.g. 25 g lysine and 25 g arginine infused in 4 h) can be given during and after the infusion of the radiopharmaceutical, in order to reduce the kidney uptake (37-39).

After [111]In, the next radionuclide investigated was [90]Y, emitting ß-particles with a high maximum energy (2.27 MeV) and a long maximum particle range (>10 mm). The first SST analogue radiolabeled with [90]Y and applied for PRRT in animals and patients was [[90]Y-DOTA,Tyr[3]]octreotide, in which, in comparison with octreotide, the phenylalanine residue at position 3 has been replaced with tyrosine; this makes the compound more hydrophilic and increases the affinity for $sst_2$, leading to higher uptake in $sst_2$-positive tumors both in preclinical studies and in patients (29, 40).

We compared the radiotherapeutic effect of different doses [[90]Y-DOTA,Tyr[3]]octreotide in rats bearing pancreatic CA20948 tumors of different size in the flank (41). After the highest dose, i.e. 370 MBq [[90]Y-DOTA,Tyr[3]]octreotide, 50% complete response was reached for the small tumors (< 1 cm$^2$), whereas only growth delay was found in the very large tumors (> 12 cm$^2$). Medium sized tumors (about 8 cm$^2$), however, showed 100% cure after this same dose of [[90]Y-DOTA,Tyr[3]]octreotide. So, in this study a difference is found in the radiotherapeutic effects in CA20948 tumors of different size. In larger tumors more clonogenic, presumably hypoxic, cells will be present, thereby limiting radiocurability. The small tumors on the other hand, will not absorb all energy emitted by [90]Y, thereby decreasing tumor curability.

Significant tumor growth delay was also found in rats bearing small CA20948 tumors after radionuclide therapy with up to 3 x 740 MBq [[64]Cu-TETA,Tyr[3]]octreotide or [[64]Cu-TETA,Tyr[3]]octreotate (42, 43), the first compound given either fractionated or as a single dose. Dose fractionation in 2 doses induced significantly increased tumor-growth inhibition compared with rats given a single dose. However, in this study the single 555 MBq dose was bound to twice the amount of peptide compared to the two 278 MBq doses. So, partial saturation of the receptors using the single high dose and therefore relatively lower uptake of radioactivity in the tumor may have contributed to these findings. Tumor growth inhibition in the same model was also found after treatment of CA20948 tumor-bearing rats with [[90]Y-DTPA-benzyl-acetamido-Tyr[3]]octreotide (44). Using 370 MBq/kg of [[90]Y-DOTA,Tyr[3]]octreotide the same group observed even complete tumor reduction in 5 out of 7 rats (45). These preclinical studies show the great promise of $^{90}$Y-labeled octreotide analogs for the clinical studies (see below).

The next analogue investigated in preclinical radionuclide therapy studies was [$^{177}$Lu-DOTA,Tyr$^3$]octreotate. $^{177}$Lu emits gamma radiation with a suitable energy for imaging and therapeutic β-particles with low to medium energy (maximum 0.50 MeV), so the same compound can be used for both imaging and dosimetry and radionuclide therapy, thus obviating the need for a pretherapeutic diagnostic study. The approximate range of the β-particles is 20 cell diameters, whereas the range of those emitted by $^{90}$Y is 150 cell diameters. Less "cross-fire" induced radiation damage in the renal glomeruli can therefore be expected with $^{177}$Lu. Also, in comparison with $^{90}$Y, a higher percentage of the $^{177}$Lu radiation energy will be absorbed in very small tumors and (micro)metastases.

We investigated the anti-tumor effects of [$^{177}$Lu-DOTA,Tyr$^3$]octreotate in various models, including a rat liver micrometastatic model, mimicking disseminated disease, and a solid tumor model. [$^{177}$Lu-DOTA,Tyr$^3$]octreotate showed anti-tumoral effects in the rat liver tumor metastases model leading to significant better survival in the treated rats (46).

In the radionuclide therapy studies using $^{177}$Lu-labeled octreotate for therapy in solid tumors, 100% cure was found in the groups of rats bearing small (≤ 1 cm$^2$) CA20948 tumors after two repeated doses of 277.5 MBq or after a single dose of 555 MBq [$^{177}$Lu-DOTA,Tyr$^3$]octreotate (estimated tumor dose 60 Gy) (31). After therapy with the same doses of [$^{177}$Lu-DOTA,Tyr$^3$]octreotide, that has a lower tumor uptake than the octreotate analog, these data were 50% and 60% cure in rats bearing small tumors. In rats bearing larger (≥ 1 cm$^2$, range 1.4 – 10 cm$^2$) tumors, 40 and 50% cure were found in the groups that received one or two 277.5 MBq injections of [$^{177}$Lu-DOTA,Tyr$^3$]octreotate, respectively (31). However, in another study in a different rat pancreatic tumor model (AR42J), in which a more favourable tumor dose was reached after 555 MBq [$^{177}$Lu-DOTA,Tyr$^3$]octreotate (140 Gy), all rats but one were cured irrespective the size of their tumor (unpublished result). So, [$^{177}$Lu-DOTA,Tyr$^3$]octreotate showed excellent therapeutic results in the rats bearing small to big tumors, the findings for the small tumors were in accordance with those of an earlier study (47).

In a different set of experiments the combination of [$^{177}$Lu-DOTA,Tyr$^3$]octreotate and [$^{90}$Y-DOTA,Tyr$^3$]octreotide (at a constant total dose) was studied in rats that each bore both a small (0.1 cm$^2$) and a large tumor (8 cm$^2$), to mimic the clinical situation, in which large tumors and small metastases are usually present in the same patient. The rats treated with the combination of 50% [$^{177}$Lu-DOTA,Tyr$^3$]octreotate plus 50% [$^{90}$Y-DOTA,Tyr$^3$]octreotide showed a longer survival (area under survival curve: 146 days) than those treated with 100% $^{90}$Y-DOTATOC (57 days) or [$^{177}$Lu-DOTA,Tyr$^3$]octreotate (50 days) (48). This underscores the great promise of

[177]Lu- and [90]Y-labeled SST analogues for radionuclide therapy and the potential of the combination of these radionuclides with different $\beta$-energies and particle ranges to achieve higher cure rates in the presence of tumors of different size.

We conclude that [[177]Lu-DOTA,Tyr[3]]octreotate is a very promising SST analog for radionuclide therapy in patients suffering from sst$_2$-expressing tumors. In patients with tumors of different size, including small metastases, also combinations of radionuclides are interesting, e.g. [90]Y and [177]Lu, to obtain the widest range of tumor curability.

**Clinical Data**

Various multicentre phase 1 and phase 2 PRRT trials have been performed using these [90]Y- and [177]Lu-labeled SST analogues. Otte et al. (49-51) described a study in which patients received four or more single doses of [[90]Y-DOTA,Tyr[3]]octreotide with ascending activity at intervals of approximately 6 weeks (cumulative dose 6120+/-1347 MBq/m$^2$) with the aim of performing an intra-patient dose escalation study. In total, 127 single treatments were given. In eight of these 127 single treatments, total doses of $\geq$ 3700 MBq were administered. In an effort to prevent renal toxicity, two patients received Hartmann-Hepa 8% amino acids (including lysine and arginine) solution during all therapy cycles, while 13 patients did so during some but not all therapy cycles; in 14 patients no solution was administered during the therapy cycles. Of the 29 patients, 24 patients showed no severe renal or haematological toxicity (toxicity < or = grade 2 according to the National Cancer Institute grading criteria). These 24 patients received a cumulative dose of $\leq$ 7400 MBq/m$^2$. Five patients developed renal and/or haematological toxicity. All of these five patients received a cumulative dose of > 7400 MBq/m$^2$ and had received no Hartmann-Hepa 8% solution during the therapy cycles. Four of the five patients developed renal toxicity; two of these patients showed stable renal insufficiency and two required haemodialysis. Two of the five patients exhibited anaemia (both grade 3) and thrombocytopenia (grade 2 and 4, respectively). Twenty of the 29 patients had a disease stabilization, two a partial remission, four a reduction of tumor mass <50% and three a progression of tumor growth.

Waldherr et al. reported several phase 2 studies in patients with neuroendocrine tumors (52-54). The patients received four or more doses of [[90]Y-DOTA,Tyr[3]]octreotide with ascending activity at intervals of approximately 6 weeks. Observed renal or haematological toxicity was $\leq$ grade 2 according to the National Cancer Institute grading criteria. The cumulative dose was $\leq$ 7,400 MBq/m$^2$. Complete and partial responses obtained in different studies amounted to 24% in 400 patients treated at the University Hospital Basel in Switzerland. In addition, distinct protocols were compared: in the above mentioned studies, patients received four injections of 1,850 MBq/m$^2$ (at intervals of ca. 6 weeks), while in another study two injections of

3,700 MBq/m$^2$ were administered at an interval of 8 weeks. Interestingly, the results from the last study were the most impressive. A higher percentage of complete responses plus partial remissions (24% after four injections versus 34% after two injections) was found, while side-effects were not significantly different. The study indicates that treatment interval and dosing may play a role in the outcome of PRRT studies (54).

The European Institute of Oncology group in Milan, Italy reported on 256 patients, mostly recruited in two distinct protocols, using [$^{90}$Y-DOTA,Tyr$^3$]octreotide with and without the administration of kidney-protecting agents (38, 55-57). About eighty percent of the patients presented with progressive disease before the start of therapy. No major acute reactions were observed up to an activity of 5.55 GBq per cycle. The patients received three or more equal i.v. injections of [$^{90}$Y-DOTA,Tyr$^3$]octreotide, starting with 1.1 GBq per cycle in escalating dosage (0.37 GBq increment per cycle) in subsequent groups. A typical administration consisted of 130 $\mu$g of DOTATOC labeled with 4.81 GBq of $^{90}$Y. Cumulative activity administered to each patient ranged between 7.4 and 21.3 GBq. Reversible grade 3 haematological toxicity on white blood cells and/or platelets was found in 43% of patients injected with 5.18 GBq, which was defined as the maximum tolerated dose per cycle. None of the patients developed acute or delayed kidney nephropathy. Complete or partial reductions in tumor mass were found in 27% of patients.

Another study with [$^{90}$Y-DOTA,Tyr$^3$]octreotide (OctreoTher, $^{90}$Y-SMT-487) was the phase I Novartis study performed in Rotterdam, Brussels and Tampa, which aimed to define the maximum tolerated single- and four-cycle doses of [$^{90}$Y-DOTA,Tyr$^3$]octreotide (58, 59). The cumulative radiation dose to kidneys was limited to 27 Gy, according to standard MIRD. Forty-seven patients were enrolled; all received amino acids concomitant with [$^{90}$Y-DOTA,Tyr$^3$]octreotide for kidney protection. At baseline, 81% of the patients had progressive disease, and 19% had stable disease. With cycle doses ranging from 1.3 to 10.8 GBq and cumulative doses from 1.7 to 27 GBq, the maximum tolerated dose was not reached. Three patients had dose-limiting toxicity: one had liver toxicity grade 3, one had thrombocytopenia grade 4 and one had myelodysplastic syndrome. After a median follow-up of 19 months, minor remissions were seen in 18% of these patients, and partial responses in 10%. An important observation in this study was a clear dose-response relation: the percentage reduction in tumor volume increased with increasing tumor radiation dose (up to about 600 Gy) (60). Prior chemotherapy predisposed to haematological toxicity. Renal toxicity was mild in these patients, with individualized dosimetry and amino acid infusion for kidney protection. A phase-II Novartis multicentre trial of [$^{90}$Y-DOTA,Tyr$^3$]octreotide treatment in patients with SCLC and breast cancer started recently. The first results from one of the institutions showed that [$^{90}$Y-DOTA,Tyr$^3$]octreotide is effective in improving the clinical status, based on a semiquantitative scoring system, of patients with receptor-positive tumors (61). In 67% of these patients a favourable clinical response was found. Three treatment cycles of 4.4 GBq [$^{90}$Y-DOTA,Tyr$^3$]octreotide each were given to 21 patients, amino acids were given to reduce renal uptake. From the same study is was also concluded that patients with diffuse SST receptor-positive hepatic metastases can be treated with this administered activity with only a small chance of developing mild acute or subacute hepatic radiation injury (62).

Despite the differences in the protocols used, the rate of complete plus partial responses seen in the various aforementioned studies [$^{90}$Y-DOTA,Tyr$^3$]octreotide studies (10-34%) consistently exceeds that obtained with [$^{111}$In-DTPA]octreotide (see above).

Another application of [$^{90}$Y-DOTA,Tyr$^3$]octreotide is PRRT of SST receptor-positive malignant gliomas by local injection instead of intravenous administration. Small peptides, such as [$^{90}$Y-DOTA,Tyr$^3$]octreotide, have the potential to target infiltrative disease within normal brain tissue. Schumacher et al. (63) treated patients with progressive gliomas of WHO grades II and III and extensively debulked low-grade gliomas with varying fractions of [$^{90}$Y-DOTA,Tyr$^3$]octreotide. The radiolabeled peptide was injected locally into the resection cavity or into solid tumor. The activity per single injection ranged from 555 to 1,875 MBq, and the cumulative activity from 555 to 7,030 MBq, yielding dose estimates from 76 to 312 Gy. In the progressive gliomas, lasting responses were obtained for at least 13-45 months without the need for steroids. Based on these observations, the feasibility of local radiotherapy following extensive debulking was also assessed, and such therapy was found to be well tolerated. It was concluded that targeted beta-particle irradiation based on diffusible small peptidic vectors appears to be a promising modality for the treatment of malignant gliomas.

New is the use of [$^{177}$Lu-DOTA,Tyr$^3$]octreotate, which shows the highest tumor uptake of all tested octreotide analogues so far in patients with neuro-endocrine tumors, with excellent tumor to kidney ratios (39). The effects of [$^{177}$Lu-DOTA,Tyr$^3$]octreotate therapy have been reported in 67 patients with neuroendocrine gastroenteropancreatic (GEP) tumors (64). Patients were treated with dosages of 3,700, 5,550 or 7,400 MBq [$^{177}$Lu-DOTA,Tyr$^3$]octreotate, up to a final cumulative dose of 22.2-29.6 GBq with treatment intervals of 6-9 weeks. Three months after the final administration, complete or partial remissions were found in 30% of the patients, and a minor response in 12 %. Tumor response was positively correlated with high uptake on the OctreoScan, limited hepatic tumor mass and high Karnofsky Performance Score. The side-effects of treatment with [$^{177}$Lu-DOTA,Tyr$^3$]octreotate were few and mostly transient, with mild bone marrow depression as the most common finding. One patient developed myelodysplastic syndrome (MDS) several months after his last treatment, most probably caused by previous chemotherapy. Kidney function and pituitary function did not deteriorate, except in one patient with compromised kidney function at start who had renal insufficiency 1 year after the last treatment. Other side-effects can be ascribed to the radiation dose to the testes in men: this dose leads to significantly lower serum testosterone and inhibin-B levels which in turn give rise to higher serum LH and FSH concentrations, thereby substantiating that the pituitary function is unimpaired. It is the current practice to adopt an expectant attitude when dealing with patients with GEP tumors owing to the limited efficacy of chemotherapy protocols. However, given the high success rate of therapy with radiolabeled SST analogues emitting $\beta$-particles, such as [$^{177}$Lu-DOTA,Tyr$^3$]octreotate, and the absence of serious side-effects, their use can be advocated in patients with GEP tumors without waiting for tumor progression. The results obtained with [$^{177}$Lu-DOTA,Tyr$^3$]octreotate are very encouraging, yet a direct, randomised comparison with [$^{90}$Y-DOTA,Tyr$^3$]octreotide treatment is lacking. Also, the reported percentages of tumor remission after [$^{90}$Y-DOTA,Tyr$^3$]octreotide treatment vary (see above). This may have several causes: 1. The administered doses and dosing schemes differ: some studies use dose-escalating schemes, whereas others use fixed doses; 2. There are several patient and tumor characteristics that determine treatment outcome, such as amount of uptake on the octreoscan, the estimated total tumor burden, and the extent of liver involvement. Therefore, differences in patient selection may play an important role in determining treatment outcome. For these reasons we planned a multicenter randomised trial comparing the effects of [$^{90}$Y-DOTA,Tyr$^3$]octreotide treatment with those of [$^{177}$Lu-DOTA,Tyr$^3$]octreotate. Because in animal experiments $^{90}$Y-labeled SST analogues are more effective for larger tumors, and $^{177}$Lu-labeled SST analogues are more effective for smaller tumors, whereas their combination was found to be most effective (see above), a third treatment arm will consist of sequential [$^{90}$Y-DOTA,Tyr$^3$]octreotide and [$^{177}$Lu-DOTA,Tyr$^3$]octreotate treatment. Apart from the combination of analogues labeled with different radionuclides, future directions to improve this therapy may also include efforts to upregulate the SST receptor expression on the tumors, as well as studies to the effects of the use of radiosensitizers.

**Lanreotide**
Virgolini et al. developed an $^{111}$In-/$^{90}$Y- labeled SST analog, [DOTA]lanreotide, for tumor diagnosis and therapy (65-68). They described that $^{111}$In-/$^{90}$Y-labeled [DOTA]lanreotide bound with high affinity to a number of primary human tumors. [$^{111}$In-DOTA]lanreotide bound with high affinity to hsst$_2$, hsst$_3$, hsst$_4$, and hsst$_5$ and with lower affinity to hsst$_1$ expressed on COS7 cells, making it a universal receptor binder (69). In Sprague Dawley rats, [$^{90}$Y-DOTA]lanreotide was rapidly cleared from the circulation and concentrated in SST receptor-positive tissues, such as pancreas or pituitary. It was concluded that this radiolabeled peptide can be used for SST receptor-mediated diagnosis as well as systemic radiotherapy of human tumors.

However, Reubi et al. found in vitro in cell lines transfected with the different SSTR subtypes that whereas [Y-DOTA]lanreotide had a good affinity for the

$sst_5$, it had a low affinity for $sst_3$ ($IC_{50}$ 290 nM) and $sst_4$ ($IC_{50}$ > 10000 nM) (30). Thereby, they challenged the concept that lanreotide is a universal binder to the different SST receptors. Froidevaux et al. (70) concluded from their comparison study of among other things [DOTA,Tyr³]octreotide and [DOTA]lanreotide in rats that radiolabeled [DOTA,Tyr³]octreotide has more potential for clinical application than [DOTA]lanreotide.

**Clinical Data**

Lanreotide was the second analogue labeled with ⁹⁰Y that was used for clinical PRRT studies. Virgolini et al. (66) reported on the biodistribution, safety and radiation absorbed dose of [¹¹¹In-DOTA]lanreotide. The tumor localizing capacity of [¹¹¹In-DOTA]lanreotide was initially investigated in 10 patients. The radiolabeled analog was then administered to 14 cancer patients evaluated for possible radiotherapy with [⁹⁰Y-DOTA]lanreotide. A comparison with [¹¹¹In-DTPA]octreotide scintigraphy was performed in 8 of the patients. The mean radiation absorbed dose amounted to 1.2 (range 0.21-5.8) mGy/MBq for primary tumors and/or metastases. The ¹¹¹In-DOTA-lanreotide radiation absorbed tumor dose was significantly higher (ratio 2.25 +/- 0.60, p < 0.01) when directly compared with ¹¹¹In-octreotide.

Leimer et al. (71) reported the case of a gastrinoma patient with liver metastases who was treated with [⁹⁰Y-DOTA]lanreotide. After four infusions of [⁹⁰Y-DOTA]lanreotide (each 1 GBq, approximately 30 nmol) over a 6-months period, the [¹¹¹In-DOTA]lanreotide scintigraphy of the liver had returned to a nearly normal condition and a decreased uptake by the recurrent gastrinoma was calculated. The imaging results were well correlated with a 25% regression of the liver metastases as indicated by CT.

[⁹⁰Y-DOTA]lanreotide treatment was further studied at different centres in the MAURITIUS trial (67). In this study, cumulative treatment doses of up to 8.58 GBq [⁹⁰Y-DOTA]lanreotide were given as a short-term intravenous infusion. Treatment results in 154 patients indicated minor responses in 14%. No severe acute or chronic haematological toxicity or changes in renal or liver function parameters due to [⁹⁰Y-DOTA]lanreotide were reported. In two-thirds of patients with neuroendocrine tumor lesions, [⁹⁰Y-DOTA,Tyr³]octreotide showed a higher tumor uptake than [⁹⁰Y-DOTA]lanreotide, which can be explained by the lower affinity of [⁹⁰Y-DOTA]lanreotide for $sst_2$.

# CONCLUSION

This chapter shows that radionuclide therapy with radiolabeled SST analogs, like [DOTA,Tyr³]octreotide and [DOTA,Tyr³]octreotate is a most promising new treatment modality for patients bearing $sst_2$-positive tumors. In PRRT the goal is to deliver an effective radiation dose to the tumor without causing undesired effects in healthy tissues. Improvements in the success of radionuclide therapy also depend on the optimization of radiation doses to tumor versus normal organs in individual patients. This requires patient-specific measurements and application of models to estimate the radiation dose. It is important to collect these essential data and to perform the analyses required to optimize the treatment approaches in order to ensure the best medical care for patients. In order to improve upon current results, combined modalities such as PRRT plus surgery are of interest. The use of radiolabeled peptides may be considered in an adjuvant setting to eradicate occult metastases, possibly originating from tumor spill, after surgery of receptor-positive tumors. Starting therapy at this early stage is favourable because of the small tumor burden. PRRT can, however, also be applied for cancer recurrence at a later stage. In patients with non-resectable tumor(s), PRRT can be applied as a first step to reduce tumor size or number, followed by surgery when possible. In conclusion, radiolabeled peptides have started a new era in nuclear oncology, not only for diagnosis but also for radionuclide therapy.

## REFERENCES

1. Kvols, L. K., Moertel, C. G., O'Connell, M. J., Schutt, A. J., Rubin, J., and Hahn, R. G. Treatment of the malignant carcinoid syndrome. Evaluation of a long-acting somatostatin analogue. N Engl J Med, *315:* 663-666, 1986.
2. Eriksson, B. and Oberg, K. Summing up 15 years of somatostatin analog therapy in neuroendocrine tumors: future outlook. Ann Oncol, *10:* S31-38, 1999.
3. Lamberts, S. W., Krenning, E. P., and Reubi, J. C. The role of somatostatin and its analogs in the diagnosis and treatment of tumors. Endocr Rev, *12:* 450-482, 1991.
4. Lamberts, S. W., Reubi, J. C., and Krenning, E. P. Somatostatin analogs in the treatment of acromegaly. Endocrinol Metab Clin North Am, *21:* 737-752, 1992.
5. Patel, Y. C., Greenwood, M. T., Panetta, R., Demchyshyn, L., Niznik, H., and Srikant, C. B. The somatostatin receptor family. Life Sci, *57:* 1249-1265, 1995.
6. Patel, Y. C. Somatostatin and its receptor family. Front Neuroendocrinol, *20:* 157-198, 1999.
7. Schonbrunn, A. Somatostatin receptors present knowledge and future directions. Ann Oncol, *10:* S17-21, 1999.
8. Krenning, E. P., Kwekkeboom, D. J., Oei, H. Y., de Jong, R. J., Dop, F. J., de Herder, W. W., Reubi, J. C., and Lamberts, S. W. Somatostatin receptor scintigraphy in carcinoids, gastrinomas and Cushing's syndrome. Digestion, *55:* 54-59, 1994.
9. Krenning, E. P., Kwekkeboom, D. J., Oei, H. Y., de Jong, R. J., Dop, F. J., Reubi, J. C., and Lamberts, S. W. Somatostatin-receptor scintigraphy in gastroenteropancreatic tumors. An overview of European results. Ann N Y Acad Sci, *733:* 416-424, 1994.
10. Krenning, E. P., Kwekkeboom, D. J., Bakker, W. H., Breeman, W. A., Kooij, P. P., Oei, H. Y., van Hagen, M., Postema, P. T., de Jong, M., Reubi, J. C., and et al. Somatostatin receptor scintigraphy with [111In-DTPA-D-Phe1]- and [123I- Tyr3]-octreotide: the Rotterdam experience with more than 1000 patients. Eur J Nucl Med, *20:* 716-731, 1993.
11. Krenning, E. P., Bakker, W. H., Breeman, W. A., Koper, J. W., Kooij, P. P., Ausema, L., Lameris, J. S., Reubi, J. C., and Lamberts, S. W. Localisation of endocrine-related tumours with radioiodinated analogue of somatostatin. Lancet, *1:* 242-244, 1989.

246

12.    Krenning, E. P., Bakker, W. H., Kooij, P. P., Breeman, W. A., Oei, H. Y., de
       Jong, M., Reubi, J. C., Visser, T. J., Bruns, C., Kwekkeboom, D. J., and et al.
       Somatostatin receptor scintigraphy with indium-111-DTPA-D-Phe-1- octreotide in
       man: metabolism, dosimetry and comparison with iodine-123- Tyr-3-octreotide. J Nucl
       Med, *33:* 652-658, 1992.
13.    Bakker, W. H., Albert, R., Bruns, C., Breeman, W. A., Hofland, L. J., Marbach, P.,
       Pless, J., Pralet, D., Stolz, B., Koper, J. W., and et al. [111In-DTPA-D-Phe1]-
       octreotide, a potential radiopharmaceutical for imaging of somatostatin receptor-
       positive tumors: synthesis, radiolabeling and in vitro validation. Life Sci, *49:* 1583-
       1591, 1991.
14.    Andersson, P., Forssell-Aronsson, E., Johanson, V., Wangberg, B., Nilsson, O.,
       Fjalling, M., and Ahlman, H. Internalization of indium-111 into human
       neuroendocrine tumor cells after incubation with indium-111-DTPA-D-Phe1-
       octreotide. J Nucl Med, *37:* 2002-2006, 1996.
15.    De Jong, M., Bernard, B. F., De Bruin, E., Van Gameren, A., Bakker, W. H., Visser,
       T. J., Macke, H. R., and Krenning, E. P. Internalization of radiolabelled
       [DTPA0]octreotide and [DOTA0,Tyr3]octreotide: peptides for somatostatin receptor-
       targeted scintigraphy and radionuclide therapy. Nucl Med Commun, *19:* 283-288,
       1998.
16.    Hofland, L. J., van Koetsveld, P. M., Waaijers, M., and Lamberts, S. W. Internalisation
       of isotope-coupled somatostatin analogues. Digestion, *57:* 2-6, 1996.
17.    Duncan, J. R., Stephenson, M. T., Wu, H. P., and Anderson, C. J. Indium-111-
       diethylenetriaminepentaacetic acid-octreotide is delivered in vivo to pancreatic, tumor
       cell, renal, and hepatocyte lysosomes. Cancer Res, *57:* 659-671, 1997.
18.    Hornick, C. A., Anthony, C. T., Hughey, S., Gebhardt, B. M., Espenan, G. D., and
       Woltering, E. A. Progressive nuclear translocation of somatostatin analogs. J Nucl
       Med, *41:* 1256-1263, 2000.
19.    Wang, M., Caruano, A. L., Lewis, M. R., Meyer, L. A., VanderWaal, R. P., and
       Anderson, C. J. Subcellular localization of radiolabeled somatostatin analogues:
       implications for targeted radiotherapy of cancer. Cancer Res, *63:* 6864-6869, 2003.
20.    Xue, L. Y., Butler, N. J., Makrigiorgos, G. M., Adelstein, S. J., and Kassis, A. I.
       Bystander effect produced by radiolabeled tumor cells in vivo. Proc Natl Acad Sci U S
       A, *99:* 13765-13770, 2002.
21.    Bernard, B. F., Krenning, E., Breeman, W. A., Visser, T. J., Bakker, W. H., Srinivasan,
       A., and de Jong, M. Use of the rat pancreatic CA20948 cell line for the comparison of
       radiolabelled peptides for receptor-targeted scintigraphy and radionuclide therapy.
       Nucl Med Commun, *21:* 1079-1085., 2000.
22.    Slooter, G. D., Breeman, W. A., Marquet, R. L., Krenning, E. P., and van Eijck, C. H.
       Anti-proliferative effect of radiolabelled octreotide in a metastases model in rat liver.
       Int J Cancer, *81:* 767-771, 1999.
23.    De Jong, M., Breeman, W. A., Bernard, H. F., Kooij, P. P., Slooter, G. D., Van Eijck,
       C. H., Kwekkeboom, D. J., Valkema, R., Macke, H. R., and Krenning, E. P. Therapy
       of neuroendocrine tumors with radiolabeled somatostatin- analogues. Q J Nucl Med,
       *43:* 356-366, 1999.
24.    Valkema, R., De Jong, M., Bakker, W. H., Breeman, W. A., Kooij, P. P., Lugtenburg,
       P. J., De Jong, F. H., Christiansen, A., Kam, B. L., De Herder, W. W., Stridsberg, M.,
       Lindemans, J., Ensing, G., and Krenning, E. P. Phase I study of peptide receptor
       radionuclide therapy with [In- DTPA]octreotide: the Rotterdam experience. Semin
       Nucl Med, *32:* 110-122., 2002.
25.    Fjalling, M., Andersson, P., Forssell-Aronsson, E., Gretarsdottir, J., Johansson, V.,
       Tisell, L. E., Wangberg, B., Nilsson, O., Berg, G., Michanek, A., Lindstedt, G., and
       Ahlman, H. Systemic radionuclide therapy using indium-111-DTPA-D-Phe1-
       octreotide in midgut carcinoid syndrome. J Nucl Med, *37:* 1519-1521, 1996.
26.    Anthony, L. B., Woltering, E. A., Espenan, G. D., Cronin, M. D., Maloney, T. J., and
       McCarthy, K. E. Indium-111-pentetreotide prolongs survival in gastroenteropancreatic
       malignancies. Semin Nucl Med, *32:* 123-132., 2002.

27.  McCarthy, K. E., Woltering, E. A., and Anthony, L. B. In situ radiotherapy with 111In-pentetreotide. State of the art and perspectives. Q J Nucl Med, *44:* 88-95, 2000.

28.  McCarthy, K. E., Woltering, E. A., Espenan, G. D., Cronin, M., Maloney, T. J., and Anthony, L. B. In situ radiotherapy with 111In-pentetreotide: initial observations and future directions. Cancer J Sci Am, *4:* 94-102, 1998.

29.  de Jong, M., Breeman, W. A., Bakker, W. H., Kooij, P. P., Bernard, B. F., Hofland, L. J., Visser, T. J., Srinivasan, A., Schmidt, M. A., Erion, J. L., Bugaj, J. E., Macke, H. R., and Krenning, E. P. Comparison of (111)In-labeled somatostatin analogues for tumor scintigraphy and radionuclide therapy. Cancer Res, *58:* 437-441, 1998.

30.  Reubi, J. C., Schar, J. C., Waser, B., Wenger, S., Heppeler, A., Schmitt, J. S., and Macke, H. R. Affinity profiles for human somatostatin receptor subtypes SST1-SST5 of somatostatin radiotracers selected for scintigraphic and radiotherapeutic use. Eur J Nucl Med, *27:* 273-282, 2000.

31.  de Jong, M., Breeman, W. A., Bernard, B. F., Bakker, W. H., Schaar, M., van Gameren, A., Bugaj, J. E., Erion, J., Schmidt, M., Srinivasan, A., and Krenning, E. P. [177Lu-DOTA(0),Tyr3] octreotate for somatostatin receptor-targeted radionuclide therapy. Int J Cancer, *92:* 628-633., 2001.

32.  Schmitt, A., Bernhardt, P., Nilsson, O., Ahlman, H., Kolby, L., Schmitt, J., and Forssel-Aronsson, E. Biodistribution and dosimetry of 177Lu-labeled [DOTA0,Tyr3]octreotate in male nude mice with human small cell lung cancer. Cancer Biother Radiopharm, *18:* 593-599, 2003.

33.  Schmitt, J. S., Wild, D., Ginj, M., Reubi, J. C., Waser, B., De Jong, M., Bernard, H. F., Krenning, E. P., and Maecke, H. R. DOTA-NOC, a high affinity ligand of the somatostatin receptor subtypes 2,3 and 5 for radiotherapy. J Labelled Cpd Radiopharm, *44:* s697-s699, 2001.

34.  Wild, D., Schmitt, J. S., Ginj, M., Macke, H. R., Bernard, B. F., Krenning, E., De Jong, M., Wenger, S., and Reubi, J. C. DOTA-NOC, a high-affinity ligand of somatostatin receptor subtypes 2, 3 and 5 for labelling with various radiometals. Eur J Nucl Med Mol Imaging, *30:* 1338-1347, 2003.

35.  de Jong, M., Rolleman, E. J., Bernard, B. F., Visser, T. J., Bakker, W. H., Breeman, W. A., and Krenning, E. P. Inhibition of renal uptake of indium-111-DTPA-octreotide in vivo. J Nucl Med, *37:* 1388-1392, 1996.

36.  Bernard, B. F., Krenning, E. P., Breeman, W. A., Rolleman, E. J., Bakker, W. H., Visser, T. J., Macke, H., and de Jong, M. D-lysine reduction of indium-111 octreotide and yttrium-90 octreotide renal uptake. J Nucl Med, *38:* 1929-1933, 1997.

37.  Rolleman, E. J., Valkema, R., de Jong, M., Kooij, P. P., and Krenning, E. P. Safe and effective inhibition of renal uptake of radiolabelled octreotide by a combination of lysine and arginine. Eur J Nucl Med Mol Imaging, *30:* 9-15, 2003.

38.  Bodei, L., Cremonesi, M., Zoboli, S., Grana, C., Bartolomei, M., Rocca, P., Caracciolo, M., Macke, H. R., Chinol, M., and Paganelli, G. Receptor-mediated radionuclide therapy with 90Y-DOTATOC in association with amino acid infusion: a phase I study. Eur J Nucl Med Mol Imaging, *30:* 207-216, 2003.

39.  Kwekkeboom, D. J., Bakker, W. H., Kooij, P. P. M., Konijnenberg, M. W., Srinivasan, A., Erion, J. L., Schmidt, M. A., Bugaj, J. E., de Jong, M., and Krenning, E. P. [177Lu-DOTA0,Tyr3]octreotate: comparison with [111In-DTPA0]octreotide in patients. Eur J Nucl Med, *28:* 1319-1325, 2001.

40.  Kwekkeboom, D. J., Kooij, P. P., Bakker, W. H., Macke, H. R., and Krenning, E. P. Comparison of 111In-DOTA-Tyr3-octreotide and 111In-DTPA-octreotide in the same patients: biodistribution, kinetics, organ and tumor uptake. J Nucl Med, *40:* 762-767., 1999.

41.  de Jong, M., Breeman, W. A., Bernard, B. F., Bakker, W. H., Visser, T. J., Kooij, P. P., van Gameren, A., and Krenning, E. P. Tumor Response After [(90)Y-DOTA(0),Tyr(3)]Octreotide Radionuclide Therapy in a Transplantable Rat Tumor Model Is Dependent on Tumor Size. J Nucl Med, *42:* 1841-1846., 2001.

42.  Anderson, C. J., Jones, L. A., Bass, L. A., Sherman, E. L., McCarthy, D. W., Cutler, P. D., Lanahan, M. V., Cristel, M. E., Lewis, J. S., and Schwarz, S. W. Radiotherapy,

toxicity and dosimetry of copper-64-TETA-octreotide in tumor-bearing rats. J Nucl Med, *39:* 1944-1951, 1998.

43.  Lewis, J. S., Lewis, M. R., Cutler, P. D., Srinivasan, A., Schmidt, M. A., Schwarz, S. W., Morris, M. M., Miller, J. P., and Anderson, C. J. Radiotherapy and dosimetry of 64Cu-TETA-Tyr3-octreotate in a somatostatin receptor-positive, tumor-bearing rat model. Clin Cancer Res, *5:* 3608-3616, 1999.

44.  Stolz, B., Smith-Jones, P., Albert, R., Tolcsvai, L., Briner, U., Ruser, G., Macke, H., Weckbecker, G., and Bruns, C. Somatostatin analogues for somatostatin-receptor-mediated radiotherapy of cancer. Digestion, *57:* 17-21., 1996.

45.  Stolz, B., Weckbecker, G., Smith-Jones, P. M., Albert, R., Raulf, F., and Bruns, C. The somatostatin receptor-targeted radiotherapeutic [90Y-DOTA-DPhe1, Tyr3]octreotide (90Y-SMT 487) eradicates experimental rat pancreatic CA 20948 tumours. Eur J Nucl Med, *25:* 668-674, 1998.

46.  Breeman, W. A., Mearadji, A., Capello, A., Bernard, B. F., van Eijck, C. H., Krenning, E. P., and de Jong, M. Anti-tumor effect and increased survival after treatment with [177Lu-DOTA0,Tyr3]octreotate in a rat liver micrometastases model. Int J Cancer, *104:* 376-379, 2003.

47.  Erion, J. L., Bugaj, J. E., Schmidt, M. A., Wilhelm, R. R., and Srinivasan, A. High radiotherapeutic efficacy of [Lu-177]-DOTA-Y3-octreotate in a rat tumor model. J Nucl Med, *40:* 223p, 1999.

48.  De Jong, M., Bernard, H. F., Breeman, W. A. P., van Gameren, A., and Krenning, E. P. Combination of $^{90}$Y- and $^{177}$Lu-labeled somatostatin analogs is superior for radionuclide therapy compared to $^{90}$Y- or $^{177}$Lu-labeled analogs only. J Nucl Med, *43:* 123-124P, 2002.

49.  Otte, A., Herrmann, R., Heppeler, A., Behe, M., Jermann, E., Powell, P., Maecke, H. R., and Muller, J. Yttrium-90 DOTATOC: first clinical results. Eur J Nucl Med, *26:* 1439-1447, 1999.

50.  Otte, A., Herrmann, R., Macke, H. R., and Muller-Brand, J. [Yttrium 90 DOTATOC: a new somatostatin analog for cancer therapy of neuroendocrine tumors]. Schweiz Rundsch Med Prax, *88:* 1263-1268, 1999.

51.  Otte, A., Mueller-Brand, J., Dellas, S., Nitzsche, E. U., Herrmann, R., and Maecke, H. R. Yttrium-90-labelled somatostatin-analogue for cancer treatment [letter]. Lancet, *351:* 417-418, 1998.

52.  Waldherr, C., Pless, M., Maecke, H. R., Haldemann, A., and Mueller-Brand, J. The clinical value of [90Y-DOTA]-D-Phe1-Tyr3-octreotide (90Y-DOTATOC) in the treatment of neuroendocrine tumours: a clinical phase II study. Ann Oncol, *12:* 941-945., 2001.

53.  Waldherr, C., Pless, M., Maecke, H. R., Schumacher, T., Crazzolara, A., Nitzsche, E. U., Haldemann, A., and Mueller-Brand, J. Tumor response and clinical benefit in neuroendocrine tumors after 7.4 GBq $^{90}$Y-DOTATOC. J Nucl Med, *in press*, 2002.

54.  Waldherr, C., Schumacher, T., Maecke, H. R., Schirp, U., Forrer, F., Nitzsche, E. U., and Mueller-Brand, J. Does tumor response depend on the number of treatment sessions at constant injected dose using 90Yttrium-DOTATOC in neuroendocrine tumors? Eur J Nucl Med, *29:* S100, 2002.

55.  Paganelli, G., Zoboli, S., Cremonesi, M., Bodei, L., Ferrari, M., Grana, C., Bartolomei, M., Orsi, F., De Cicco, C., Macke, H. R., Chinol, M., and de Braud, F. Receptor-mediated radiotherapy with 90Y-DOTA-D-Phe1-Tyr3-octreotide. Eur J Nucl Med, *28:* 426-434., 2001.

56.  Paganelli, G., Zoboli, S., Cremonesi, M., Macke, H. R., and Chinol, M. Receptor-mediated radionuclide therapy with 90Y-DOTA-D-Phe1-Tyr3- Octreotide: preliminary report in cancer patients. Cancer Biother Radiopharm, *14:* 477-483., 1999.

57.  Chinol, M., Bodei, L., Cremonesi, M., and Paganelli, G. Receptor-mediated radiotherapy with Y-DOTA-DPhe-Tyr-octreotide: the experience of the European Institute of Oncology Group. Semin Nucl Med, *32:* 141-147., 2002.

58.  Smith, M. C., Liu, J., Chen, T., Schran, H., Yeh, C. M., Jamar, F., Valkema, R., Bakker, W., Kvols, L., Krenning, E., and Pauwels, S. OctreoTher: ongoing early

clinical development of a somatostatin- receptor-targeted radionuclide antineoplastic therapy. Digestion, *62:* 69-72., 2000.

59. Valkema, R., Kvols, L., Jamar, F., Bakker, W. H., Smith, C., Krenning, E. P., and Pauwels, S. Phase 1 study of therapy with 90Y-SMT487 (OctreoTher) in patients with somatostatin receptor-positive tumors. J Nucl Med, *43:* 33P, 2002.

60. Jonard, P., Jamar, F., Walrand, S., Collart, J. P., Valkema, R., Bakker, W. H., Labar, D., Smith, C., Kvols, L., Krenning, E. P., and Pauwels, S. Tumor dosimetry based on PET $^{86}$Y-DOTA-Tyr$^3$-octreotide (SMT487) and CT-scan predicts tumor response to $^{90}$Y-SMT487 (OctreoTher). J Nucl Med, *41:* 111P, 2000.

61. Bushnell, D., O'Dorisio, T., Menda, Y., Carlisle, T., Zehr, P., Connolly, M., Karwal, M., Miller, S., Parker, S., and Bouterfa, H. Evaluating the clinical effectiveness of 90Y-SMT 487 in patients with neuroendocrine tumors. J Nucl Med, *44:* 1556-1560, 2003.

62. Bushnell, D., Menda, Y., Madsen, M., O'Dorisio, T., Carlisle, T., Zehr, P., Ponto, L., Karwal, M., Parker, S., Ponto, J., Connolly, M., and Bouterfa, H. Assessment of hepatic toxicity from treatment with 90Y-SMT 487 (OctreoTher in patients with diffuse somatostatin receptor positive liver metastases. Cancer Biother Radiopharm, *18:* 581-588, 2003.

63. Schumacher, T., Hofer, S., Eichhorn, K., Wasner, M., Zimmerer, S., Freitag, P., Probst, A., Gratzl, O., Reubi, J. C., Maecke, R., Mueller-Brand, J., and Merlo, A. Local injection of the 90Y-labelled peptidic vector DOTATOC to control gliomas of WHO grades II and III: an extended pilot study. Eur J Nucl Med Mol Imaging, *29:* 486-493., 2002.

64. Kwekkeboom, D. J., Bakker, W. H., Kam, B. L., Teunissen, J. J., Kooij, P. P., De Herder, W. W., Feelders, R. A., Van Eijck, C. H., De Jong, M., Srinivasan, A., Erion, J. L., and Krenning, E. P. Treatment of patients with gastro-entero-pancreatic (GEP) tumours with the novel radiolabelled somatostatin analogue [(177)Lu-DOTA(0),Tyr(3)]octreotate. Eur J Nucl Med Mol Imaging, *30:* 417-422., 2003.

65. Virgolini, I., Traub, T., Novotny, C., Leimer, M., Fuger, B., Li, S. R., Patri, P., Pangerl, T., Angelberger, P., Raderer, M., Burggasser, G., Andreae, F, Kurtaran, A., and Dudczak, R. Experience with indium-111 and yttrium-90-labeled somatostatin analogs. Curr Pharm Des, *8:* 1781-1807, 2002.

66. Virgolini, I., Szilvasi, I., Kurtaran, A., Angelberger, P., Raderer, M., Havlik, E., Vorbeck, F., Bischof, C., Leimer, M., Dorner, G., Kletter, K., Niederle, B., Scheithauer, W., and Smith-Jones, P. Indium-111-DOTA-lanreotide: biodistribution, safety and radiation absorbed dose in tumor patients. J Nucl Med, *39:* 1928-1936., 1998.

67. Virgolini, I., Britton, K., Buscombe, J., Moncayo, R., Paganelli, G., and Riva, P. In- and Y-DOTA-lanreotide: results and implications of the MAURITIUS trial. Semin Nucl Med, *32:* 148-155., 2002.

68. Virgolini, I., Kurtaran, A., Angelberger, P., Raderer, M., Havlik, E., and Smith-Jones, P. "MAURITIUS": tumour dose in patients with advanced carcinoma. Ital J Gastroenterol Hepatol, *31 Suppl 2:* S227-230., 1999.

69. Smith-Jones, P. M., Bischof, C., Leimer, M., Gludovacz, D., Angelberger, P., Pangerl, T., Peck-Radosavljevic, M., Hamilton, G., Kaserer, K., Kofler, A., Schlangbauer-Wadl, H., Traub, T., and Virgolini, I. DOTA-lanreotide: a novel somatostatin analog for tumor diagnosis and therapy. Endocrinology, *140:* 5136-5148., 1999.

70. Froidevaux, S., Heppeler, A., Eberle, A. N., Meier, A. M., Hausler, M., Beglinger, C., Behe, M., Powell, P., and Macke, H. R. Preclinical comparison in AR4-2J tumor-bearing mice of four radiolabeled 1,4,7,10-tetraazacyclododecane-1,4,7,10-tetraacetic acid- somatostatin analogs for tumor diagnosis and internal radiotherapy. Endocrinology, *141:* 3304-3312., 2000.

71. Leimer, M., Kurtaran, A., Smith-Jones, P., Raderer, M., Havlik, E., Angelberger, P., Vorbeck, F., Niederle, B., Herold, C., and Virgolini, I. Response to treatment with yttrium 90-DOTA-lanreotide of a patient with metastatic gastrinoma. J Nucl Med, *39:* 2090-2094., 1998.

# 16
# SOMATOSTATIN ANALOGUE THERAPY OF NEURO-ENDOCRINE GASTRO-ENTERO PANCREATIC TUMORS

Kjell Öberg
*Department of Medicine, Uppsala University Hospital, S-751 85 Uppsala, Sweden*

## INTRODUCTION

Neuroendocrine gastrointestinal tumors derive from the neuroendocrine cell system and have widely divergent clinical presentations. These tumors are rare, the most common type, the carcinoid, occurs in 2.8-21 per million the incidence of carcinoid syndrome is about 0.5/100.000 and for endocrine pancreatic tumors 0.4/100.000. (Norheim et al., 1987, Eriksson et al., 1990) The neuroendocrine gastrointestinal tumors can be devided into several categories. One grouping is in foregut, midgut and hindgut carcinoids, where foregut tumors include thymic, bronchial, gastric, pancreatic and duodenal neuroendocrine tumours, whereas midgut tumors contain ileal, appendecial and tumors of the proximal colon. Hindgut carcinoids contain tumors of the rest of colon, sigmoideum and rectum (Williams and Sandler, 1962). This classification was originally done for carcinoid tumors but is sometimes used for all types of neuroendocrine GEP-tumors. A second grouping is carcinoid tumors versus endocrine pancreatic tumors and a third alternative is functioning, tumors versus non-functioning tumors. Functional tumors usually present with hormone related clinical symptoms, whereas non-functioning tumors although producing different secretory products, they are not giving rise to any clinical symptoms (Öberg, 1996). All these classifications, which have been based on anatomical localizations or clinical symptoms, have been questioned lately and to eliminate the present confusion, the term "carcinoid" might in the future only be used to designate a classical midgut neuroendocrine tumor with the carcinoid syndrome. Other tumors should be assigned the term

**Key words**: Neuroendocrine GEP-tumors, carcinoids, somatostatin analogue, somatostatin receptors, octreotide, lanreotide, combination therapy acromegaly, prolactinomas, Cushing's disease, clinically nonfunctioning adenomas

"neuroendocrine tumor", followed by their primary location, e.g. neuro-endocrine lung, gastric, duodenal, pancreatic, colonic and rectal tumors. The dominant hormone production may sometimes be included, e.g. somatostatin (SST) producing neuroendocrine duodenal tumor etc...

Multiple endocrine neoplasia type 1 (MEN-1) is a familial disorder, inherited as an autosomal dominant trait with variable penetrance pattern. A specific genetic lesion has been described for MEN-1 with deletion of chromosome 11q13 and mutations of MEN-1 gene. About 80% of affected MEN-1 members develop endocrine pancreatic tumors and furthermore, about 30% of all gastrin producing neuroendocrine gastrointestinal tumors are related to MEN-1 syndrome. Some of these patients also develop lung, duodenal and gastric carcinoids (Larsson et al., 1988; Skogseid et al., 1991).

Neuroendocrine gut and pancreatic tumors present various clinical symptoms related to hormone production. The most common clinical syndrome is the carcinoid syndrome, which may be present in patients with midgut carcinoid tumors with liver metastases, as well as in some patients with foregut carcinoid tumors. The syndrome consists of flushing, diarrhea, carcinoid heart disease with right heart failure, bronchial constriction and elevated levels of the serotonin metabolite 5-hydroxy-indole acetic acid. The tumors also release tachykinins, bradykinins and prostaglandins participating in the clinical syndrome. 40% of all carcinoid tumors are located in the midgut and of these a majority are malignant (Feldman, 1987). Pancreatic endocrine tumors are classified as functioning, if they are associated with the clinical syndrome related to hormone production and are considered non-functioning if they are not associated with the clinical symptoms of hormone release. The latter category constitutes around 30% of all endocrine pancreatic tumors and include tumor secreting pancreatic polypeptide, chromogranin-A, peptide-YY and neurotensin (Eriksson et al., 1990). The two most frequent clinical syndromes related to endocrine pancreatic tumors are the Zollinger-Ellison syndrome, caused by gastrin overproduction and the hypoglycemic syndrome, which is related to high insulin/proinsulin release. The Zollinger-Ellison syndrome, or gastrinoma syndrome, can also be confined to gastrin production by duodenal carcinoids (about 40%) and more than 70% of gastrin producing tumors are malignant with early lymph node involvement. Clinical symptoms related to gastrin production are gastritis, diarrhea and malabsorption but severe gastric ulcer disease is rare today due to extensive use of $H_2$-receptor blockers and proton pump inhibitors (Jensen, 1983). The second most frequent endocrine pancreatic tumor is insulinoma or hyperglycemic syndrome related to increased production and release of insulin and pro-insulin. Eighty percent of these tumors are benign solitary tumors in the pancreas and the patient can be cured by surgery. Ten to twenty percent of the tumors are malignant with distant metastases and produce very often hormones other than insulin/proinsulin, such as gastrin, ACTH and glucagon. Typical symptoms related to insulin overproduction are signs of neuroglucopenia and increased catecholamine release (Fajans and Vinik, 1989). Other functioning endocrine pancreatic tumors are VIP producing tumors causing the so called Verner-Morrison syndrome, or WDHA-syndrome, accompanied by extensive diarrhea, hypokalemia and achlorhydria. In such patients the stool volume might reach more than 10 litres per day and they are often requiring intensive care. The tumors are confined to the pancreas but sometimes also in the lung and sympathic ganglia (Long et al., 1981). Another rare clinical syndrome is the glucagonoma syndrome with the typical necrolytic migratory erythema, diabetic glucose tolerance, anaemia weight loss and thromboembolism, all signs related to glucagon production and the effects of glucagon (Stacpoole, 1981). SST-producing tumors can be both functional and non-functional. Syndromes with gall bladder dysfunction, gall stones and diabetic glucose tolerance, malabsorption and diarrhea, are sometimes related to increased levels of circulating somatostatin (Krejs et al., 1979). However, an increased level of plasma somatostatin is quite often found in patients with the so called "non-functioning islet cell tumors" (Eckhauser et al., 1986).

Forgut carcinoids comprise 15% of all carcinoid tumors and include thymic and lung carcinoids. They can in principle present with any known endocrine syndrome such as Cushing's Syndrome, acromegaly, carcinoid syndrome and SIADH. Duodenal carcinoids sometimes secrete gastrin, somatostatin and glucagon and produce related symptoms (Wilander et al., 1989). The majority of gastric carcinoids are related to chronic atrophic gastritis (CAG) and achlorhydria and due to increased gastrin production from antral G-cells, proliferation of ECL-cells with secondary secretion of histamine. These tumors are sometimes related to a different type of carcinoid flushing (Rindi et al., 1993). Midgut carcinoids present the carcinoid syndrome in 60-80% of the cases. Serotonin, bradykinins, tachykinins and prostaglandins are mediators of the syndrome (Feldman, 1987). Hindgut carcinoids constitute 20% of the carcinoids and belong to the group of non-functioning neuroendocrine tumors. Despite the production of hormones, such as chromogranin-A, PYY, HCG-α and -β subunits, they produce no related clinical symptoms (Spread et al., 1994). Appendicial carcinoids belonging to the group of midgut carcinoids are benign and belong to the group of non-functioning tumors (Öberg, 1996).

Neuroendocrine gastrointestinal tumors usually present with metastatic inoperable disease and more or less severe hormonal symptoms. A multimodal therapeutic approach is then warranted and the aims of the treatment is to control hormone symptoms, reduce circulating hormone levels, prevent further tumor growth and possibly also achieve tumor reduction, prolonged survival and improved quality of life. Surgery has been a cornerstone in the management of the neuroendocrine tumors but due to high frequency of malignant disease, surgery is seldom curative but can be of benefit due to debulking and by-passing procedures (Makridis et al., 1990). Other means of tumor reduction is embolization and chemo-embolization of liver metastases (Carrasco et al., 1983), cryotherapy and radio-frequency ablation. Selective

254

cases can also be subjected to liver transplantation. External irradiation can be used to alleviate symptoms from bone and brain metastases and more recently, tumor targeted radioactive treatment with the [111]Indium-DTPA-Octreotide and [90]Ytrium-DOTA-Octreotide has been performed (Otte et al, 1998). In earlier trials also [131]Iodine and [125]Iodine-MIBG has been used for treatment of particularly carcinoid tumors (Hoefnagel et al., 1987). Chemotherapy, i.e. combinations with streptozotocin, can produce objective responses in endocrine pancreatic tumors but is of little benefit in carcinoid tumors (Moertel et al., 1994). Alpha inteferon has been used in both EPT´s and carcinoids with responses in 40-50% of the patients (Eriksson and Öberg, 1991).


## SOMATOSTATIN ANALOGUE THERAPY

SST analogues, octreotide, being the first analogue available for clinical use, have become increasingly important in the management of patients with neuroendocrine gut and pancreatic tumors. Today, octreotide and lanreotide, both octapeptides (Fig. 1), are registered in most countries for the control of symptoms associated with metastatic carcinoid tumors as well as VIP and glucagon producing endocrine pancreatic tumors. The rational for the effect of these agents has been elucidated by the demonstration of SST receptors (SSTR) in 80-90% of tumors by immunohistochemistry, autoradiography and

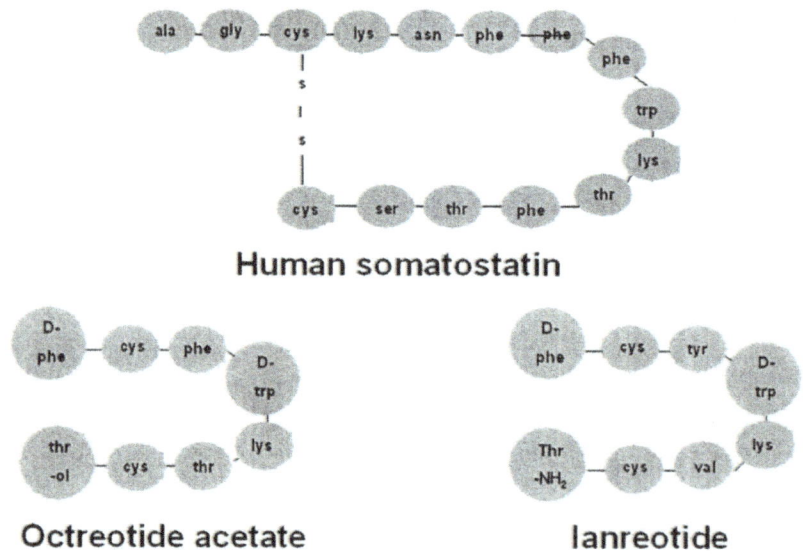

*Figure 1: Structure of human SST- 14 and the SST analogues octreotide and lanreotide.*

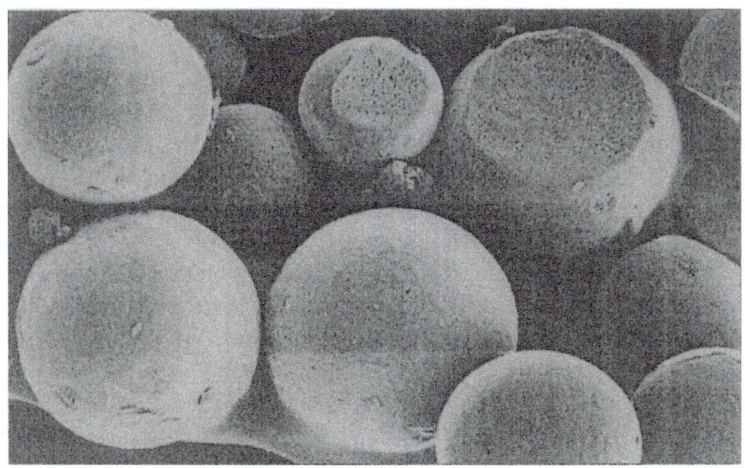

*Figure 2: Microspheres of biodegradable polymers (Sandostatin-LAR).*

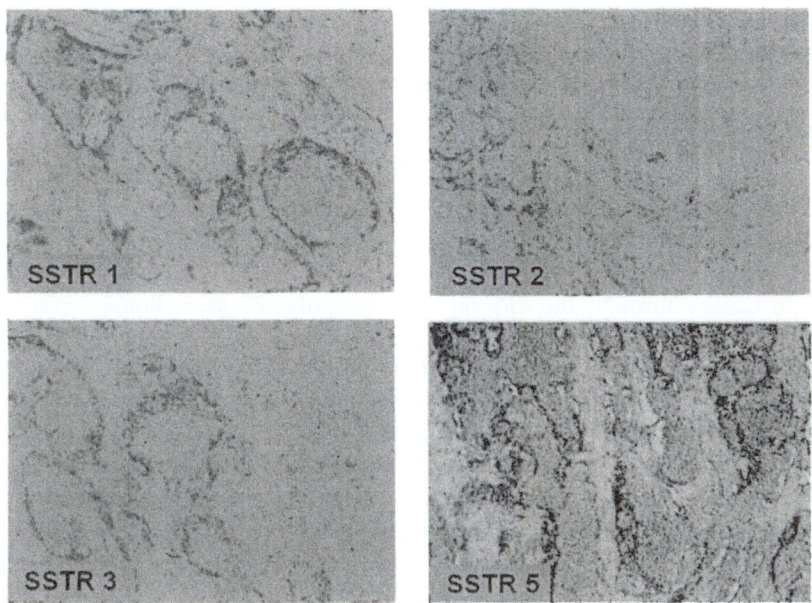

*Figure 3: Immunohistochemistry with four different specific somatostatin receptor antibodies (sst1, 2, 3, 5) in a midgut carcinoid tumor. Note the uneven distribution of the different receptor subtypes and particularly the low expression of sst2 in large areas of the tumor.*

octreotide scintigraphy, the latter method having emerged as a routine diagnostic tool for staging of patients and also predictive test for sensitivity to treatment with somatostatin analogues (Reubi et al., 1990; Krenning et al., 1993; Tiensuu-Janson et al., 1994). The SST analogues in clinical practice today, octreotide and lanreotide, both bind with high affinity to somatostatin receptor 2 and 5 and with lower affinity to receptor 3 (see chapter 6). SST and its analogues exert their cellular action through interaction with its specific cell and transmembrane receptor, which belong to the super family of G-protein coupled membrane receptors. It has been shown that SSTRs are linked to intracellular signaling pathways through pertussis toxin sensitive and pertussis toxin insensitive $G_i/G_0$ alpha G-protein subunits (Patel, 1999).

The beneficial effect of native somatostatin, which binds to all the five different subtypes of SST receptors, in blocking the carcinoid flush induced by pentagastrin, reducing circulating levels of serotonin and controlling other symptoms associated with the carcinoid syndrome, were described by Thulin et al and Frölich et al in 1978. However, the clinical use of native SST-14 was hampered by its short half-life of only two minutes, which necessitated continuous intravenous infusion and rebound phenomena that occur after withdrawal of infusion. In 1982, Bauer et al reported on the synthesis of the long-acting octapeptide analogue, octreotide (SMS-201-995), which retain the four amino acid sequence presumed to be essential for the biological activity and was cyclized with a distal disulfide bridge to prevent degradation, with a half-life of two hours that could be given subcutaneous 2-3 times daily (Fig. 1).

Both Sandostatin and lanreotide have been incorporated in bio-degradable microspheres, giving slow release formulations. Sandostatin-LAR is capsulated in poly (D,L-lactide-CO-glykolide-glucose) (Fig. 2). Slow release of octreotide acetate occurs through the cleavage of the polymer ester linkage primarily through tissue fluid hydrolysis. Administration is by intramuscular depot injection every 28 days in contrast to the subcutaneous octreotide acetate every eight hours. Lanreotide is also available in a slow release formulation. Lanreotide-PR obtained by incorporating lanreotide into polyactide-polyglycolid co-polymer microspheres. The duration of efficacy is two weeks after intramuscular injection of 30 mg. Both these formulations have been shown to reliably control symptoms in patients with the carcinoid syndrome.

## EXPRESSION OF SOMATOSTATIN RECEPTORS

All the five subtypes of SSTRs are expressed in neuroendocrine GEP-tumors. However, expression varies in every tumor, but receptor subtypes 2 and 5 are predominantly expressed in most tumors. There are, however, patients with carcinoid tumors that express all five subtypes of somatostatin receptors, including receptors 1, 2, 3 and 5. Insulin producing tumors present lower expression of SSTRs. Only about 50% are positive on octreoscan. Nowadays, the expression of SSTR receptor proteins can be visualized by specific antibodies to the different subtypes (Fig. 3). In this patient with a carcinoid of midgut type, all five receptor subtypes were expressed in the same tumor but in different areas of the tumor. Note the low expression of $sst_2$.

## TREATMENT WITH SOMATOSTATIN ANALOGUES

SST analogues, octreotide being the first available for clinical use, have become increasingly important in the management of patients with neuroendocrine gut and pancreatic tumors. The efficacy of somatostatin analogue therapy could be divided into subjective, biochemical and tumor responses (tumor size reduction). A complete remission is rarely obtained but partial remisson is more than 50% reduction of clinical symptoms, biochemical markers or tumor size; stable disease is less than 50% reduction of these parameters but also less than 25% increase of the same parameters. Progressive disease is more than 25% increase of clinical symptoms, biochemical markers or tumor size (according to WHO).

### Carcinoids

There is by now a large number of reports on longterm treatment with octreotide in patients with carcinoid tumors. In the first trial reported by Kvols et al (1986), octreotide given subcutaneously (150 µg t.i.d.) elicited symptomatic responses in 88% and biochemical responses in 72% of 25 patients with carcinoid tumors. The median duration of biochemical response was 12 months. Fifteen of the patients in this study had received prior chemotherapy. In 1989 Gorden et al, performed the first meta-analysis of the reported cases in the literature. Eighty cases of carcinoid tumors had been treated of which 62 had been treated for 1 to 18 months with 50-500 µg 2-3 times a day subcutaneously. Symptoms had improved in 54/59 patients (92%) and were unchanged in 5 cases. A biochemical response had been achieved in 48/71 patients (66%) and stabilization in 17 (24%). Reduction of tumor size was noted in 4/48 carcinoids. Unchanged tumor size was noted in 41/48 (85%) and tumor progression in one. This report did not offer information about duration of disease from diagnosis to start of octreotide, previous treatments or the duration of response to octreotide. However, 31/80 patients developed recurrent symptoms and some of them responded to increased doses. This illustrates a loss of therapeutic response or tachyphylaxis (Wynick et al., 1989), which can be circumvented temporarily by a dose increase. Compilation of our own data in 228 patients with carcinoid tumors, symptomatic improvement was obtained in 64%, biochemical response in 66% (complete and partial remissions), radiological response in 6%, stable disease in 38% and progressive disease in 56%. These patients had received subcutaneous octreotide 100-1500 µg/day, see Table 1 (Öberg, 2001; Eriksson and Öberg, 1999). In 1995, Harris and Redfern performed a meta-analysis of data compiled from 62 published studies on patients with carcinoids in order to examine the relationship between the dose of octreotide and the clinical efficacy evaluated by analyzing U-5-HIAA, flushing and diarrhea in patients.

Six dose ranges of octreotide were assessed and ranged from 100-3,000 µg/day.

**Table 1.** *Analysis of responses to somatostatin analogue therapy in different cohorts of carcinoid patients*

| Response | Standard doses (s.c.) 100-1500 µg/d N=228 | High dose (s.c.) 9-15 mg/d N=42 | Slow release (i.m.) 20-30 mg/2-4 w N=52 |
|---|---|---|---|
| Symptomatic | 64% | 50% | 54% |
| Biochemical | | | |
| CR + PR | 66% | 72% | 48% |
| SD | 25% | 25% | 40% |
| PD | 9% | 13% | 12% |
| Tumor size | | | |
| CR | - | 2% | - |
| PR | 5% | 11% | 5% |
| SD | 38% | 47% | 76% |
| PD | 56% | 39% | 19% |

CR = complete remission; PR = partial response; SD = stable disease; PD = progressive disease

The maximum clinical response to octreotide occurred in a patient treated with 300-375 µg/day, some further improvement in doses up to 1,000 µg/day. Doses of octreotide above 10,000 µg showed no additional clinical benefit. The author's conclusion from this study is that there was a significant patient to patient variation in the sensitivity to octreotide treatment and that it is important to titrate the dose of the analogue in each patient, until adequate symptom or biochemical control are achieved. Low dose octreotide therapy (50 µg t.i.d.) resulted in significantly lower biochemical response rates (30%) compared to the regular dose of analogue treatment (Öberg, et al., 1986). Carcinoid crisis can be prevented by subcutaneous or intravenous somatostatin analogues pre-, intra- and postoperatively (Kvols, et al., 1985).

**Endocrine Pancreatic Tumors**

**Insulinomas**
When octreotide (50 µg twice a day) was given preoperatively, it had inconsistent effects on insulin and glucose levels (Maton, et al, 1989). In some patients, possibly because of more profound suppression of GH and glucagon than tumor produced insulin/proinsulin, the hyperglycemia was worsened. Patients with metastatic insulinomas given octreotide over longterm improved symptomatically and biochemically in 50% of cases which correspond to later report of only 50% expression of somatostatin receptor type 2 (Scarpignato, 1995; Lamberts et al., 1991). In clinical practice, somatostatin analogue

treatment in insulinomas should not be done on an out-patient basis, since the response to this treatment is hard to predict.

## Gastrinomas

More than 50 patients with gastrin producing tumors have been treated with doses of octreotide 100-1500 µg/day, most of them on short term basis. Of these 90% had a good clinical response with control of gastric hypersecretion, pain and diarrhea (Maton, et al., 1989). A significant drop in serum gastrin and basal acid secretion was obtained. In a long term study by Ruszniewski and co-workers, the maximal acid output decreased during 9-12 months of treatment suggesting an anti-trophic effect (i.e. reduction of parietal cell mass) of octreotide. The efficacy of octreotide in lowering the number of antral G-cells and its effect on long-lasting hypergastrinemia (i.e. ECL-cell hyperplasia) are currently being evaluated. However, octreotide or other somatostatin analogues has an undefined role in gastrinomas because of their efficient and more convenient oral treatment with proton pump inhibitors. With the recent development of slow release formulations of somatostatin analogues, there may be subgroups of malignant gastrinomas that could benefit from this treatment.

## VIPomas

In patients with VIP producing tumors, not curable by surgery, octreotide is now being regarded as the treatment of choice (Maton et al., 1989; Scarpignato, 1995). Symptomatic improvement occurs in more than 80% of patients at doses of 100-450 µg/day. However, in some patients the beneficial effect lasted only a few days requiring increases in dose. Biochemical responses could be noted in about 80% of patients. Symptomatic relief was not always related to the reduction in plasma concentration of VIP, indicating that octreotide has a direct effect on the gut. In some cases, octreotide has been shown to change the molecular form of circulating VIP, possibly into less bioactive forms. Octreotide and other somatostatin analogues can be combined with chemotherapy for longterm management of VIP producing malignant tumors (Wood et al., 1985).

## Glucagonomas

In about 90% of patients with glucagonomas who presented a skin rash when treatment was initiated, octreotide cleared it in a few days. Other symptoms in the syndrome such as weight loss and diarrhea can improve but octreotide has varying effects on diabetes. Plasma glucagon was reduced in 60% of patients (Scarpignato, 1995). The symptomatic response is independent of plasma glucagon concentrations suggesting a direct effect of octreotide on the skin. Also in glucagonomas, octreotide can change circulating molecular forms of glucagon indicating inhibition of posttranslational processing of pre-proglucagon and thereby reducing circulating bioactive glucagon (Jockenhovel et al., 1994).

## Somatostatinomas

The small number of patients with somatostatin producing tumors has been treated with octreotide and some of them have improved symptomatically and biochemically, which correlated with the presence of somatostatin receptors in the tumors evidenced by octreoscan (Angeletti et al., 1998).

## ANTIPROLIFERATIVE EFFECTS OF OCTREOTIDE

The inhibition of tumor growth in patients with carcinoids and endocrine pancreatic tumors has been reported to be low in most studies. In a study by Saltz et al., 1993, in patients with neuroendocrine tumors treated with octreotide (150-250 μg t.i.d), no regression was documented. However, octreotide stabilized the size assessed by CT in 50% of the patients for a median duration of five months. In a German multicenter trial, 52 patients with different forms of neuroendocrine malignancies and CT documented tumor progression were treated with octreotide, 200 μg t.i.d. Stabilization of tumor growth was achieved in 19/52 patients (36%) for a median duration of 18 months (Arnold et al., 1996). In the meta analysis of Maton and Gorden, tumor regression in endocrine pancreatic tumors was noticed in 8/46 patients (17%), no change in 18/46 (39%) and progression in 20/46 (44%). Even though a reduction of tumor size is rarely seen with standard octreotide treatment, a stabilization of further tumor growth suggests that octreotide has an antiproliferative effect. A possible mechanism of the anti-neoplastic action of SST analogues might be direct anti-mitotic effects via SSTRs on tumor cells. Suppression of the release of trophic hormones (gut peptides) indirect inhibition of growth factors (IGF-1, EGF, PDGF), inhibition of angiogenesis, induction of apoptosis and possibly modulation of the immune response (reviewed in Scarpignato and Pelosini, 2001).

In summary, standard octreotide treatment of neuroendocrine gastrointestinal tumors gives symptomatic or subjective responses in 30-75% of the patients and the response rate appears to be dose-dependant. Biochemical responses are achieved in 30-60% of patients and are also dose-dependant. A significant tumor reduction is only reported in < 10% of the patients but stabilization can be achieved in another 35-50% of patients. Several in vitro-studies indicate that the anti-proliferative effect might be dose-related and might be mediated through receptor subtypes 2 and 5 (Taylor et al., 1988; Weckbecker 1992; Buscail et al., 1995). The inhibition of hormone synthesis and release is also mediated via receptor subtype 2 and tumor growth may continue despite ongoing anti-secretory effects, suggesting that these actions might be mediated through different mechanisms. Octreotide scintigraphy (octreoscan) should be performed in all patients with neuroendocrine tumors, not only for diagnosis and staging of disease but also might be used as a predictive test before initiation of treatment (Tiensuu-Janson et al, 1994; Shi et al., 1998). Another test suggested is the Sandostatin suppression test, where the hypersecreted hormones are measured one hour before and at 30 minute intervals for three hours after administration of single 0.05 mg doses of octreotide after an overnight fast (Shi et al., 1998).

## HIGH DOSE SOMATOSTATIN ANALOGUE THERAPY

A few studies have addressed the potential value of high dose SST analogue treatment in neuroendocrine gastro-intestinal tumors. A dose-related tumor response has been demonstrated in a variety of tumor models with increasing SST analogue treatment. In a study performed by our group, we treated 19 patients with advanced neuroendocrine GEP-tumors (13 carcinoids and 6 EPT) with a mean duration of disease before start of SST analogue therapy of 56.7 months and 19 months, respectively. All except four were pretreated with standard doses of octreotide. Octreotide scintigraphy was positive in 17/18 patients before initiation of treatment with lanreotide in escalating doses up to 12 mg/day, divided into four injections/day. The highest tolerable dose was maintained for one year or until progression. Biochemical responses were achieved in 11/19 patients (58%). One patient (5%) showed a partial tumor response, whereas stabilization was obtained in 12/19 patients (70%) (Eriksson et al., 1997). Biopsies taken before and during high dose treatment were analyzed with regard to mitotic and apoptotic index indices. Patients with biochemical responses and stable tumor disease showed an increase in the number of apoptotic cells in the tumor with time and also decreasing Ki-67 index (Imam et al., 1997). We have also a preliminary observation that high dose treatment reduces the VEGF-expression, both at mRNA and protein level in the same patients, which might indicate an anti-angiogenetic effect. Apoptosis may have been mediated by binding of lanreotide to somatostatin receptor type 3 (Sharma et al., 1996). Moreover in this trial, positron emission tomography using the tracer [11]C-l-DOPA showed that lanreotide inhibited exocytosis more strongly than inhibition of synthesis of the peptide (Bergström et al., 1996).

Anthony et al. (1993) treated 13 patients refractory to standard dose of octreotide with octreotide 6 mg/day and achieved a partial tumor response in 4/13 (31%) and stabilization in 2/13 (15%). In the same report, high dose lanreotide 9 mg/day given to 13 patients with mixed diagnoses, among six midgut carcinoids three obtained partial tumor response. Faiss and co-workers (1997) treated 30 patients with metastatic neuroendocrine GI-tumors, with 15 mg/day of lanreotide for one year, one complete and one partial tumor response was achieved. When compiling our own data on high dose octreotide and lanreotide treatment (3-12 mg/day), symptomatic improvement was seen in 50%, biochemical responses in 72%, and radiological responses in 13% (Table I).

To summarize, high dose treatment can induce antitumor responses in 10-15% of neuroendocrine GEP-tumors. This figure is somewhat higher than standard dose treatment and taken together with the observed induction of apoptosis and possibly anti-angiogenesis indicate that high dose octreotide/lanreotide can produce additional antiproliferative effects in patients failing on standard doses of SST analogues.

## CONTINUOUS INFUSION OF SOMATOSTATIN ANALOGUES

Several studies in acromegaly have indicated that continuous infusion of octreotide has an advantage over intermittent subcutaneous injections. One can achieve more pronounced control of growth hormone (GH) and insulin like growth factors (IGF-1) levels, clinical and biochemical control can be achieved at lower doses and the adverse effects may be less (Harris et al., 1995). To explore this, we performed a European multicenter trial treating 35 patients with the carcinoid syndrome (19 had failed standard doses of octreotide) with RC-160 (vapreotide, octastatin) at a dose of 1.5 mg/day, as a continuous subcutaneous infusion via micropump for 3-6 months (Eriksson et al., 1996). This was the first clinical study where vapreotide was used and was of particular interest since *in vitro*-studies had suggested that vapreotide might have a stronger antiproliferative effect than both octreotide and lanreotide (Hofland et al., 1994). In this trial subjective improvement and disease stabilization were observed in 60% of patients. However, the biochemical response rate was rather low (23%) and there was no tumor response. One observation in this trial was the low frequency of side effects which were considerably lower than with subcutaneous intermittent administration of SST analogues.

### Slow Release Formulations

One of the most important improvements in SST analogue treatment is the development of slow release formulations. Sandostatin-LAR and Lanreotide-PR are such formulations, in which octreotide and lanreotide respectively, have been incorporated into microspheres of bio-degradable polymers (Lancranjan et al., 1995; Heron et al., 1993). Lanreotide prolonged release was used by Ruszniewski and co-workers (1996) in 39 patients with carcinoid tumors. Patients received a dose of 30 mg every two weeks and they reported subjected responses in about 55%, biochemical responses in 42%, but no tumor responses after six months of treatment. Somewhat higher subjective (73%) and biochemical (55%) response rates were obtained in a German study, including 11 patients (Scherübl et al., 1994). We conducted a European multicenter trial and included 55 patients (48 carcinoids and 7 EPT) that were treated with lanreotide prolonged release 30 mg intramuscularly every two weeks for six months. Symptomatic improvement was observed in 38% of carcinoids, 66% of gastrinomas, and 1% of VIP-oma. Biochemical responses were obtained in 47% and tumor responses in 7% of the patients. Stabilization of tumor growth was achieved in 80% of the patients. In this trial, quality of life was studied using the validated instrument QLQ C30 and the assessment showed already after one month, significant improvement of emotional and cognitive function, and overall health as well as sleeping disorders and diarrhea. This was the first trial to demonstrate that longacting slow release SST analogue treatment improves the quality of life in patients with neuroendocrine tumors (Wymenga et al., 1999).

In a trial by Rubin and co-workers (1999), octreotide acetate long-acting formulation was compared with subcutaneous octreotide acetate in malignant carcinoid syndrome in an open labelled study. A total of 93 patients were included in the study and the patients received double blinded octreotide-LAR at 10, 20 and 30 mg every four weeks versus open labelled subcutaneous octreotide every eight hours. The authors conclude that octreotide-LAR controls the symptoms of carcinoid syndromes at least as well as subcutaneous octreotide (0.3-0.9 mg/24h). The starting dose of 20 mg of octreotide-LAR intramuscular is recommended. Supplementary subcutaneous octreotide is needed for approximately two weeks after initiation of octreotide-LAR treatment. Occasional supplementary subcutaneous injections may be required for possibly 2-3 months until steady-state octreotide levels of LAR are achieved.

One study, from Ricci and co-workers, suggests a difference in response between lanreotide and octreotide in patients with metastatic neuroendocrine tumors. In their study, 15 patients (8 endocrine pancreatic tumors, 7 midgut carcinoid tumors) were treated with lanreotide-PR 30 mg every two weeks for a median time of eight months. These patients were then treated with octreotide-LAR at a dose of 20 mg every four weeks. An objective partial response was documented in one patient (7%), no change in six (40%), and progressive disease in eight patients (53%). The partial response was observed in one patient with non-functioning endocrine pancreatic tumor with progressive liver and lymph node metastases after six months of intramuscular lanreotide treatment. The median duration of disease stabilization was 7.5 months (range 6-12 + months). The overall biochemical response rate was 41% and the overall symptomatic response rate was 82%. The authors suggested that the lack of cross-resistance between depot lanreotide and octreotide might be a result of a more efficient occupancy of SST receptor due to differences in the pharmaco-kinetic profile of the two drugs.

## COMBINATION THERAPY WITH SOMATOSTATIN ANALOGUES

The combination of SST analogues with other agents is an interesting area for future studies. We have previously reported that a combination of octreotide and interferon-alpha produced biochemical responses in 75% of patients resistant to either interferon-alpha alone or regular doses of octreotide (Janson et al., 1992). *In vitro* and *in vivo*-studies of BON-1 tumors indicate that a combination of these two compounds has a stronger anti-proliferative effect than interferon or octreotide alone. In a study by Frank and co-workers (1999), 22 patients were treated with a combination of Sandostatin 200 µg subcutaneously three times daily and alpha interferon 5 million IU three times/week. Overall tumor growth was inhibited in 13/22 patients (59%). Tumor inhibition was associated with significantly longer survival, a mean survival was 23 months in the nine patients who did not respond versus 90 months in the 13 who did respond. Eleven of the responders survived for over four years, compared with none of the non-responders. Others have also reported dramatic responses of the combination of α-IFN and somatostatin analogue in carcinoids (Joensuu et al., 1992).

An Italian trial reported on 58 patients with metastatic neuroendocrine tumors who were initially treated with octreotide at a dose of 500 μg t.i.d. The dose was increased to 1,000 μg t.i.d. and this regimen was continued until disease progression. Twenty-five patients with progressive disease during octreotide treatment received concomitant chemotherapy, Dabarbacin 200 mg/m$^2$, 5-FU 250 mg/m$^2$, epidoxorubicin 25 mg/m$^2$, administered intravenously daily for three days every three weeks. In 27 patients the disease stabilized for at least six months. A partial remission was achieved in two patients, which lasted 10 and 14 months, respectively. The median survival of the entire group was 22 months, however, in this trial no additive or synergistic effect was obtained by adding chemotherapy (Di Bartolomeo et al., 1996).

Weckbecker and co-workers have demonstrated that Doxorubicin in combination with octreotide showed a significant synergistic effect in *in vitro* and *in vivo*-studies in mice. That observation has to be further explored. SSTR-targeted chemotherapy represents an appealing approach to treatment of SSTR expressing tumor by synthesizing conjugates of SST analogues and cytotoxic drugs (such as methotrexate or doxorubicin) (Radulovic et al., 1992; Nagy et al., 1986).

## ADVERSE REACTIONS TO SOMATOSTATIN ANALOGUE THERAPY

The most common side-effects of SST analogues are generally mild and include nausea, transient abdominal cramps, flatulence, diarrhea and local reactions at the injection site. Most of these minor side-effects resolve with time. In 20-50% of the patients, gallstones are formed *de novo* but these remained virtually asymptomatic (Trendle et al., 1997). Rare adverse events to SST analogue treatment include hypocalcemia, bradycardia, acute pancreatitis, hepatitis, jaundice, transitory ischemic attacks and a negative inotropic effect of the analogue.

### Impact on Survival

There are no randomized controlled trials where SST analogues have been compared with other forms of therapy in neuroendocrine GEP-tumors. Therefore all survival data are compared with historical reports. In a study of

80 patients with metastatic carcinoid tumors where patients received 0.15-0.5 mg octreotide subcutaneous three times daily, overall a median survival from diagnosis of metastatic disease was 8.8 years compared with 1.8 years in historical controls (Anthony et al., 1996).

Kvols and Reubi (1993) reported a median survival of about three years in midgut carcinoid tumors with the carcinoid syndrome and extensive liver metastases. This was from start of octreotide treatment, compared to historical data from the same center where cytotoxic treatment, streptozotocin + 5-FU, generated a median survival of 12 months in the same category of patients. Survival data from our own center indicated for patients with carcinoid syndrome and metastatic disease treated with alpha interferon, a median survival of 48 months from start of treatment. All these data have to be taken with caution since no randomized trials have been reported so far.

**Future Aspects**

SST analogues can relieve symptoms, reduce circulating hormone levels and stabilize tumor growth in about 50% of patients with malignant neuroendocrine tumors. There is no doubt that SST analogue therapy has significantly improved the quality of life in patients with malignant neuroendocrine tumors, but the true impact on survival has not been demonstrated as yet. Promising future areas of SST analogue treatment include high dose slow release treatment and the tumor targeted radioactive somatostatin analogue therapy and also tumor targeted cytotoxic SST analogue therapy. The combination of somatostatin analogues with interferon and other agents is also worthy of investigation. New SST analogues, both non-peptide and peptide analogues, are currently developed. Analogues that see all five subtypes of somatostatin receptors are ready for clinical trials. Also subtype-specific analogues for the five different subtypes are ready for the first trials. In the future, patients may be treated with the "panreceptor" analogue or a cocktail of one, two or three different subtype-specific analogues depending on the expression pattern of somatostatin receptors within the tumors (Shimon et al., 1997). The cross-talk between different SSTRs, as well as with other G-protein coupled receptors (e.g., dopamine receptors), might be further areas to explore for treatment and diagnosis (Rocheville et al., 2000). The anti-angiogenetic effects of somatostatin analogues have to be evaluated and also in combinations with other angiogenetic agents (Woltering et al., 1997).

# REFERENCES

1. Angeletti, S., Corleto, V.D., and Schiallachi, O. (1998). Use of the somatostatin analogue octreotide to localise and manage somatostatin-producing tumours. Gut 42, 792-794.
2. Anthony, L., Johnson, D., and Hande, K. (1993). Somatostatin analogue phase I trials in neuroendocrine neoplasms. Acta Oncol 32, 217-223.
3. Anthony, L.B., Skyr, Y., Wim, S.D., and Oates, J.A. (1996). Octreotide acelate (OA) in the management of malignant carcinoid syndrome (CS). Ann Oncol 7 (suppl 5), 47 (abs 219P).
4. Arnold, R., Trautmann, M.E., and Creutzfeldt, W. (1996). Somatostatin analogue and inhibition of tumor growth in metastatic endocrine gastroenteropancreatic tumors. Gut 38, 430-438.
5. Bauer, W., Briner, U., and Doepfner W. (1982). A very potent and selective octapeptide analogue of somatostatin with prolonged action. Life Sci 31, 1133-1140.
6. Bergström, M., Eriksson, B., and Öberg K. (1996). In vivo demonstration of enzyme activity in endocrine pancreatic tumors –decarboxylation of $^{11}$C-dopa to $^{11}$C-dopamine. J Nucl med 36, 32-37.
7. Buscail, L., Esteve, J.P., and Saint-Laurent, N. (1995). Inhibition of cell proliferation by the somatostatin analogue RC-160 is mediated by somatostatin receptor subtypes SSTR2 and SSTR5 through different mechanisms. Proc Natl Acad Sci 92, 1580-1584.
8. Carrasco, C.H., Chuang, V., and Wallace, S. (1983). Apudoma metastatic to the liver: Treatment by hepatic artery embolization. Radiology 149, 79-83.
9. Di Bartolomeo, M., Bajetta, E., Buzzoni, R., Mariani, L., Carnaghi, C., Somma, L., Zilembo, N., Di Leo, A., and the ITMO Association (1996). Clinical efficacy of octreotide in the treatment of metastatic neuroendocrine tumors. Cancer 77, 402-408.
10. Eckhauser, F.E., Cheung, P.S., and Vinik, A.I. (1986). Nonfunctioning malignant neuroendocrine tumors of the pancreas. Surgery 100, 978.
11. Eriksson, B., Arnberg, H., and Lindgren, P.G. (1990). Neuroendocrine pancreatic tumors: Clinical presentation, biochemical and histopathological tumors: Clinical presentation, biochemical and histopathological findings in 84 patients. J Int Med 228, 103-113.
12. Eriksson, B., Janson ET, and Bax, N.D.S. (1996). The use of new somatostatin analogues, lanreotide and octastatin, in neuroendocrine gastrointestinal tumors. Digestion 57, 77-80.
13. Eriksson, B., Renstrup, J., Imam, H., and Öberg, K. (1997). High-dose treatment with lanreotide of patients with advanced neuroendocrine gastrointestinal tumors: Clinical and biological effects. Ann Oncol 8, 1041-1044.
14. Eriksson, B., and Öberg, K. (1999). Summing up 15 years of somatostatin analog therapy in neuroendocrine tumors. Future outlook. Ann Oncol 10 (Suppl 2), S1-8.
15. Faiss, S., and Wiedenmann, B. (1997). Dose-dependent and antiproliferative effects of somatostatin. J Endocrinol Invest 20, 68-70.
16. Fajans, S.S., and Vinik, A.I. (1989). Insulin-producing islet cell tumors. Endocrinol Metab Clin North Am 18, 45.
17. Feldman, J.M. (1987). Carcinoid tumors and syndrome. Semin Oncol 14, 237.
18. Frank, M., Klose, K.J., Wied, M., Ishaque, N., Schade-Brittinger C., and Arnold, R. (1999). Combination therapy with octreotide and α-interferon: Effect

of tumor growth in metastatic endocrine gastroenteropancreatic tumors. Am J Gastroenterol 94, 1382-1387.

19. Fröhlich, J.C., Bloomgarden, Z.T., and Oates, J.A. (1978). The carcinoid flush: Provocation by pentagastrin and inhibition by somatostatin. N Engl J Med 299, 1055-1057.

20. Gorden, P., Comi, R.J., Maton, P.N., and Go V.L.W. (1989). Somatostatin and somatostatin analogue (SMS 201-995) in treatment of hormone-secreting tumors of the pituitary and gastrointestinal tract and non-neoplastic diseases of the gut. Ann Intern Med 110, 35-50.

21. Harris, A., Kokoris, S., and Ezzat, S. (1995). Continuous versus intermittent sub-cutaneous infusion of octreotide in the treatment of acromegaly. J Clin Pharmacol 35, 59-71.

22. Harris, A., and Redfern, J.S. (1995). Octreotide treatment of carcinoid syndrome: Analysis of published dose-titration data. Aliment of published dose-titration data. Aliment Pharmacol Ther 9, 387-394.

23. Heron, I., Thomas, F., and Dero, M. (1993). Pharmacokinetics and efficacy of a long-acting formulation of the new somatostatin analog BIM-23014 in patients with acromegaly. J clin Endocrinol Metabol 76, 721-727.

24. Hoefnagel, C.A., den Hartog Jager, F.C., and Taal, B.G. (1987). The role of [125]I-MIBG in the diagnosis and therapy of carcinoids. Eur J Nucl Med 13, 187.

25. Hofland, L.J., Koetsveld, and Waaijers, M. (1994). Relative potencies of the somato-statin analogs octreotide, BIM-23014, and RC-160 on the inhibition of hormone release by cultures human endocrine tumor cells and normal rat anterior pituitary cells. Endocrinol 134, 301-306.

26. Imam, H., Eriksson, B., and Lukinius A. (1997). Induction of apoptosis in neuro-endocrine tumors of the digestive system during treatment with somatostatin analogs. Acta Oncol 36, 607-614.

27. Janson, E.T., Ahlström, H., Andersson, T., and Öberg, K (1992). Octreotide and interferon-α: A new combination for the treatment of malignant carcinoid tumors. Eur J Cancer 28A, 1647-1650.

28. Jensen, R.T. (1983). Zollinger-Ellison syndrome: Current concepts and management. Ann Int Med 98, 59-75.

29. Jockenhovel, S., Lederbogen S., and Olbricht, T. (1994). The long-acting somatostatin analogue: Symptomatic and peptide response. Gut 26, 127-133.

30. Joensuu, H., Kätkä, K. and Kujari, H. (1992). Dramatic response of a metastatic carcinoid tumour to a combination of interferon and octreotide. Acta Endocrinol (Copenh) 126, 184-185.

31. Krejs, G.J., Ocri, L., and Conlon, M. (1979). Somatostatinoma syndrome (biochemical, morphological, and clinical features). N Engl J Med 301, 285.

32. Krenning, E.P., Kwekkeboom, D.J., and Bakker, W.H. (1993). Somatostatin receptor scintigraphy with ([111]In-DTPA-D-Phe[1]) and ([123]Tyr[3])-octreotide. The Rotterdam experience with more than 1000 patients. Eur J Nucl Med 20, 716-731.

33. Kvols, L.K., Marsh, H.M., and Moertel, C.G. (1985). Rapid reversal of carcinoid crisis with a somatostatin analouge. N Engl J Med 313, 1229-1230.

34. Kvols, L.K., Moertel, C.G., and O'Connell M.J. (1986). Treatment of the malignant carcinoid syndrome. Evaluation of a long-acting somatostatin analogue. N Engl J Med 315, 663-666.

35. Kvols, L.K., and Reubi, J.C. (1993). Metastatic carcinoid tumors and the malignant carcinoid syndrome. Acta Oncol 32:2, 197-201.

36. Lamberts, S.W.J., Krenning, E.P., and Reubi, J.C. (1991). The role of somatostatin and its analogs in the diagnosis and treatment of tumors. Endocrinol Rev 12, 450-482.

37. Lancranjan, I., Bruns, C., and Grass, P. (1995). Sandostatin-LAR: Pharmacokinetics, pharmacodynamics, efficacy, and tolerability in acromegalic patients. Metabolism 44, 18-26.
38. Larsson, C., Skogseid B., and Öberg, K. (1988). Multiple endocrine neoplasia type 1 gene maps to chromosome 11 and is lost in insulinoma. Nature 332, 85-87.
39. Long, R.G., Bryant, M.G., and Mitchell, S.J. (1981). Clinicopathological study of pancreatic and ganglioneuroblastoma tumors secreting vasoactive intestinal polypeptide (VIPomas). Br J Med J 282, 1767.
40. Makridis, C., Öberg, K., and Juhlin, C. (1990). Surgical tretment of midgut carcinoid tumors. World J Surg 14, 377.
41. Maton, P.N., Gardner, J.D., and Jensen, R.T. (1989). Use of long-acting somatostatin analogue SMS 201-995 in patients with pancreatic islet cell tumors. Dig Dis Sci 34, 285-291.
42. Moertel, C.G., Johnson, C.M., and McKusick, M.A. (1994). The management of patients with advanced carcinoid tumours and islet cell carcinomas. Ann Intern Med 120, 302-309.
43. Nagy, A., Schally, A.V., Halmos, G., Armatis, P., Cai, R.Z., and Csernus, V. (1986). Synthesis and biological evaluation of cytotoxic analogues of somatostatin containing doxorubicin and its potent derivative 2-pyrrolinodoxorubicin. Br J Surg 83, 456-460.
44. Norheim, I., Öberg, K., and Theodorsson-Norheim, E. (1987). Malignant carcinoid tumors. An analysis of 103 patients with regard to tumor localization, hormone production, and survival. Ann Surg 206, 115-125.
45. Otte, A., Mueller-brand, J., and Dellas, S. (1998), Yttrium-90-labelled somatostatin analogue for cancer treatment. Lancet 351, 417-418.
46. Patel, Y.C. (1999). Somatostatin and its receptor family. Frontiers in Neuroendo-crinology 20, 157-198.
47. Radulovic, S., Nagy, A., Szoke, B., and Schally, A. (1992). Cytotoxic analogue of somatostatin containing methotrexate inhibits growth of MIA PaCA-2 human pancreatic xenografts in nude mice. Cancer lett 62, 263-271,
48. Reubi, J.C., Kvols, L.K., and Waser, B. (1990). Detection of somatostatin receptors in surgical and percutaneous needle biopsy samples and islet cell carcinomas. Cancer Res 50, 5969-5977.
49. Ricci, S., Anonuzzo, A., Galli, L., Ferdeghini, M., Bodei, L., Orlandini, C., and Conte, P.F. (2000). Octreotide acelate long-acting release in patients with metastatic neuroendocrine tumors pretreated with lanreotide. Ann Oncol 11, 1127-1130.
50. Rindi, G., Luinetti, O., and Cornaggia, M. (1993). Three subtypes of gastric argyrophil carcinoid and the gastric neuroendocrine carcinoma. A clinicopathological study. Gastroenterol 104, 994-1006.
51. Rocheville, M., Lange, D.C., Kumar, U., Patel, S.C., Patel, R.C., and Patel, Y.C., (2000). Receptors for dopamine and somatostatin: Formalin of hetero-oligomers with enhanced functional activity. Science 288, 154-157.
52. Rubin, J., Ajani, J., Schinner, W., Wenok, A.P., Bukowski, R., Pommier, R., Salz, L., Danona, P., and Anthony, L. (1999). Octreotide acetate long-acting formulation versus open-label subcutaneous octreotide acetate in malignant carcinoid syndrome. J Clin Oncol 17, 600-606.
53. Ruszniewski, P., Ramdani, A., and Cadiot G. (1993). Long-term treatment with octreotide in patients with the Zollinger-Ellison syndrome. Eur J Clin Invest 23, 296-301.
54. Ruszniewski, P., Ducreux, M., and Chayvialle, J.A. (1996). Treatment of the carcinoid syndrome with the long-acting somatostatin analogue lanrcotide: A prospective study in 39 patients. Gut 39, 279-283.

55. Saltz, L., Trochanowski, B., and Buckley, M. (1993). Octreotide as an antineoplastic agent in the treatment of functional and non-functional neuroendocrine tumors. Cancer 72, 244-248.
56. Scarpignato, C. (1995). Somatostatin analogues in the management of endocrine tumors of the pancreas. In Mignon M, Jensen RT (eds): Endocrine Tumors of the Pancreas. Basel: Karger, 385-414.
57. Scarpignato, C., and Pelosini, I. (2001). Somatostatin analogs for cancer treatment and diagnosis – an overview; In: Somatostatin Analogs in Cancer Management. Ed. C. Scarpignato (Karger). Chemotherapy 47 (Suppl 2), 1-29.
58. Scherübl, H., Wiedenmann, B. and Riecken, E.O. (1994). Treatment of the carcinoid syndrome with a depot formulation of the somatostatin analogue lanreotide. Eur J Cancer 6, 1590-1591.
59. Sharma, K., Patel, X.C., and Srikant, C.B. (1996). Subtype-selective induction of wild-type p53 and apoptosis, but not cell cycle arrest, by human somatostatin receptor 3. Mol Endocrinol 10, 1688-1696.
60. Shi, W., Buchanan, K.D., Johnston, C.F., Larkin, C., Ong, Y.L., Ferguson, R., and Laird, J. (1998). The octreotide suppression test and ([111]In-DTPA-D-Ph[1])-octreotide scintigraphy in neuroendocrine tumors correlate with responsiveness to somatostatin analogue treatment. Clin Endocrinol 48, 303-309.
61. Shimon, I., Taylor, J., and Weiss, M.H. (1997). Somatostatin receptor (SSTR) subtype-selective analogues differentially suppress in vitro growth hormone and prolactin in human pituitary adenomas. Novel potential therapy for functional pituitary tumors. J clin Invest 100, 2386-2392.
62. Skogseid, B., Eriksson, B., and Lundqvist, G. (1991). Multiple endocrine neoplasia type 1: A ten-year preospective screening study in four kindreds. J Clin Endocrinol Metab 73, 281-287.
63. Spread, C., Berkel, H., and Jewell L. (1994). Colon carcinoid tumors. A population-based study. Dis Colon Rectum 37, 482-491.
64. Stacpoole, P.W. (1981). The glucagonoma syndrome: Clinical features, diagnosis, and treatment. endocr Rev 2, 347.
65. Taylor, J.E., Bogden, A.E., Moreau, J.P., and Coy D.H. (1988). In vitro inhibition of human small-cell carcinoma (NCL-H69) growth by somatostatin analogue. Biochem Biophys Res Commun 153, 81-86.
66. Tiensuu Janson, E., Westlin, J.E., and Eriksson, B. (1994). ([111]In-DTPA-D-Phe[1])-Octreotide scintigraphy in patients with carcinoid tumors: The predictive value for somatostatin analog treatment. Eur J Endocrinol 131, 577-581.
67. Trendle, M.C., Moertel, C.G., and Kvols, L.K. (1997). Incidence and morbidity ofcholelithiasis in patients receiving chronic octreotide for metastatic carcinoid and malignant islet cell tumours. Cancer 79, 830-834.
68. Thulin, L., Samnegård, H., and Tyden, G. (1978). Efficacy of somatostatin in a patient with carcinoid syndrome. Lancet 2, 43.
69. Weckbecker, G., Liu, R., Tolcsvai, L., and Bruns, C. (1992). Antiproliferative effectsof the somatostatin analogue octreotide (SMS 201-995) on ZR-75-1 human breast cancer cells in vivo and in vitro. Cancer Res 52, 4973-4978.
70. Wilander, E., Lundqvist, M., and Öberg, K. (1989). Gastrointestinal carcinoid tumours. Prog Histochem Cytochem 19, 1-85.
71. Williams, E.D., and Sandler, M. (1963). The classification of carcinoid tumours. The Lancet 1, 238-239.
72. Woltering, E.A., Watson, J.C., Alperin-Lea, C, Sharma, C., Keenan, E., Kurdzawa, D.L., and Barrie, R. (1997). Somatostatin analogues: Angiogenesis inhibitors with novel mechanisms of action. Invest New Drugs 15, 77-86.
73. Wood S.M., Kraenzlin,M.E., Adrian, T.E., and Bloom, S.R (1985) Treatment of patients with pancreatic endocrine tumors using a new long-acting somatostatin analogue: Symptomatic and peptide response. Gut 26, 438-444.

74. Wymenga, A.N.M., Eriksson, B., and Salmela, P.I. (1999). Efficacy and safety of lanreotide prolonged release in patients with gastrointestinal neuro-endocrine tumors with hormone related symptoms. J Clin Oncol *4*, 1111-1117.
75. Wynick, D., Andersson, J.V.,Williams, S.J. , and Bloom, S.R. (1989). Resistance of metastatic pancreatic endocrine tumours after long-term treatment with the somatostatin analogue octreotide (SMS 201-995). Clin Endocrinol *30*, 385-388.
76. Öberg, K., Norheim, I., Lundqvist, G., and Wide, L. (1986). Treatment of the carcinoid syndrome with SMS 201-995, a somatostatin analouge. Scand J Gastroenterol *119*, 191-192.
77. Öberg, K., and Eriksson, B. (1991). The role of interferons in the management of carcinoid tumors. Acta Oncol *30*, 519-522.
78. Öberg, K. (1996). Neuroendocrine tumors. Ann Oncol *7*, 453-463.
79. Öberg, K. (2001). Established clinical use of octreotide and lancreotide in oncology. Chemotherapy *47*, (suppl 2), 40-53.

# 17
# VASCULAR EFFECTS OF SOMATOSTATIN

Pekka Häyry and Einari Aavik
*Rational Drug Design Programme, Biomedicum and Transplantation Laboratory, University of Helsinki and Helsinki University Central Hospital, POBox 21 (Haartmaninkatu 3), FIN 00014 University of Helsinki, Finland*

## INTRODUCTION

Somatostatin (SST) was first identified in sheep hypothalamus as a peptide that inhibited the release of growth hormone. Later it was found that SST is produced throughout the central nervous system and in several peripheral organs. SST-like molecules are found in all vertebrates and in some invertebrate species, as well as in the plant kingdom. Thus SST represents a phylogenetically ancient multigene family of peptides with two bioactive products in higher species, SST-14 and SST-28, a congener of SST-14 extended at the N-terminus (Patel, 1999).

SST acts not only as a neurotransmitter but has diverse paracrine/autocrine regulatory effects on several tissues, including pancreatic islet tissue, gastrointestinal tract, immune cells and tissues and may, thereby, be linked in the pathophysiology of various diseases including diabetes mellitus, Alzheimer's Disease and diseases regulated by immune/inflammatory effects (Patel, 1999). This article focuses on the vascular effects of SST.

Abbreviations: SST - somatostatin, SSTR - somatostatin receptor, PTCA - percutaneous transluminal coronary angioplasty, IGF-1 - insulin-like growth factor – 1, IGF-1R - insulin-like growth factor - 1 receptor, SMC - smooth muscle cell, EC - endothelial cell, EGF - epidermal growth factor, NF-κB - nuclear factor - kappa B, PTP - phosphotyrosine phosphatase, PDGF - platelet derived growth factor, PDGFR - platelet derived growth factor, 6-keto-PGF1-alpha - 6-keto-prostaglandin F(1alpha).

## VASCULAR EFFECTS OF SOMATOSTATIN AND ANALOGS

The discovery of the vascular effects of SST should be accredited to Foegh, Ramwell and their collaborators at Georgetown University. They used an air-drying model of rat carotid artery and demonstrated that out of the 5 synthetic SST-like peptides, which were administrated at a dose range of 20-100 ug/kg/d s.c. 2 days prior to and 5 days after endothelial injury, BIM23014 (lanreotide, Angiopeptin) and its closely-related congener BIM23034, inhibited the generation of myointimal hyperplasia and thymidine incorporation to the carotid artery (Lundergan et al., 1989). Because all five analogs equally inhibited the release of growth hormone, but only angiopeptin and BIM23034 were inhibitory to myointimal proliferation, the results suggested that certain SST analogs inhibit myointimal proliferation by a unique mechanism not related to growth hormone regulation.

### Myointimal hyperplasia
Myointimal hyperplasia is a common manifestation of vascular disease regardless how the disease has been inflicted. Such diseases include classical atherosclerosis, where the lesions are mostly focal, proximal and often confined to characteristic areas of stress in the arterial tree, particularly in coronaries, carotids and lower parts of aorta and iliac arteries. The etiological factors of atherosclerosis are a multitude and not within the scope of this review (Ross, 1999; Schwartz and Murry, 1998). Another pattern of myointimal hyperplasia is observed in chronically-rejecting allografts where the manifestations in the intra-allograft arteries are concentric and generalized, leading ultimately to insufficient blood supply, fibrosis and loss of the transplant (Hayry et al., 1993a) .

Myointimal hyperplasia, leading to (re-)occlusion of the vessel, is also a common complication after vascular surgical procedures. Such procedures include restenosis after percutaneous transluminal angioplasty (PTCA), endarterctomies, vascular stenting and coronary bypass procedures, particularly when venous autografts are used. The failure rate in these operations is significant and there is currently not effective prophylaxis to prevent reocclusion. Thus the observation of the Foegh-Ramwell group (Lundergan et al., 1989) turned out to be of considerable interest, as it opened the possibility to generate SST agonists for the prevention of restenosis and chronic allograft rejection.

### SSTR FAMILIES
Five distinct SSTR genes have been described that encode receptor proteins of the 7-transmembrane domain class within the superfamily of G-protein coupled receptors. These five receptors show equally high affinity to SST and corticostatins, SST-28 being somewhat more selective than SST-14 for SSTR5 (for ref., see Patel, 1999).

Both structural and functional information indicate that the receptors can be divided into two groups: one group that includes SSTR2, SSTR3 and SSTR5 and the other one that includes SSTR1 and SSTR4. These two groups can be differentiated both by their selectivity in binding to synthetic SST analogs, as well as on the basis of their amino-acid homologies (see Patel, 1999; Patel et al., 1995; Reisine and Bell, 1995).

The best investigated peptide analogs in vascular biology, SMS201-995, (octreotide, Sandostatin), and BIM23014 (lanreotide, Angiopeptin) show high selectivity to SSTR2 (3) and 5, whereas CH 275 (Liapakis et al., 1996; Chen et al., 1999) is the only artificial analog with high selectivity to SSTR1 and 4 and low or no selectivity to SSTR2, 3 and 5 (Patel and Srikant, 1994; Reisine and Bell, 1995; Patel, 1999). Most recently Rohrer et al. (1998) have synthesized non-peptide ligands having selective agonistic affinity to all of these five receptors, and providing thereby a powerful tool to investigate the physiological functions of the receptor subtypes separately. To our knowledge only two (subtype-selective) antagonists have been identified (Bass et al., 1996; Reubi et al., 2000).

## SSTR AGONIST LIGANDS IN EXPERIMENTAL STUDIES ON MYOINTIMAL HYPERPLASIA

After their initial observation in the rat, Foegh and Ramwell confirmed their result using arterial balloon catheter injury in hypercholesterolemic rabbits. Angiopeptin, 20 ug/kg/d significantly inhibited myointimal thickening of infra-renal aorta as well as of common and external iliac arteries. As angiopeptin treatment did not significantly modify the plasma-lipid levels, they concluded that the inhibitory effect is unrelated to plasma-lipid concentrations (Howell et al., 1993). Later they also showed that angiopeptin inhibits transplant arteriosclerosis in cardiac, renal and vascular allografts and, concomitantly, abrogated the increase of IGF-1 in the vascular wall following immune and mechanical injury (Foegh and Ramwell, 1995). In our own study on non-immunosuppressed rat aortic allografts, administration of angiopeptin at a dose rate of 80 ug/kg/d with osmotic minipumps, inhibited cell proliferation and intimal hyperplasia and concomitantly, several locally-produced peptide growth factors, i.e., EGF, IGF-1 and PDGF (Hayry et al., 1993b). On the other hand, no effect was demonstrated on the proliferation of in vitro propagated primary smooth muscle cell lines in this study.

The anti-proliferative effect of SST analogs in rodent vessels, were confirmed by the group of Bruns. Sandostatin administrated with osmotic minipumps at a rate of up of 10 ug/kg/hour for 16 days, inhibited myointimal hyperplasia after rat carotid angioplasty (Weckbecker et al., 1997), intimal hyperplasia in mouse carotid artery allografts at a dose rate of 50 ug/kg/h and chronic vasculopathy in partially immunosuppressed rat kidney allografts at a dose rate of up to 30 mg/kg, initiated on day -2 and continued to day 40 after transplantation

(Schuurman et al., 1996). Their experience in rodent models has recently been summarized (Bruns et al., 2000).

The results in large animals were equally promising. Santoian et al. (Santoian et al., 1993) performed overstretched balloon injury in normolipidemic pig coronary arteries receiving 50 ug/kg Angiopeptin one hour before and at the time of balloon injury, and administrated Angiopeptin thereafter at a rate of 100 ug/kg/d s.q. for 14 days. Control animals were inflicted with balloon injury only. Compared to controls, Angiopeptin significantly inhibited intimal hyperplasia and increased residual lumen area. Subsequently Hong et al. (1997) administrated Angiopeptin to pigs with a Palmaz-Schatz stent in the left anterior descending coronary artery. One group received local saline infusion on site of the stent via local delivery catheter, the second group received 200 ug of Angiopeptin locally after the stent was placed in the artery, the third group was treated with miniosmotic pump, 200 ug/kg/d systemically and the fourth group received combined local and systemic treatment. Systemic treatment produced least neointimal hyperplasia and significantly reduced stent restenosis compared to the control group.

## RESULTS OF CLINICAL TRIALS

Thus the results of all experimental studies were favourable and decisions were reached to design a clinical study. In these studies, restenosis after trans-luminal PTCA was investigated. In the first Scandinavian study (Eriksen et al., 1995), 112 patients were randomized in a double-blind study to receive continuous subcutaneous angiopeptin, 750 ug/d, or placebo as infusion from the day before PTCA and during the following four days. 80 patients had a successful PTCA and 75 of these patients with 94 lesions underwent angiography 6 months thereafter. All 112 patients underwent a 12-month clinical follow-up examination. The risk factors were equal in both groups. The hierarchial 12-month event rate (death, myocardial infarction, coronary artery bypass grafting and repeated PTCA) was reduced from 34% to 25% (p=0.30) by angiopeptin, by intent-to treat analysis. Restenosis (> or = 50% diameter stenosis) was significantly reduced in patients treated with Angiopeptin (12% vs. 40%, p=0.003). Late lumen loss was also significantly reduced after Angiopeptin treatment (0.12 ± 0.46 mm vs. 0.52 ± 0.64 mm, p=0.03). Thus, though the results were marginal, it looked as if Angiopeptin infusion for 5 days tended to decrease the clinical events and restenosis after PTCA.

In the second multicenter study (Emanuelsson et al., 1995), the patients received a continuous infusion of either placebo or angiopeptin for 6 to 24 hours before PTCA and for 4 days after PTCA. In total, 553 patients with 742 lesions were randomized. Angiopeptin decreased the clinical events during the 12-month follow-up from 36% in the placebo group to 28% in the angiopeptin treated patients (p=0.046). In contrast, no significant effect was seen in angiographic variables using computerized coronary angiography.

Finally, the efficacy of Sandostatin treatment was also investigated in a placebo-controlled trial (von Essen et al., 1997). Altogether 274 patients received either Sandostatin 500 ug every 8 hour or placebo starting one hour before angioplasty continuing thereafter for three weeks. The minimal lumen diameters before and after angioplasty and at 6-month follow-up were analyzed with a digital quantitative algorithm. 217 patients completed the 6-month follow-up. The minimal lumen diameter was not any different in the Sandostatin-treated vs. placebo group (p=0.70) and the restenosis rates were identical (p=1.0). Neither was there any difference in the frequency of clinical events. Thus, under this constellation octreotide did not reduce the angiographically-determined restenosis rate or the incidence of clinical events after coronary angioplasty. These results were obviously a great disappointment. Both Henri Beafour Institute producing angiopeptin and Novartis producing Sandostatin decided not to go any further. Thus, what actually failed? One possibility was that the ligands, having agonistic activity to the SSTR2,3,5 family were not actually the right ligands to target to in the vascular wall. Another possibility is that also the mode of administration was wrong.

## EXPRESSION OF SSTR SUBTYPES IN THE VASCULAR WALL

Radioactive ligand binding studies particularly by the group of Reubi, have clearly demonstrated that SSTRs are actually expressed in the vascular wall. In these studies SSTRs were identified initially with receptor autoradiography using $[^{125}I\text{-}Tyr^3]$ octreotide or $[Leu^8, D\text{-}Trp^{22}, {}^{125}I\text{-}Tyr^{25}]$ SST-28 as radioligands. They first demonstrated (Reubi et al., 1994b) that SSTRs are present in high density in most intramural veins but not in arteries of intestines in Crohn's Disease or ulcerative colitis. Later (Reubi et al., 1994a; Reubi et al., 1996b) they confirmed this result in the vessels of a variety of human tumors.

In order to identify specificially the receptor subtypes in human tumors and in peritumoral vessels, they used in situ hybridization techniques. The majority of gastrointestinal vessels expressed SSTR2 mRNA whereas SSTR1 and SSTR3 expression was less frequent. Needless to say, this technology was far from being quantitative, but they concluded that at least in colorectal carcinomas the peritumoral veins express high density SSTRs, probably of the SSTR2 type (Reubi et al., 1996a). Another dimension was revealed in studies targeting models of arterial injury. Here, the credit should be particularly given to the group at the University of British Columbia. In their first study (Chen et al., 1997), transaortic balloon injury of the rat iliac artery was performed and the rats were sacrificed sequentially at 48 hours, 1 week and 1 month post injury, perfusion fixed and stained with antibodies against SSTR2, 3 and 5. SSTR2 was identified in the intimal surfaces of normal and injured vessels. SSTR2 immune reactivity was more prominent at 1 week post injury. There was no immunostaining with SSTR3 and SSTR5 antibodies.

In our own study (Khare et al., 1999), we investigated the expression pattern of all five SSTRs in rat thoracic aorta in resting state and at 15 min, 3, 7, and 14 days after balloon endothelial denudation, using semiquantitative reversed transcriptase polymerase chain reaction (RT-PCR) and immunocytochemistry for the expression of proteins. All five SSTRs were expressed in rat aorta both as mRNA and protein and displayed a time-dependent, subtype selective response to endothelial denudation. mRNA for SSTR1 and 2 increased acutely, SSTR1 far more than SSTR2, on days 3 and 7, coincident with smooth muscle cell proliferation, and declined to basal levels on day 14. SSTR3 and 4 mRNAs displayed a different pattern with a delayed, more gradual increase, beginning on days 3-7 and continuing thereafter. SSTR5 mRNA was expressed at a very low level only (Figure 1). In immunohistochemistry, SSTR antigens were localized predominantly in the

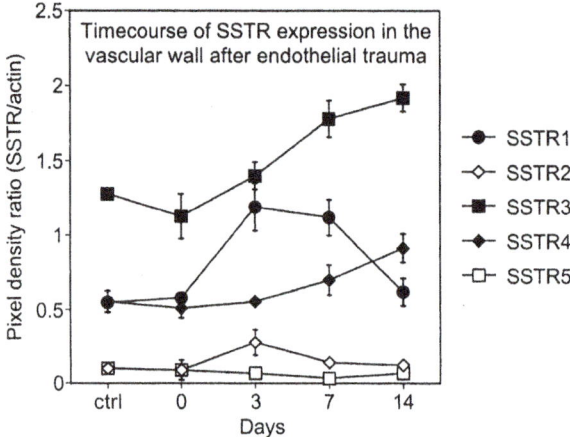

*Figure 1. Time course of the expression of SSTR1-5 mRNA in semiquantitative RT-PCR after carotid denudation in the rat. From Khare et al., Faseb J 13:387-94.*

SMC that were present in the media and had migrated into the intima, or occasionally in the endothelial cells. Antigen expression correlated mostly with receptor mRNA expression, with some exceptions. Notably, only SSTR1 and 4 proteins were prominently expressed in the media and neointima, co-localizing with the smooth muscle cells, but SSTR2,3 and 5 proteins were expressed very weakly, if at all (Table 1). We concluded that there are dynamic changes in SSTR expression after vascular injury and that in the view of the prominent and early expression of SSTR1, this might be the optimal subtype to target to for the inhibition of myointimal proliferation and SSTR3 and 4 for migration and remodelling.

**Table 1.** *Intensity of immunoreactivity in the immunohistochemical staining of the SSTR subtype expression in denuded rat carotid at different time points after injury*

| Subtype | Control | 3 days | 7 days | 14 days | 60 days |
|---|---|---|---|---|---|
| **ADVENTITIA** | | | | | |
| SSTR1 | - | - | - | - | - |
| SSTR2 | - | - | - | - | - |
| SSTR3 | - | - | - | - | - |
| SSTR4 | - | - | - | - | - |
| SSTR5 | - | - | - | - | - |
| | | | | | |
| **MEDIA** | | | | | |
| SSTR1 | +/- | +/- | + | + | - |
| SSTR2 | + | + | - | - | - |
| SSTR3 | - | - | +/- | - | + |
| SSTR4 | +/- | +/- | +/- | - | + |
| SSTR5 | - | +/- | - | - | - |
| | | | | | |
| **INTIMA** | | | | | |
| SSTR1 | - | - | + | + (EC) | - |
| SSTR2 | - | - | - | - | - |
| SSTR3 | - | - | - | - | + |
| SSTR4 | - | - | +/- | + | + |
| SSTR5 | - | - | - | - | - |

*All readings were made blind. - indicates no reactivity, +/- possible positive reaction, + positive reaction. EC indicates that immunoreactivity was also observed in the endothelial cells in addition to smooth muscle cells (Khare et al, FASEB J. 1999, 13:387-394).*

In their second study, the University of British Columbia group performed balloon endothelial injury to rat left common and iliac arteries with and without circumferential arterial dissection (Curtis et al., 2000a) . The receptor expression was quantitated 1 and 2 months after injury by RT-PCR and by immunohistochemistry . Normal rat iliac arteries expressed only SSTR2 and 3 and there was a significant upregulation of SSTR 2 messenger RNA in injured arteries compared to controls. Earlier time points were not investigated.

In their third study, they concentrated upon human vessels (Curtis et al., 2000b). Samples of normal veins and arteries as well as atherosclerotic arteries expressed predominantly SSTR1 both when using semi-quantitative RT-PCR and immunohistochemistry. No evidence of SSTR3 or 5 expression was detected in normal or diseased vessels. In subsequent in vitro studies, human endothelial cell line and human umbilical vein endothelial cells were shown to express SSTR1 and 4 and exposure of these cells to 10 nM SST or 10 nM SSTR-1-specific agonist, CH275, resulted in alterations to the acting cytoskeleton as characterized by a loss of actin-stress fibres coupled with an increase in lamellipodia formation at the plasma membrane. Primary smooth muscle cell cultures or -lines were not investigated.

These results, demonstrating the predominant expression of SSTR1 (and 4) in the vascular (arterial) wall, are also consistent with our (unpublished) studies on primate denudation injury using a recently-described baboon model (Du Toit et al., 2001), but clearly contradictory to the earlier findings after rat balloon injury by Curtis (Curtis et al., 2000a). A likely explanation is that in their rat study, the University of British Columbia group did not sample the arteries shortly after injury, when the differentials in gene expression levels were clear and prominent. Thus, taken together, there seems to be a consensus that the SSTR1,4 family is expressed in the vascular wall and that the expression of SSTR1 (and possibly 4) acutely increases after injury and in atherosclerotic arteries. The expression pattern also suggests that targeting in previous clinical studies to SSTR1 and 5 in the prevention of vascular complications after PTCA, might have been a wrong concept. Instead, one should target to SSTR1 (and 4) to get an optimal response in the prevention of intimal hyperplasia.

## PREVENTION OF INTIMAL HYPERPLASIA USING LIGANDS SELECTIVE TO THE SSTR1/4 FAMILY

Until recently only few results, and all of them in vitro results, have been available where ligands targeting to other subtypes than SSTR2,5 have been used to investigate the effects of these ligands on vascular cells. As already referred to, Curtis et al. (2000b) demonstrated an inhibitory effect of the SSTR1,4 agonist CH275 on stimulated human endothelial cells in vitro. In another report, Lauder et al. (1997) demonstrated an in vitro effect of SSTR5 receptor selective agonist, L-362,855 on basic FGF-induced regeneration of vascular smooth muscle cells and of CHO cells stably transfected with SSTR5; the effect was 100-times more potent than that of SST. Such studies are, however, difficult to correlate with the situation in vivo.

We have recently synthesized CH275 and compared the efficacy of this SSTR1/4 selective ligand to native SST14 in the prevention of intimal hyperplasia after carotid denudation in the rat. Several end-points were quantitated 14 days after injury using a large dose response scale from 10 to 500 ug/kg/d as infusion or as daily injections . When the drugs were given as one injection per day, the SSTR1/4 selective peptide inhibited significantly more prominently and more parametres in intimal hyperplasia compared to native SST-14 (unpublished). On the other hand, when the drugs were administered as continuous infusion for 14 days, only a very minor effect was observed with SST14 and none with CH275. We interpret the lack of efficacy when given as infusion to be due to internalization of the receptor via continuous exposure to ligand. Thus also the mode of administration in clinical angiopeptin studies might have been wrong.

Interesting in this context are also the results of Rohrer et al. (1998) on the functional activity of SSTR subtype-selective non-peptide SST analogs on cultured cells, providing additional - although indirect - evidence for targeting to SST1,4 family in the prophylaxis and treatment of intimal hyperplasia. As referred to earlier, they synthetized five non-peptide agonist ligands selective to any one of the five SSTRs. The proper function of the receptor was documented with cAMP in CHO K1 cells, growth hormone release was investigated from primary cultures of rat pituitary cells and glucagon and insulin release from mouse pancreatic islet cell preparations. Native SST14 inhibited growth hormone release, glucagon release and insulin release in micromolar concentrations. Compounds selective for SSTR2 and 5 were inhibitory to all three parameters of SST function, whereas compounds selective to SSTR1,3 and 4 did not inhibit growth hormone, glucagon or insulin release in this experimental setting. Thus these results mirror the receptor expression studies in arteries, and suggest that the "classical" somatostatin functions on peripheral organs are mediated by a set of receptors which are not prominently expressed in the vascular wall; instead, none of the "classical" functions were mediated by those receptors which are prominently expressed, namely SST1,3 and 4 (Rohrer et al., 1998).

## SPECULATIONS ON THE MECHANISM OF SOMATOSTATIN EFFECTS ON THE VASCULAR WALL

Shared within the superfamily of G-protein coupled receptors, the SSTRs have been reported to modulate many cellular effector proteins like adenylate cyclase, phospolipase C, calcium channels, potassium channels, NA+/H+ exchanger, protein tyrosine phosphatases, phospholipase A2, mitogen-activated protein kinase (Patel, 1999) . The receptors are coupled to several intracellular transduction cascades, both via pertusis toxin sensitive (Gi/G0) and -insensitive G-proteins (Gq, G14, G16). It is likely that the different SSTR subtypes employ different transduction systems, depending on the cell type concerned. As these events are discussed in detail elsewhere in this book, we will confine ourselves on the mechanism of action of SST and its analogs on cell types in the vascular wall.

### Cell Types in the Vascular Wall and Clonality of the Lesion

The vascular wall comprises of adventitial fibroblast-type cells and additional cell types of connective tissue, smooth muscle cells confined primarily in the media and expressing the characteristic "contractile" phenotype and endothelial cells of the inner layer of the vessel and the vasa vasorum within the wall itself. In contrast to rodents or normolipidemic rabbits that do not poses intima, human vessels also contain an intmal layer which develops concomitantly with age (Schwartz et al., 1995) and this seems to be the constellation also with primates (Du Toit et al., 2001).

Contrary to general belief, the vascular wall cells are not in static condition, but evidence exists that they are replaced during the process of ageing. This may be particularly the case in (neo)intima formation. This was first documented (but already forgotten) by Benditt (Benditt and Benditt, 1973), demonstrating that the lesions in human arteries are "clonal", resembling to a certain extent neoplastic lesions. The finding of Benditt also suggested that the cells in various atherosclerotic lesions share a common ancesteral origin, and are not necessarily derived from the juxtaposed media layer as commonly believed (Thyberg et al., 1990).

The origin of neointimal cells has been documented best in allograft models. In chronic rejection, the allograft vessel endothelial cells seem to retain the donor phenotype, whereas the cells in the neointima (at least in rodent models), are recipient-derived (Plissonnier et al., 1995). Thus they are definitely not derived from the vascular media of the transplant vessels. This is also likely, though not proven, in the case of the neointimal cells after endothelial injury. The number of cells (nuclei) in the intima after endothelial denudation in rat, cannot be explained via depletion of media cells and/or by proliferation of cells in the adventitia and the media (Du Toit et al., 2001). A likely source of the neointimal cells is the pluripotent stem cell, which would also explain the "clonality" observed by Benditt (Benditt and Benditt, 1973). The mechanisms whereby SST regulates these events of neointimal formation are still a matter of speculation.

**Growth Factors as Mediators of Somatostatin Effects in the Vascular Wall**

The observation of the Foegh-Ramwell group (Lundergan et al., 1989) that out of the five synthetic SST-like peptides only two, angiopeptin and BIM23034, inhibited neointimal formation although all of them inhibited the release of growth hormone, suggests that regulation of growth hormone is not a prime pathway of SST effects in the vascular wall. Furthermore, with the lipidemic rabbit model (Howell et al., 1993) they showed that angiopeptin treatment, though inhibitory to neointimal formation, did not significantly modify the plasma lipid levels. These observations suggest that the effects of SST on intimal hyperplasia are not systemic, but mediated rather by local effects to the cells in the vascular wall.

In our own study with rat aortic allografts (Hayry et al., 1993b) we found that angiopeptin treatment of the recipient rat at a flow rate of 80 mg/kg/d with osmotic minipumps, significantly inhibiting proliferation of smooth muscle cells and their appearance in the intima. The treatment did not reduce the expression of interleukin 1, thromboxane B2 or 6-keto-PGF1-apha proteins but, instead, significantly reduced the EGF, IGF-1 and PDGF protein levels. The finding suggests that some of the effects of SST peptides may be mediated via the regulation of these ligands and/or their receptors. In another study (Mennander et al., 1993) we were not able to demonstrate any effect of angiopeptin on the intensity of perivascular inflammation or proliferation of inflammatory cells, focusing the likely effects of SST to the smooth muscle cells and/or their precursors and/or the endothelial cells. More generally the result suggested that SST peptides regulate the vascular effects via interacting with the local autocrine or paracrine mechanisms regulating cell growth and/or the mechanisms recruiting the precursors of smooth myscle cells into the intima of the injured vessel. The regulation of the autocrine/ paracrine effects of two growth factors, IGF-1 and PGDF, by SST peptides has been of particular interest.

IGF-1 is a major growth factor regulating cell replication in the vascular wall. Blocking of IGF-1 binding to receptor by non-degradable IGF-1 mimetic (JB3, composed of D- rather than L-amino acids), significantly inhibited cell replication and intimal thickening in denuded rat carotid compared to scrambled peptide (Hayry et al., 1995). To the contrary, JB3 had no effect on the replication of primary smooth muscle cell lines in vitro (Hayry et al., 1995). The same peptide equally inhibited retinal neovascularization in mice in a model where proliferative retinopathy has been well-characterized for VEGF-dependence (Smith et al., 1999).

Several studies clearly suggest that SST regulates the IGF-1 axis in the vascular wall. Angiopeptin at a dose level of only 10 ug/kg twice daily, inhibited the increase of immunoreactive IGF-1 on days 1, 2 and 4 after balloon catheterization of iliac arteries and aorta in rabbit (Howell et al., 1994). In human coronary arteries, IGF-1 is expressed on a higher level in diseased vessels compared to normal controls and both Sandostatin and Angiopeptin inhibited the IGF-1 and basic FGF-induced human coronary artery smooth muscle cell proliferation in vitro (Grant et al., 1994) . During the Scandinavian clinical trial for the prevention of restenosis after PTCA, Frystyk et al. (1996) found that lanreotide administration decreased circulating serum total and free IGF-1 (and increased IGF-binding protein 1) on day 4 postoperatively, though the difference to the placebo group was numerically not that great. Finally, treatment with Sandostatin decreased IGF-1 mRNA expression in normal rat arteries by 70%, and prevented the induction of IGF-1 gene after balloon injury (Yumi et al., 1997), though the upregulation of platelet-derived growth factor A-gene was not affected. Because in this study there was no chance in plasma growth hormone, IGF-1 and glucagon, their results suggested that the octreotide effect was selective to the IGF-1 axis and mediated locally.

Another set of paracrine events in the vascular wall, possibly regulated by SST peptides (Hayry et al., 1993b) is the PDGF axis. PDGF A and B and the receptors PDGFR-$\alpha$ and -$\beta$ have been shown to be intimately involved in the generation of arterial fibroproliferative dysplasias. This was first demonstrated by the late Russell Ross and his collaborators when blocking the PDGF action with proper monoclonal antibodies not only in rodents (Ferns et al., 1991) but also in subhuman primates (Kraiss et al., 1993). In our own studies we used small molecular weight antagonists preventing the autophosphorylation of the PDGF receptor, CGP53716 later STI573. Administration of the antagonist at a dose rate of 50 ug/kg/d inhibited intimal hyperplasia after rat carotid injury (Myllarniemi et al., 1997) and significantly reduced allograft arteriosclerosis in suboptimally immunosuppressed rat cardiac allografts (Sihvola et al., 1999). Considering that angiopeptin administration significantly reduced the PDGF-expression after aorta transplantation (Häyry et al., 1993b), it is possible that the PDGF-PDGFR axis may be another site of SST regulation in the vascular wall.

### Somatostatin Effects on Intracellular Events *in vivo*

Very little information exists on the effects of SST peptides on intracellular events in the vascular wall. Induction of two proto-oncogenes, c-fos and c-jun, is one of the earliest events associated with balloon denudation in rabbit (Bauters et al., 1992) and rat (Miano et al., 1990), occurring within an hour of the procedure. Angiopeptin induced an about 40% reduction in c-fos and c-jun expression compared to saline treated controls when as little as 20 ug/kg/d of angiopeptin was administered to the rabbits, but the treatment was initiated at 24 h before the operation (Bauters et al., 1994).

Octreotide seems to induce phosphotyrosine phosphatase (PTP) activity and prevents the reduction of serine/threonine phosphatase (PP2A) activity upon arterial injury in rats. Enhanced protein phosphatase activity increases the pool of nonphosphorylated transcription factors (specifically fos/jun dimer activator protein AP1 and NF-κB) and thus, yields a two-fold reduction in smooth muscle cell proliferation at day 3 post balloon injury (Yamashita et al., 1999).

Hence, taken together, these observations are concordant with the paradigm that one major mechanism of SST action in the vascular wall is the regulation of paracrine and intracrine events of various growth factors and their receptors, necessary for the replication and locomotion of smooth muscle cells or their precursors.

### *In vitro* Effects of Somatostatins

Alderton et al. (1998) used *in vitro* "denudation" of rat aortic smooth muscle cell monolayers by wounding a confluent cell layer with a perspex comb and producing parallel cell free areas on coverslips. "Post denudation" the cells were grown in serum free medium and basic FGF (10 ng/ml) was used to stimulate the regrowth of these cells. SST14 caused concentration-dependent [0.1-1000 nmol] inhibition of basic FGF-stimulated regrowth of smooth muscle cells, whereas angiopeptin displayed weak agonistic activity only, inhibiting basic FGF-stimulated regrowth at concentrations greater than 100 nmol. The findings suggest that although both families of SSTRs are growth inhibitory, the anti-proliferative effect of the SSTR2,5 family on smooth muscle cells is relatively weak and that the strong anti-proliferative effect on native SST14 may be attributed to SSTRs 1 and 4 rather than to SSTR2 and 5. The previously reported discrepancy on the lack of inhibition of SST on the replication of rat media-derived smooth muscle cells *in vitro*, contrasting the anti-proliferative effect and inhibition intimal thickening *in vivo* (Hayry et al., 1993b), should also be explained. This is difficult to explain unless the paradigm that the intimal cells derive from the media (see; for example Thyberg et al., 1990), is basically wrong.

Different *in vitro* results have been obtained when arterial explants and not propagated vascular smooth muscle cells have been used in the studies. Sakamoto et al. (1998) investigated the effects of Sandostatin in canine carotid and coronary arterial explants. After 6 days of culture, smooth mucle cells grew in 33% and 58% of the explants of injured canine coronary tissues, respectively; in contrast, smooth muscle cell outgrowth was not observed from any of the explants on normal canine arterial tissue. Sandostatin completely inhibited smooth muscle cell outgrowth in injured canine carotid arterial tissue at a concentration of $10^{-6}$ mol. Sarkar et al. (1999) used media transfer and coculture to investigate the effects of various SST isoforms and analogs on the proliferation of smooth muscle cells, mitogenesis of serum restimulated quiecent smooth muscle cells and on arterial explants according to Vargas (Vargas et al., 1997). Sandostatin, Angiopeptin and native SST-14 had no effect on the replication or growth of canine or rat smooth muscle cell lines; however, in the explants, Angiopeptin (75 umol/l) significantly inhibited DNA synthesis within the first 24 hours. Similar results were observed with rat aorta explants on fibronectin surface with Sandostatin (Danesi et al., 1997).

Taken together, these experiments demonstrate that cultured media-derived smooth muscle cell lines are relatively resistant to the inhibitory effects of SST, but that the inhibitory effects may be readily demonstrated if vascular explants, particularly deriving from denuded vessels, are used. Thus the explants may contain cells which have better growth potential than do media cells of resting vessels, or lines derived from media cells, and that the number of such cells in the explants is increased by a prior denudation procedure. These cells may well represent tissue specific stem cells in the vascular wall and/or pluripotent stem cells migrating to the wall as a consequence of previous *in vivo* injury.

## TIMING OF ADMINISTRATION

In acute injury models a variety of growth factors and their receptors are expressed shortly after the injury, concomitantly with the replication and migration of smooth muscle cells and their precursors into the intima. Concomitantly the expression of SSTR1 and 4 is upregulated. Thus, it is of interest to investigate what is the proper timing for SST administration for the prevention of injury and how does the timing regulate the putative regulators of SST effects.

Few such studies exist. Foegh et al. (1994) performed aorta and common and external iliac artery balloon injuries in the rabbit. Angiopeptin (up to 200 ug/kg/d) was administrated from day 1 before injury to days 1, 5 or 21 days after injury. Administration of angiopeptin reduced intimal thickening by approximately 50%, as evaluated at 21 days after injury. Treatment initiated at the time of injury was found to be effective, but delaying treatment for 8, 18 or 27 hours post injury abrogated the inhibitory effect on myointimal thickening. Thus administration of angiopeptin for two days was as efficacious as for 3 weeks, but delay in treatment for as little as 8 hours after injury abrogated the effect. This clearly speaks for the importance of early events and suggests that SST inhibits the expression of early genes causally related to vascular injury response and thereby triggering vascular cell proliferation.

In another study Bauters et al. (1994) used aortic balloon denudation in rabbits to optimize the administration schedule. Control animals received twice daily injections of saline beginning 24 h before balloon denudation, one group received angiopeptin 10 ug/kg/d beginning 24 h before denudation and the third group the same dose of angiopeptin beginning one hour after balloon denudation. The treatment was carried to 28 days when histological evaluation was performed. The degree of neointimal thickening was significantly less in the group where treatment was began on -24 hours, compared to controls, whereas the delay in the initiation of the treatment to 8 hours post injury largely abolished the angiopeptin effect. This result is remarkably similar to that of Foegh et al. (1994).

## PARADIGM FOR DRUG DESIGN

This literature survey demonstrates that only a limited repertoire of SSTRs is expressed in the arterial wall, particularly in human carotid and coronary arteries and in aorta, the sites of lesion formation in atherosclerosis and in preclinical models of vascular injury. The receptors present in noninjured vascular wall are SSTR1 and 4 rather than SSTR2 and 5. The SSTR1 and 4 subtypes are strongly upregulated as a consequence of injury in rat and SSTR1 is the predominant (only) receptor in atherosclerotic human arteries. Although all SSTRs may display some growth regulatory potential, this potential seems to reside in the SSTR1,4 rather than the SSTR2,5 family. In addition, considering the differential in the expression of the SSTR1,4 in the injured arterial wall vs., for example, in the alimentary track, the results suggest that targeting to SSTR1 (and 4) instead of SSTR2,5, would be a basis for rational drug design in vasculoprotective SST analogs.

Second, it seems plausible to suggest that the current paradigm of the intimal cells deriving from the vascular media is wrong. A more plausible explanation is that the intimal cells derive from distant pluripotent stem cells. It also appears that the growth inhibitory effect of SST is directed to early events in the induction of repair response to vascular injury. It seems likely that at least some of these effects are mediated via interference with the paracrine and autocrine events of several growth factors, including IGF-1 and PDGF, and on the signalling cascades downstream of these receptors.

In terms of acute injury, it seems mandatory that SST treatment is initiated immediately after injury or, preferably, before injury has been performed. It is not sure how long this treatment should be continued but this may depend on the species involved. In the rat, the re-endothelialization occurs slowly (compared to primates including humans) and it may be that treatment should be carried until the endothelial lining has been completely re-established.

## ACKNOWLEDGEMENTS

This study was supported by grants from the Technology Development Centre (TEKES), the Academy of Finland and Helsinki University Hospital Research Funds, Helsinki, Finland.

## REFERENCES

1. Alderton, F., Lauder, H., Feniuk, W., Fan, T. P., and Humphrey, P. P. (1998). Differential effects of somatostatin and angiopeptin on cell proliferation. Br J Pharmacol 124, 323-30.
2. Bass, R. T., Buckwalter, B. L., Patel, B. P., Pausch, M. H., Price, L. A., Strnad, J., and Hadcock, J. R. (1996). Identification and characterization of novel somatostatin antagonists. Mol Pharmacol 50, 709-15.
3. Bauters, C., de Groote, P., Adamantidis, M., Delcayre, C., Hamon, M., Lablanche, J. M., Bertrand, M. E., Dupuis, B., and Swynghedauw, B. (1992). Proto-oncogene expression in rabbit aorta after wall injury. First marker of the cellular process leading to restenosis after angioplasty? Eur Heart J 13, 556-9.
4. Bauters, C., Van Belle, E., Wernert, N., Delcayre, C., Thomas, F., Dupuis, B., Lablanche, J. M., Bertrand, M. E., and Swynghedauw, B. (1994). Angiopeptin inhibits oncogene induction in rabbit aorta after balloon denudation. Circulation 89, 2327-31.
5. Benditt, E. P., and Benditt, J. M. (1973). Evidence for a monoclonal origin of human atherosclerotic plaques. Proc Natl Acad Sci U S A 70, 1753-6.
6. Bruns, C., Shi, V., Hoyer, D., Schuurman, H., and Weckbecker, G. (2000). Somatostatin receptors and the potential use of Sandostatin to interfere with vascular remodelling. Eur J Endocrinol 143, S3-7.
7. Chen, J. C., Hsiang, Y. N., and Buchan, A. M. (1997). Somatostatin receptor expression in rat iliac arteries after balloon injury. J Invest Surg 10, 17-23.
8. Chen, L., Hoeger, C., Rivier, J., Fitzpatrick, V. D., Vandlen, R. L., and Tashjian, A. H. (1999). Structural basis for the binding specificity of a SSTR1-selective analog of somatostatin. Biochem Biophys Res Commun 258, 689-94.
9. Curtis, S. B., Chen, J. C., Winkelaar, G., Turnbull, R. G., Hewitt, J., Buchan, A. M., and Hsiang, Y. N. (2000a). Effect of endothelial and adventitial injury on somatostatin receptor expression. Surgery 127, 577-83.
10. Curtis, S. B., Hewitt, J., Yakubovitz, S., Anzarut, A., Hsiang, Y. N., and Buchan, A. M. (2000b). Somatostatin receptor subtype expression and function in human vascular tissue. Am J Physiol Heart Circ Physiol 278, H1815-22.
11. Danesi, R., Agen, C., Benelli, U., Paolo, A. D., Nardini, D., Bocci, G., Basolo, F., Campagni, A., and Tacca, M. D. (1997). Inhibition of experimental angiogenesis by

286

the somatostatin analogue octreotide acetate (SMS 201-995). Clin Cancer Res 3, 265-72.

12. Du Toit, D., Aavik, E., Taskinen, E., Myburgh, E., Aaltola, E., Aimonen, M., Aavik, S., van Wyk, J., and Häyry, P. (2001). Structure of carotid artery in baboon and rat and differences in their response to endothelial denudation angioplasty. 2001 33, 63-78.

13. Emanuelsson, H., Beatt, K. J., Bagger, J. P., Balcon, R., Heikkila, J., Piessens, J., Schaeffer, M., Suryapranata, H., and Foegh, M. (1995). Long-term effects of angiopeptin treatment in coronary angioplasty. Reduction of clinical events but not angiographic restenosis. European Angiopeptin Study Group. Circulation 91, 1689-96.

14. Eriksen, U. H., Amtorp, O., Bagger, J. P., Emanuelsson, H., Foegh, M., Henningsen, P., Saunamaki, K., Schaeffer, M., Thayssen, P., and Orskov, H. (1995). Randomized double-blind Scandinavian trial of angiopeptin versus placebo for the prevention of clinical events and restenosis after coronary balloon angioplasty. Am Heart J 130, 1-8.

15. Ferns, G. A., Raines, E. W., Sprugel, K. H., Motani, A. S., Reidy, M. A., and Ross, R. (1991). Inhibition of neointimal smooth muscle accumulation after angioplasty by an antibody to PDGF. Science 253, 1129-32.

16. Foegh, M. L., Asotra, S., Conte, J. V., Howell, M., Kagan, E., Verma, K., and Ramwell, P. W. (1994). Early inhibition of myointimal proliferation by angiopeptin after balloon catheter injury in the rabbit. J Vasc Surg 19, 1084-91.

17. Foegh, M. L., and Ramwell, P. W. (1995). Angiopeptin: experimental and clinical studies of inhibition of myointimal proliferation. Kidney Int Suppl 52, S18-22.

18. Frystyk, J., Skjaerbaek, C., Alexander, N., Emanuelsson, H., Suryapranata, H., Beyer, H., Foegh, M., and Orskov, H. (1996). Lanreotide reduces serum free and total insulin-like growth factor-I after angioplasty. Circulation 94, 2465-71.

19. Grant, M. B., Wargovich, T. J., Ellis, E. A., Caballero, S., Mansour, M., and Pepine, C. J. (1994). Localization of insulin-like growth factor I and inhibition of coronary smooth muscle cell growth by somatostatin analogues in human coronary smooth muscle cells. A potential treatment for restenosis? Circulation 89, 1511-7.

20. Hayry, P., Isoniemi, H., Yilmaz, S., Mennander, A., Lemstrom, K., Raisanen-Sokolowski, A., Koskinen, P., Ustinov, J., Lautenschlager, I., and Taskinen, E. (1993a). Chronic allograft rejection. Immunol Rev 134, 33-81.

21. Hayry, P., Myllarniemi, M., Aavik, E., Alatalo, S., Aho, P., Yilmaz, S., Raisanen-Sokolowski, A., Cozzone, G., Jameson, B. A., and Baserga, R. (1995). Stabile D-peptide analog of insulin-like growth factor-1 inhibits smooth muscle cell proliferation after carotid ballooning injury in the rat. Faseb J 9, 1336-44.

22. Hayry, P., Raisanen, A., Ustinov, J., Mennander, A., and Paavonen, T. (1993b). Somatostatin analog lanreotide inhibits myocyte replication and several growth factors in allograft arteriosclerosis. Faseb J 7, 1055-60.

23. Hong, M. K., Kent, K. M., Tio, F. O., Foegh, M., Kornowski, R., Bramwell, O., Cathapermal, S. S., and Leon, M. B. (1997). Single-dose intramuscular administration of sustained-release Angiopeptin reduces neointimal hyperplasia in a porcine coronary in-stent restenosis model. Coron Artery Dis 8, 101-4.

24. Howell, M., Orskov, H., Frystyk, J., Flyvbjerg, A., Gronbaek, H., and Foegh, M. (1994). Lanreotide, a somatostatin analogue, reduces insulin-like growth factor I accumulation in proliferating aortic tissue in rabbits in vivo. A preliminary study. Eur J Endocrinol 130, 422-5.

25. Howell, M. H., Adams, M. M., Wolfe, M. S., Foegh, M. L., and Ramwell, P. W. (1993). Angiopeptin inhibition of myointimal hyperplasia after balloon angioplasty of large arteries in hypercholesterolaemic rabbits. Clin Sci (Colch) 85, 183-8.

26. Khare, S., Kumar, U., Sasi, R., Puebla, L., Calderon, L., Lemstrom, K., Hayry, P., and Patel, A. Y. (1999). Differential regulation of somatostatin receptor types 1-5 in rat aorta after angioplasty. Faseb J 13, 387-94.

27. Kraiss, L. W., Raines, E. W., Wilcox, J. N., Seifert, R. A., Barrett, T. B., Kirkman, T. R., Hart, C. E., Bowen-Pope, D. F., Ross, R., and Clowes, A. W. (1993). Regional expression of the platelet-derived growth factor and its receptors in a primate graft model of vessel wall assembly. J Clin Invest 92, 338-48.

28. Lauder, H., Sellers, L. A., Fan, T. P., Feniuk, W., and Humphrey, P. P. (1997). Somatostatin sst5 inhibition of receptor mediated regeneration of rat aortic vascular smooth muscle cells. Br J Pharmacol 122, 663-70.

29. Liapakis, G., Hoeger, C., Rivier, J., and Reisine, T. (1996). Development of a selective agonist at the somatostatin receptor subtype sstr1. J Pharmacol Exp Ther 276, 1089-94.

30. Lundergan, C., Foegh, M. L., Vargas, R., Eufemio, M., Bormes, G. W., Kot, P. A., and Ramwell, P. W. (1989). Inhibition of myointimal proliferation of the rat carotid artery by the peptides, angiopeptin and BIM 23034. Atherosclerosis 80, 49-55.

31. Mennander, A., Raisanen, A., Paavonen, T., and Hayry, P. (1993). Chronic rejection in the rat aortic allograft. V. Mechanism of the angiopeptin (BIM23014C) effect on the generation of allograft arteriosclerosis. Transplantation 55, 124-8.

32. Miano, J. M., Tota, R. R., Vlasic, N., Danishefsky, K. J., and Stemerman, M. B. (1990). Early proto-oncogene expression in rat aortic smooth muscle cells following endothelial removal. Am J Pathol 137, 761-5.

33. Myllarniemi, M., Calderon, L., Lemstrom, K., Buchdunger, E., and Hayry, P. (1997). Inhibition of platelet-derived growth factor receptor tyrosine kinase inhibits vascular smooth muscle cell migration and proliferation. Faseb J 11, 1119-26.

34. Patel, Y. C. (1999). Somatostatin and its receptor family. Front Neuroendocrinol 20, 157-98.

35. Patel, Y. C., Greenwood, M. T., Panetta, R., Demchyshyn, L., Niznik, H., and Srikant, C. B. (1995). The somatostatin receptor family. Life Sci 57, 1249-65.

36. Patel, Y. C., and Srikant, C. B. (1994). Subtype selectivity of peptide analogs for all five cloned human somatostatin receptors (hsstr 1-5). Endocrinology 135, 2814-7.

37. Plissonnier, D., Nochy, D., Poncet, P., Mandet, C., Hinglais, N., Bariety, J., and Michel, J. B. (1995). Sequential immunological targeting of chronic experimental arterial allograft. Transplantation 60, 414-24.

38. Reisine, T., and Bell, G. I. (1995). Molecular biology of somatostatin receptors. Endocr Rev 16, 427-42.

39. Reubi, J. C., Horisberger, U., and Laissue, J. (1994a). High density of somatostatin receptors in veins surrounding human cancer tissue: role in tumor-host interaction? Int J Cancer 56, 681-8.

40. Reubi, J. C., Mazzucchelli, L., and Laissue, J. A. (1994b). Intestinal vessels express a high density of somatostatin receptors in human inflammatory bowel disease. Gastroenterology 106, 951-9.

41. Reubi, J. C., Schaer, J. C., Laissue, J. A., and Waser, B. (1996a). Somatostatin receptors and their subtypes in human tumors and in peritumoral vessels. Metabolism 45, 39-41.

42. Reubi, J. C., Schaer, J. C., Wenger, S., Hoeger, C., Erchegyi, J., Waser, B., and Rivier, J. (2000). SST3-selective potent peptidic somatostatin receptor antagonists. Proc Natl Acad Sci U S A 97, 13973-8.

43. Reubi, J. C., Waser, B., Laissue, J. A., and Gebbers, J. O. (1996b). Somatostatin and vasoactive intestinal peptide receptors in human mesenchymal tumors: in vitro identification. Cancer Res 56, 1922-31.

44. Rohrer, S. P., Birzin, E. T., Mosley, R. T., Berk, S. C., Hutchins, S. M., Shen, D. M., Xiong, Y., Hayes, E. C., Parmar, R. M., Foor, F., Mitra, S. W., Degrado, S. J., Shu, M., Klopp, J. M., Cai, S. J., Blake, A., Chan, W. W., Pasternak, A., Yang, L., Patchett, A. A., Smith, R. G., Chapman, K. T., and Schaeffer, J. M. (1998). Rapid identification of subtype-selective agonists of the somatostatin receptor through combinatorial chemistry. Science 282, 737-40.

45. Ross, R. (1999). Atherosclerosis--an inflammatory disease. N Engl J Med 340, 115-26.

46. Sakamoto, H., Sakamaki, T., Kanda, T., Ito, Y., Sumino, H., Masuda, H., Ohyama, Y., Ono, Z., Kurabayashi, M., Kobayashi, I., and Nagai, R. (1998). The somatostatin analog, octreotide, inhibits in vitro outgrowth of smooth muscle cells from canine coronary and carotid atherosclerotic plaque tissues. Res Commun Mol Pathol Pharmacol 101, 25-34.

47. Santoian, E. D., Schneider, J. E., Gravanis, M. B., Foegh, M., Tarazona, N., Cipolla, G. D., and King, S. B. (1993). Angiopeptin inhibits intimal hyperplasia after angioplasty in porcine coronary arteries. Circulation 88, 11-4.

288

48. Sarkar, R., Dickinson, C. J., and Stanley, J. C. (1999). Effects of somatostatin, somatostatin analogs, and endothelial cell somatostatin gene transfer on smooth muscle cell proliferation in vitro. J Vasc Surg 29, 685-93.
49. Schuurman, H. J., Beckmann, N., Briner, U., Bruns, C., Bruttel, K., Tanner, M., Tolcsvai, L., and Weckbecker, G. (1996). Magnetic resonance imaging in assessment of rejection of a kidney allograft in the rat: effect of the somatostatin analogue SMS 201-995. Transplant Proc 28, 3272-5.
50. Schwartz, S. M., deBlois, D., and O'Brien, E. R. (1995). The intima. Soil for atherosclerosis and restenosis. Circ Res 77, 445-65.
51. Schwartz, S. M., and Murry, C. E. (1998). Proliferation and the monoclonal origins of atherosclerotic lesions. Annu Rev Med 49, 437-60.
52. Sihvola, R., Koskinen, P., Myllarniemi, M., Loubtchenkov, M., Hayry, P., Buchdunger, E., and Lemstrom, K. (1999). Prevention of cardiac allograft arteriosclerosis by protein tyrosine kinase inhibitor selective for platelet-derived growth factor receptor. Circulation 99, 2295-301.
53. Smith, L. E., Shen, W., Perruzzi, C., Soker, S., Kinose, F., Xu, X., Robinson, G., Driver, S., Bischoff, J., Zhang, B., Schaeffer, J. M., and Senger, D. R. (1999). Regulation of vascular endothelial growth factor-dependent retinal neovascularization by insulin-like growth factor-1 receptor. Nat Med 5, 1390-5.
54. Thyberg, J., Hedin, U., Sjolund, M., Palmberg, L., and Bottger, B. A. (1990). Regulation of differentiated properties and proliferation of arterial smooth muscle cells. Arteriosclerosis 10, 966-90.
55. Vargas, R., Wroblewska, B., Rego, A., Cathapermal, S., and Ramwell, P. W. (1997). Angiopeptin inhibits thymidine incorporation by explants of porcine coronary arteries. J Cardiovasc Pharmacol 29, 278-83.
56. von Essen, R., Ostermaier, R., Grube, E., Maurer, W., Tebbe, U., Erbel, R., Roth, M., Oel, W., Brom, J., and Weidinger, G. (1997). Effects of octreotide treatment on restenosis after coronary angioplasty: results of the VERAS study. VErringerung der Restenoserate nach Angioplastie durch ein Somatostatin-analogon. Circulation 96, 1482-7.
57. Weckbecker, G., Pally, C., Raulf, F., Beckmann, N., Schuurman, J. H., Rudin, M., and Bruns, C. (1997). The somatostatin analog octreotide as potential treatment for re-stenosis and chronic rejection. Transplant Proc 29, 2599-600.
58. Yamashita, M., Dimayuga, P., Kaul, S., Shah, P. K., Regnstrom, J., Nilsson, J., and Cercek, B. (1999). Phosphatase activity in the arterial wall after balloon injury: effect of somatostatin analog octreotide. Lab Invest 79, 935-44.
59. Yumi, K., Fagin, J. A., Yamashita, M., Fishbein, M. C., Shah, P. K., Kaul, S., Niu, W., Nilsson, J., and Cercek, B. (1997). Direct effects of somatostatin analog octreotide on insulin-like growth factor-I in the arterial wall. Lab Invest 76, 329-38.

# 18

# THE USE OF SOMATOSTATIN ANALOGUES IN DIABETIC RETINOPATHY

Lois E.H. Smith and Maria B. Grant
*Department of Ophthalmology, Children's Hospital, Harvard Medical School, Boston MA, 02115*
*Department of Pharmacology and Therapeutics, University of Florida College of Medicine, Gainesville, FL. 32610*

Diabetic retinopathy is the most common cause of blindness in the working age population in the Western world. Before the advent of intensive therapy with the goal of near normalization of blood glucose levels, greater than 90% of diabetics developed some degree of retinopathy after 20 years of diabetes (1,2). Despite the emphasis on "tight" blood glucose control supported by the Diabetes Control and Complications Trial and the United Kingdom Prospective Diabetes Study, tight control will not be possible in all patients and thus diabetic retinopathy will thus remain a clinically relevant complication. Diabetic Retinopathy occurs in two phases, non-proliferative retinopathy and proliferative retinopathy.

Non-proliferative or background retinopathy can be seen as early as one year after diagnosis of diabetes. The initial signs are vascular basement membrane thickening, pericyte loss, development of microaneurysms and acellular capillaries. These early "background" changes in the retinal vessels are not usually associated with visual loss, although there can be increased permeability leading to exudates (serum protein deposits) or to macular edema. Vascular occlusion starts in the capillary bed but eventually includes venules and arterioles. This mid-stage disease process is characterized by vascular abnormalities such as beading, larger areas of capillary non-perfusion, infarcts or cotton wool spots and areas of hemorrhage.

The proliferative phase of diabetic retinopathy occurs in a subset of diabetic patients. In this group, non-perfused retina becomes hypoxic and produces angiogenic growth factors such as vascular endothelial growth factor (VEGF), which stimulate new blood vessel growth. This proliferative retinopathy or neovascularization can eventually lead to blindness. The current standard treatment for proliferative retinopathy is ablation of the retina by photocoagulation at a defined stage of the disease, destroying the local cellular source of the growth factors but also damaging neural retina. A therapeutic approach that targets growth factor function or directly inhibits vascular growth would be preferable to the current destructive treatment.

Some of the molecular pathways and angiogenic factors that lead to new blood vessel growth have been elucidated although the interaction among factors is still in the early stages of discovery. Experiments in several model systems have indicated that somatostatin (SST) receptors (SSTR) are an attractive therapeutic target for inhibition of neovascularization. Pre-clinical and experimental clinical studies suggest that SST analogues may also be effective in directly inhibiting growth in many types of cells and in tumors (3-5). Activation of SSTR inhibits angiogenesis in several model systems (6,7) in addition to reducing neovascularization in diabetic retinopathy (8-11) and retinopathy of prematurity (12). Initially identified as a hypothalamic hormone capable of blocking growth hormone (GH) release from the anterior pituitary (13), SST, is now known to inhibit a wide range of peptides. SST analogues are currently approved for clinical use to treat conditions ranging from peptic ulcers to endocrine tumors.

A role for GH in proliferative diabetic retinopathy (PDR) was first suggested by Poulsen who described PDR regression in a post-partum woman with spontaneous pituitary infarction resulting in hypopituitarism. This led to the hypothesis that hypophysectomy might be a treatment for PDR. Subsequent controlled clinical trials demonstrated that hypophysectomy could improve PDR (14,15), and that the improvement correlated with the degree of serum GH reduction. However because of the difficulty in clinical management of diabetic individuals with panhypopitutarism, pituitary ablation was not an ideal therapeutic modality.

Inhibition of GH secretion by SST analogs has been the basis of several clinical trials involving patients with PDR (9-11). In the earliest clinical trials, SST analogs did not show consistent efficacy. However, these studies were limited to a short dosing duration and doses were often inadequate to suppress serum GH or IGF-I levels. Kirkegaard and coworkers treated patients with background retinopathy with the SST analogue, octreotide and reported no improvement in clinical outcome compared to untreated controls (16). The SST analog (BIM 23014/Lanreotide) was administered via continuous infusion pump for 4 weeks to 17 patients with non-high risk PDR. Eight of 11 subjects completed the dosing schedule with the SST analog. The retinal disease did not progress in 6 patients and showed signs of regression in 2 patients, whereas there was disease progression in half of the 6 control subjects (9). Visual acuity also improved during several months of continuous infusion with the commercially available SST analog, octreotide, in an uncontrolled clinical trial of 4 patients (10).

In a longer term study results have been much more promising. In a randomized prospective controlled study compared the effect of long-term administration of the highest tolerated dose of octreotide with conventional diabetes management in 22 patients with early proliferative and late preproliferative diabetic retinopathy. The intent of the investigators was to lower serum IGF-1 into the hypopituitary range of <75 ng/ml. Only one of 22 eyes (11 patients), compared to 9 of 24 control eyes, required photocoagulation therapy at the end of the 15-month treatment period. The time to development of high risk PDR was significantly longer for octreotide-treated patients than for control patients ($P$= 0.006). The incidence of retinal disease progression was only 27% in octreotide-treated patients and 42% in conventionally managed patients. The patients on octreotide also received thyroid replacement to avoid possible hypothyroidism associated with TSH suppression seen with chronic octreotide therapy (11). The dramatic effect of octreotide in reducing the need for laser therapy in these patients treated with exogenous thyroid hormone is particularly interesting when one considers that SSTR expression is up-regulated by T3 in the pituitary (17,18).

The effect of SST analogue on proliferative retinopathy may be mediated through a number of pathways. Several clinical studies have correlated serum IGF-I levels with retinal neovascularization. Merimee and colleagues found increased serum IGF-I levels in a subset of patients with "rapidly accelerating PDR" (19) and elevated levels of IGF-1 have been found in the vitreous of patients with PDR (20). In a subsequent prospective study, patients with elevated serum IGF-1 levels at the onset of retinal neovasularization – compared with 3 months prior – were shown to have an increased frequency of PDR (21). Other studies have failed to replicate the finding of elevated serum IGF-1 levels (22).

Animal models support a role for GH and IGF-1 in the progression of retinopathy. Genetically engineered mice that produce a GH receptor antagonist have been studied to define the role of GH in retinal neovascularization. Using these transgenic mice in a model of hypoxia-induced neovascularization designed to mimic proliferative retinopathy it was determined that GH inhibition decreased retinal neovascularization. In a parallel study of wild-type mice treated with the GH suppressing SST analogue MK-678, reduced neovascularization was observed compared to untreated mice. These experiments suggest that the inhibitory effect of SST analogues is mediated through the suppression of GH secretion or GH's effect on IGF-1. Consistent with the absence of an increased incidence of retinopathy in acromegalic patients, GH over-expressing (giant) mice did not have a significant increase in angiogenesis retinopathy compared to wild type controls (12,23,24). However, only 10 % of acromegalic patients actually have diabetes and this is usually mild disease associated with GH induced insulin resistance.

IGF-1 appears to be critical to the mechanism of SST inhibition of retinal neovascularization since neovascularization was not inhibited in mice treated with both the SST analog MK678 as well as exogenous IGF-1. Furthermore,

IGF-1 receptor antagonists can block hypoxia-induced retinal neovascularization, thereby bypassing the need to suppress GH and implicating IGF-1 as a mediator of proliferation (25). These animal studies support a role for SST analogues in the suppression of GH and IGF-1 production and the subsequent inhibition of retinal neovascularization.

The SSTR family consists of 5 major subtypes. All of the SST analogs discussed above have relative type 2 SSTR (SSTR2) specificity. Non-peptide and modified peptide analogues with greater receptor affinity and selectivity than the native hormone have been developed for clinical and experimental use (26-29). These analogues show differential affinities for each of the five SSTRs and have allowed investigators to determine functional effects mediated by individual receptor subtypes. Inhibition of GH secretion is most sensitive to SSTR2 agonists, while inhibition of insulin release is mediated primarily through SSTR5 receptors (30). The first generation of SST analogues have the highest affinity for SSTR2 but this selectivity is only 20 to 100 times greater for SSTR2 over SSTR3 and SSTR5. In contrast, several non-peptide analogues have receptor selectivity of 4 orders of magnitude or greater. Subtype-selective agonists may therefore be useful in modulating endothelial cell function without altering hormone secretion.

SST may reduce endothelial cell proliferation and neovascularization by several mechanisms, including inhibition of post-receptor signaling events of peptide growth factors such as IGF-1, VEGF, epidermal growth factor and platelet derived growth factor (6,31-34). SSTR2 and SSTR3-selective analogues inhibit retinal endothelial cell growth in vitro (33,35), and this direct inhibitory effect on retinal endothelial cell proliferation is achieved without altering "hormone" or growth factor secretion (35).

The signal transduction pathways used by SST analogues to reduce retinal endothelial cell viability have not been elucidated, but SSTR signaling has been investigated in other systems. SSTRs signal through 2 main pathways, often resulting in apoptosis or cell cycle arrest. The first pathway involves coupling through pertussis-toxin sensitive G-proteins of the $G\alpha_{i/o}$ class leading to inhibition of cyclic AMP (cAMP) production (36-38). This inhibition is thought to be the mechanism of action for the cytostatic effects of SST analogues in cells where cell proliferation is cAMP dependent (39). The second major signaling event initiated by SSTR activation leads to increased activation of the protein tyrosine phosphatases SHP-1 (40) and SHP-2 (41,42), which can inhibit signaling through tyrosine kinase receptors. SSTR stimulation can lead to both apoptosis and cell cycle arrest. Cells require cdk2/cyclin E activity and phosphorylation of Rb to traverse the $G_{1/S}$-phase checkpoint. SST analogue-induced cell cycle inhibition at this point is thought to result from induction of cdk inhibitors such as p21 and p27 (5,43). Inhibition of cdk2/cyclin E activity in response to SSTR2 stimulation has been reported in multiple cell types (5,39,41,44).

Cell cycle arrest does not precede apoptosis in cells treated with SST analogues specific for the type 3 SSTR (45). Activation of SSTR3 leads to increased synthesis of p53 and increased expression of p53-regulated genes, such as BAX, which regulate apoptosis (45,46). The combination of these data suggest that SST may promote both cell-cycle arrest and induce apoptosis independent of the modulation of systemic GH and IGF-1 levels. In addition, the cell- and receptor-specific effects indicate that SSTR signaling must be evaluated in individual cell types since there is little consensus on the mechanism of action for SST analogues at the cellular level. Retinal endothelial cells are sensitive to agonists that activate both SSTR2 and SSTR3. While both clinical and experimental evidence implicates the GH IGF-1 cascade in the efficacy of SSTR2-selective agonists, recent experimental evidence suggests a role for the SSTR3 in SST-induced endothelial cell apoptosis. A long acting, sustained-release preparation developed for the treatment of acromegaly is currently in a multi-center clinical trial for the treatment of severe non-proliferative and non high-risk proliferative diabetic retinopathy.

## REFERENCES

1.  Klein R, Klein BE, Moss SE. Relation of glycemic control to diabetic microvascular complications in diabetes mellitus. *Ann Intern Med*. 1996; 124: 90-96.
2.  Klein R, Klein BE, Moss SE. The Wisconsin epidemiological study of diabetic retinopathy: a review. *Diabetes Metab Rev*. 1989; 5: 559-570.
3.  Albini A, Florio T, Giunciuglio D, Masiello L, Carlone S, Corsaro A, Thellung S, Cai T, Noonan DM, Schettini G. Somatostatin controls Kaposi's sarcoma tumor growth through inhibition of angiogenesis. *Faseb J*. 1999; 13: 647-655.
4.  Cattaneo MG, Amoroso D, Gussoni G, Sanguini AM, Vicentini LM. A somatostatin analogue inhibits MAP kinase activation and cell proliferation in human neuroblastoma and in human small cell lung carcinoma cell lines. *FEBS Lett*. 1996; 397: 164-168.
5.  Charland S, Boucher MJ, Houde M, Rivard N. Somatostatin inhibits Akt phosphorylation and cell cycle entry, but not p42/p44 mitogen-activated protein (MAP) kinase activation in normal and tumoral pancreatic acinar cells. *Endocrinology*. 2001; 142: 121-128.
6.  Lawnicka H, Stepien H, Wyczolkowska J, Kolago B, Kunert-Radek J, Komorowski J. Effect of somatostatin and octreotide on proliferation and vascular endothelial growth factor secretion from murine endothelial cell line (HECa10) culture. *Biochem Biophys Res Commun*. 2000; 268: 567-571.
7.  Danesi R, Del Tacca M. The effects of the somatostatin analog octreotide on angiogenesis in vitro. *Metabolism*. 1996; 45: 49-50.
8.  Lee HK, Suh KI, Koh CS, Min HK, Lee JH, Chung H. Effect of SMS 201-995 in rapidly progressive diabetic retinopathy [letter]. *Diabetes Care*. 1988; 11: 441-443.
9.  McCombe M, Lightman S, Eckland DJ, Hamilton AM, Lightman SL. Effect of a long-acting somatostatin analogue (BIM23014) on proliferative diabetic retinopathy: a pilot study. *Eye*. 1991; 5: 569-575.
10. Mallet B, Vialettes B, Haroche S, Escoffier P, Gastaut P, Taubert JP, Vague P. Stabilization of severe proliferative diabetic retinopathy by long-term treatment with SMS 201-995. *Diabetes Metab*. 1992; 18: 438-444.
11. Grant MB, Mames RN, Fitzgerald C, Hazariwala KM, Cooper-DeHoff R, Caballero S, Estes KS. The efficacy of octreotide in the therapy of severe nonproliferative and early

proliferative diabetic retinopathy: a randomized controlled study. *Diabetes Care*. 2000; 23: in press.

12. Smith LE, Kopchick JJ, Chen W, Knapp J, Kinose F, Daley D, Foley E, Smith RG, Schaeffer JM. Essential role of growth hormone in ischemia-induced retinal neovascularization. *Science*. 1997; 276: 1706-1709.

13. Brazeau P, Vale W, Burgus R, Ling N, Butcher M, Rivier J, Guillemin R. Hypothalamic polypeptide that inhibits the secretion of immunoreactive pituitary growth hormone. *Science*. 1973; 179: 77-79.

14. Patterson RH, Jr. Hypophysectomy: transfrontal technique and results in the management of metastatic cancer and diabetic retinopathy. *Clin Neurosurg*. 1974; 21: 60-67.

15. Arslan M. Ultrasonic selective hypophysectomy in Cushing's disease, acromegaly and diabetic retinopathy. *Acta Otolaryngol*. 1967; 63: 252-263.

16. Kirkegaard C, Norgaard K, Snorgaard O, Bek T, Larsen M, Lund-Andersen H. Effect of one year continuous subcutaneous infusion of a somatostatin analogue, octreotide, on early retinopathy, metabolic control and thyroid function in Type I (insulin-dependent) diabetes mellitus. *Acta Endocrinol (Copenh)*. 1990; 122: 766-772.

17. de los Frailes MT, Cacicedo L, Lorenzo MJ, Fernandez G, Sanchez-Franco F. Thyroid hormone action on biosynthesis of somatostatin by fetal rat brain cells in culture. *Endocrinology*. 1988; 123: 898-904.

18. Lam KS, Wong RL. Thyroid hormones regulate the expression of somatostatin receptor subtypes in the rat pituitary. *Neuroendocrinology*. 1999; 69: 460-464.

19. Merimee TJ, Zapf J, Froesch ER. Insulin-like growth factors: studies in diabetics with and without retinopathy. *N Engl J Med*. 1983; 309: 527-530.

20. Spranger J, Buhnen J, Jansen V, Krieg M, Meyer-Schwickerath R, Blum WF, Schatz H, Pfeiffer AF. Systemic levels contribute significantly to increased intraocular IGF- I, IGF-II and IGF-BP3 [correction of IFG-BP3] in proliferative diabetic retinopathy. *Horm Metab Res*. 2000; 32: 196-200.

21. Hyer SL, Sharp PS, Brooks RA, Burrin JM, Kohner EM. A two-year follow-up study of serum insulinlike growth factor-I in diabetics with retinopathy. *Metabolism*. 1989; 38: 586-589.

22. Janssen JA, Lamberts SW. Circulating IGF-I and its protective role in the pathogenesis of diabetic angiopathy. *Clin Endocrinol (Oxf)*. 2000; 52: 1-9.

23. Ballintine EJ, Foxman S, Gorden P, Roth J. Rarity of diabetic retinopathy in patients with acromegaly. *Arch Intern Med*. 1981; 141: 1625-1627.

24. Amemiya T, Toibana M, Hashimoto M, Oseko F, Imura H. Diabetic retinopathy in acromegaly. *Ophthalmologica*. 1978; 176: 74-80.

25. Smith LE, Shen W, Perruzzi C, Soker S, Kinose F, Xu X, Robinson G, Driver S, Bischoff J, Zhang B, Schaeffer JM, Senger DR. Regulation of vascular endothelial growth factor-dependent retinal neovascularization by insulin-like growth factor-1 receptor. *Nat Med*. 1999; 5: 1390-1395.

26. Rohrer SP, Birzin ET, Mosley RT, Berk SC, Hutchins SM, Shen DM, Xiong Y, Hayes EC, Parmar RM, Foor F, Mitra SW, Degrado SJ, Shu M, Klopp JM, Cai SJ, Blake A, Chan WW, Pasternak A, Yang L, Patchett AA, Smith RG, Chapman KT, Schaeffer JM. Rapid identification of subtype-selective agonists of the somatostatin receptor through combinatorial chemistry [published erratum appears in Science 1998 Nov 27;282(5394):1646]. *Science*. 1998; 282: 737-740.

27. Gillespie TJ, Erenberg A, Kim S, Dong J, Taylor JE, Hau V, Davis TP. Novel somatostatin analogs for the treatment of acromegaly and cancer exhibit improved in vivo stability and distribution. *J Pharmacol Exp Ther*. 1998; 285: 95-104.

28. Tejeda M, Gaal D, Schwab RE, Pap A, Szuts T, Keri G. Influence of various administration routes on the antitumor efficacy ofTT-232, a novel somatostatin analog. *Anticancer Res*. 2000; 20: 1023-1027.

29. Teplan I. Peptides and antitumor activity. Development and investigation of some peptides with antitumor activity. *Acta Biol Hung*. 2000; 51: 1-29.

30. Parmar RM, Chan WW, Dashkevicz M, Hayes EC, Rohrer SP, Smith RG, Schaeffer JM, Blake AD. Nonpeptidyl somatostatin agonists demonstrate that sst2 and sst5

inhibit stimulated growth hormone secretion from rat anterior pituitary cells. *Biochem Biophys Res Commun.* 1999; 263: 276-280.

31. Cattaneo MG, Scita G, Vicentini LM. Somatostatin inhibits PDGF-stimulated Ras activation in human neuroblastoma cells. *FEBS Lett.* 1999; 459: 64-68.

32. Paques M, Massin P, Gaudric A. Growth factors and diabetic retinopathy. *Diabetes Metab.* 1997; 23: 125-130.

33. Grant MB, Caballero S, Millard WJ. Inhibition of IGF-I and b-FGF stimulated growth of human retinal endothelial cells by the somatostatin analogue, octreotide: a potential treatment for ocular neovascularization. *Regul Pept.* 1993; 48: 267-278.

34. Burghardt B, Barabas K, Marcsek Z, Flautner L, Gress TM, Varga G. Inhibitory effect of a long-acting somatostatin analogue on EGF- stimulated cell proliferation in Capan-2 cells. *J Physiol Paris.* 2000; 94: 57-62.

35. Wilson SH, Davis MI, Caballero S, Grant MB. *Modulation of retinal endothelial cell behavior by somatostatin analogues: implications for diabetic retinopathy.* in *Growth Hormone and IGF Research.* 2001. Monte Carlo, Monaco: Oxford Clinical Communications.

36. Bruno JF, Xu Y, Song J, Berelowitz M. Molecular cloning and functional expression of a brain-specific somatostatin receptor. *Proc Natl Acad Sci U S A.* 1992; 89: 11151-11155.

37. Law SF, Manning D, Reisine T. Identification of the subunits of GTP-binding proteins coupled to somatostatin receptors. *J Biol Chem.* 1991; 266: 17885-17897.

38. Patel PC, Barrie R, Hill N, Landeck S, Kurozawa D, Woltering EA. Postreceptor signal transduction mechanisms involved in octreotide-induced inhibition of angiogenesis. *Surgery.* 1994; 116: 1148-1152.

39. Medina DL, Toro MJ, Santisteban P. Somatostatin interferes with thyrotropin-induced G1-S transition mediated by cAMP-dependent protein kinase and phosphatidylinositol 3- kinase. Involvement of RhoA and cyclin E x cyclin-dependent kinase 2 complexes. *J Biol Chem.* 2000; 275: 15549-15556.

40. Thangaraju M, Sharma K, Leber B, Andrews DW, Shen SH, Srikant CB. Regulation of acidification and apoptosis by SHP-1 and Bcl-2. *J Biol Chem.* 1999; 274: 29549-29557.

41. Florio T, Thellung S, Arena S, Corsaro A, Bajetto A, Schettini G, Stork PJ. Somatostatin receptor 1 (SSTR1)-mediated inhibition of cell proliferation correlates with the activation of the MAP kinase cascade: role of the phosphotyrosine phosphatase SHP-2. *J Physiol Paris.* 2000; 94: 239-250.

42. Reardon DB, Dent P, Wood SL, Kong T, Sturgill TW. Activation in vitro of somatostatin receptor subtypes 2, 3, or 4 stimulates protein tyrosine phosphatase activity in membranes from transfected Ras-transformed NIH 3T3 cells: coexpression with catalytically inactive SHP-2 blocks responsiveness. *Mol Endocrinol.* 1997; 11: 1062-1069.

43. Pages P, Benali N, Saint-Laurent N, Esteve JP, Schally AV, Tkaczuk J, Vaysse N, Susini C, Buscail L. sst2 somatostatin receptor mediates cell cycle arrest and induction of p27(Kip1). Evidence for the role of SHP-1. *J Biol Chem.* 1999; 274: 15186-15193.

44. Srikant CB. Cell cycle dependent induction of apoptosis by somatostatin analog SMS 201-995 in AtT-20 mouse pituitary cells. *Biochem Biophys Res Commun.* 1995; 209: 400-406.

45. Sharma K, Patel YC, Srikant CB. Subtype-selective induction of wild-type p53 and apoptosis, but not cell cycle arrest, by human somatostatin receptor 3. *Mol Endocrinol.* 1996; 10: 1688-1696.

46. Sharma K, Srikant CB. Induction of wild-type p53, Bax, and acidic endonuclease during somatostatin-signaled apoptosis in MCF-7 human breast cancer cells. *Int J Cancer.* 1998; 76: 259-266.

# 19

# SOMATOSTATIN AND ITS RECEPTORS: PAST, PRESENT AND THE FUTURE

Coimbatore B. Srikant
*Fraser Laboratories, McGill University Health Centre and Royal Victoria Hospital, Montreal, Quebec, Canada H3A 1A1*

Since its discovery somatostatin (SST) has served as an important hormonal model in understanding the complexities governing the hormone synthesis, processing, secretion and function at a cellular level and its physiology and pathophysiology in health and disease. Initially the distribution, gene expression, processing and physiological actions were defined. Then the heterogeneity and pharmacology of its receptors, as well as the structure-function characteristics of synthetic agonists of somatostatin receptors (SSTRs) were elucidated. Third, cloning of SSTR subtypes and the attendant explosion of activity in characterizing the molecular biological and pharmacological properties of the SSTR subtypes contributed to a further refinement of our understanding of the biology of SST and SSTRs. Fourth, animal models lacking SST or individual SSTR subtypes generated through gene knockout strategies have enabled the assessment of their physiological importance. The preceding chapters in this book have discussed these issues in detail. The landmark developments in this field (Table 1) have helped change the initial perspective of SST as an endocrine "inhibitory hormone" into one that is an important regulator of embryonic development, neuronal patterning, cell differentiation, proliferation, apoptosis, secretion, motility, and carcinogenic and neoangiogenic processes. Here I reflect on the progress made to date in this field and consider the challenges and promises that lie ahead.

Characterization of the SST gene led to the identification of the DNA binding consensus sequences CRE and TATA in promoter region, the roles of transcription factors CREB and CBP and provided the first insights into the regulation of gene transcription by extracellular signals that impinge on cAMP (chapter 1). SST is a proven model to elucidate the biosynthetic processing of prohormones, the control of regulated secretory pathways and the role of specific prohormone convertases that influence the generation of the different active hormonal forms in a cell-specific manner (chapter 2). However, the cell specific regulatory control mechanisms in neuronal and endocrine tissues remain to be fully understood. Further research in the areas of processing and

*Table 1. Landmark Developments in Somatostatin Research*

> Discovery of SST
> Demonstration of wide anatomical distribution and multiplicity of actions
> Regulation of SST gene expression
> Identification of SST-28 and Antrin
> Elucidation of the structure and its processing of preproSST
> Molecular evolution of somatostatin genes
> Identification of cortistatin
> Identification of SST receptors and their heterogeneity
> Development of long acting / metabolically stable SST analogs and radionuclide tagged derivatives
> Development of nonpeptide SST agonists/antagonists
> Use of radionuclide-tagged SST derivatives for tumor imaging
> Clinical applications of SST analogs in endocrine diseases, neurodegeneration, cancer and angiogenesis
> Cloning of SSTR subtypes
> SSTR subtype selective antiproliferative actions
> Identification of SSTR-interacting proteins
> Gene knockout (SST, SSTR1, SSTR2 and SSTR5)
> Signaling diversity arising out of agonist-dependent oligomeric assembly of SSTRs between subtypes or with other GPCRs
> Evolutionary aspects of SSTR and identification of related receptors

diversification of hormonal secretion into regulated and constitutive pathways should not only provide a greater insight into the intricate regulation of hormonal processing but also help unravel mechanisms of cell-specific regulation and organelle-specific functions of processing enzymes.

The multiple actions of SST controlling the diverse processes including cell proliferation, apoptosis and secretion can now be explained, at least in part, by the selective manipulation of signaling processes in the cell through its receptor subtypes (chapters 9, 10). The existence of non-classical regulatory actions of SSTRs is suggested by the molecular interactions with adaptor/functional proteins containing the PDZ-domain (chapter 12). The predicted possibility that other, as yet unknown, SSTR subtypes may exist is supported by the identification of MrgX2, an orphan GPCR as a cortistatin selective, SSTR-related receptor [1, see chapter 3]. The significant body of work related to the evolutionary aspects of somatostatin genes (chapter 4) and the identification of SSTR-like receptors in organisms with simple genetic background such as the drosophila (chapter 12) should encourage further studies on the evolutionary divergence and functions of SSTR-related receptors and their agonists.

The documented physical interaction between SSTR subtypes amongst themselves or with other GPCRs indicates the existence of greater complexity in hormonal signaling. However, it is possible that the complexities controlling the in vivo hormonal efficacy may be more intricate than can be determined from in vitro expression systems of recombinant receptors. Research in this area is bound to be both exciting and challenging, should provide leads into important aspects relating to targeted delivery of specific SST agonists or antagonists to diseased organs and manipulation of receptor-mediated signaling.

A great deal has been learnt about the relevance of SSTR expression in tumors and in normal tissues including vascular structures (chapters 7, 8, 11, 17). This has led to intense research in the diagnosis and therapy of tumors using radiotherapeutic and chemotherapeutic derivatives of SST analogs such as octreotide, lanreotide and vapriotide (see chapters 13, 14, 15, 16). However, it should be borne in mind that these compounds preferentially interact with SSTR2. Nevertheless, on the basis of its relative high expression as determined by qualitative or semi-quantitative RT-PCR analysis, SSTR2 is touted as a potential prognostic marker in cancer: presence of SSTR2 in tumors may suggest a good prognosis whereas its absence might indicate poor prognosis pointing to early death (see Chapter 7). It should be pointed out that analysis of the expression pattern of SSTR subtypes remains to be confirmed at a protein level.

The multiplicity of SSTR subtypes has created an opportunity and a challenge to develop agonists and antagonists with absolute receptor subtype selectivity. The development of a vast array of agonists led to extensive studies on the pharmacological and functional specificities of SSTR subtypes (see chapter 6). Apart from the expected inter-species differences in the binding of such agonists, several of them appear to exhibit unexpected affinity for certain SSTR subtypes in heterologous systems. Some peptide and non-peptide SST ligands were found to display antagonistic activity but the varying degree of partial agonism of such compounds limit their utility in defining subtype-specific receptor functions. Much progress remains to be made in elucidating the structural constraints that govern SST-SSTR interaction. There is tremendous activity in this area utilizing such techniques as combinatorial chemistry and high throughput screening for ligand development and mutagenic studies to define the ligand binding domains of SSTR subtypes. Further insights into the structural features of the SST agonists that dictate subtype selectivity may be gained using techniques such as solid state NMR imaging as has been reported recently in the case of neurotensin which was shown to undergo rearrangement to assume a rigid linear β-strand conformation when bound to its receptor [14]. Understanding how subtle changes in the conformational constraints in the SST agonists alter their subtype-specificity should aid the design of second generation of agonists, antagonists and inverse agonists with exquisite SSTR subtype-selectivity. Such compounds should prove invaluable in defining how SSTR subtypes function individually, in synergy or in opposition to elicit the desired physiological responses in a predictable manner. Insights in to these aspects is vital for optimally exploiting the signaling propensity of coexpressed SSTR subtypes and of other GPCRs that physically and functionally respond to SSTR activation, and manipulate the coupling specificity of SSTR subtypes to diverse members of the G protein family and their role in linking these subtypes to different second messengers / signaling cascades. This is relevant in the context of highly variable subtypic expression of GPCRs particularly in tumor cells. It is therefore with great excitement that we await future developments in drug design and pharmaco-functional analysis of SSTR subtype specific agonists, antagonists and inverse agonists. The availability of subtype-specific agonists and antagonists will, no doubt, contribute to the optimal exploitation of their therapeutic potential in endocrine/neuroendocrine disorders and in the subtype-specific diagnosis and therapy of SSTR positive tumors.

Gene targeting of SST and SSTR subtypes is underway. Enforced expression of SSTR2 in pancreatic cancer cells was reported to induce SST expression which in turn enhanced death receptor-mediated apoptosis apoptotic responsiveness [2]. Given that SSTR2 is not expressed in some aggressive cancers, Susini and coworkers have postulated that SSTR2 gene therapy would be beneficial [2,3]. However, such a tenet has not been widely confirmed given that endogenous production of SST in SSTR2 positive tumor cells has not been convincingly shown. If true, loss of in vivo SSTR2 expression should be expected to favor carcinogenesis. This clearly is not the case since increased carcinogenic activity has not yet been documented in rodents following genetic ablation of SSTR2 or, for that matter SST. Moreover, no increase in SST expression has been documented in SSTR2-/-, SSTR1-/- or SSTR5-/- mice [4]. It was somewhat disappointing knockout of SST or of its receptor subtypes 1, 2 and 5 did not induce major or· lethal phenotypic change in transgenic mice. Nevertheless, mice lacking SST or SSTR subtypes have proven valuable in screening SST agonists for subtype-specific actions. Only subtle changes in the regulation of secretion of GH and gastric acid (SST-/- mice) [4, see also chapter 5], of insulin and glucagon in SSTR2-/- mice [5,6] and of insulin and GH in mice lacking SSTRs 2/5 have been observed [7,8]. SSTR2-/- mice were also found to display impaired coordination of motor activity, exploratory and emotional reactivity [9-12]. Increased subtype- and tissue-specific expression of SSTRs 1-5 without any change in cortistatin levels were evident in SST-/- mice [13]. Interestingly elevated SST-like immunoreactivity in proximal gut region was observed hinting at the likely existence of a novel SST-like gene in SST-/- mice [13]. These mice should enable creation of subsequent generations of animals through selective cross-breeding and permit the evaluation of appropriate double or multiple knockout phenotypes to assess the importance of SST and its receptors in limb bud development, organellar differentiation, neuronal patterning, regulation of neuroendocrine, endocrine and exocrine secretion as well as in cancer and neurodegenerative disorders.

In summary, these developments have established somatostatin and its receptors as ideal models to probe the biology of peptide hormones and GPCRs. Further exploration of amphioxus genome should lead to better mapping of evolutionary relationships and divergence of somatostatin and structurally related peptides and their receptors. Receptor-mediated actions of somatostatin is regulated in a complex manner not only by activation of diverse/distinct signaling pathways by SSTR subtypes but also by their selectivity for agonist binding, interaction with multiple G proteins and other accessory / adaptor proteins, but also by the molecular association of SSTR subtypes between themselves and / or other GPCRs. Development of newer generation of agonists and antagonists with improved SSTR subtype selectivity should not only increase our understanding of the cellular actions of SST, but also allow the precise evaluation of the changes in the expression of SSTR subtypes in normal development and in tumorigenisis, and permit the rationalized exploitation of subtype-selective agonists in the diagnosis and treatment of SSTR positive cancers.

## REFERENCES

1.  Robas, N, Mead, E, and Fidock, M, *MrgX2 is a high potency cortistatin receptor expressed in dorsal root ganglion.* J Biol Chem, 2003. **278**: 44400-44404.
2.  Guillermet, J, Saint-Laurent, N, Rochaix, P, Cuvillier, O, Levade, T, Schally, AV, Pradayrol, L, Buscail, L, Susini, C, and Bousquet, C, *Somatostatin receptor subtype 2 sensitizes human pancreatic cancer cells to death ligand-induced apoptosis.* Proc Natl Acad Sci U S A, 2003. **100**: 155-160.
3.  Benali, N, Cordelier, P, Calise, D, Pages, P, Rochaix, P, Nagy, A, Esteve, JP, Pour, PM, Schally, AV, Vaysse, N, Susini, C, and Buscail, L, *Inhibition of growth and metastatic progression of pancreatic carcinoma in hamster after somatostatin receptor subtype 2 (sst2) gene expression and administration of cytotoxic somatostatin analog AN-238.* Proc Natl Acad Sci U S A, 2000. **97**: 9180-9185.
4.  Ramirez, JL, Mouchantaf, R, Kumar, U, Otero Corchon, V, Rubinstein, M, Low, MJ, and Patel, YC, *Brain somatostatin receptors are up-regulated in somatostatin-deficient mice.* Mol Endocrinol, 2002. **16**: 1951-1963.
5.  Zheng, H, Bailey, A, Jiang, MH, Honda, K, Chen, HY, Trumbauer, ME, Van der Ploeg, LH, Schaeffer, JM, Leng, G, and Smith, RG, *Somatostatin receptor subtype 2 knockout mice are refractory to growth hormone-negative feedback on arcuate neurons.* Mol Endocrinol, 1997. **11**: 1709-1717.
6.  Martinez, V, Curi, AP, Torkian, B, Schaeffer, JM, Wilkinson, HA, Walsh, JH, and Tache, Y, *High basal gastric acid secretion in somatostatin receptor subtype 2 knockout mice.* Gastroenterology, 1998. **114**: 1125-1132.
7.  Strowski, MZ, Kohler, M, Chen, HY, Trumbauer, ME, Li, Z, Szalkowski, D, Gopal-Truter, S, Fisher, JK, Schaeffer, JM, Blake, AD, Zhang, BB, and Wilkinson, HA, *Somatostatin receptor subtype 5 regulates insulin secretion and glucose homeostasis.* Mol Endocrinol, 2003. **17**: 93-106.

8.  Strowski, MZ, Parmar, RM, Blake, AD, and Schaeffer, JM, *Somatostatin inhibits insulin and glucagon secretion via two receptors subtypes: an in vitro study of pancreatic islets from somatostatin receptor 2 knockout mice.* Endocrinology, 2000. **141**: 111-117.
9.  Allen, JP, Hathway, GJ, Clarke, NJ, Jowett, MI, Topps, S, Kendrick, KM, Humphrey, PP, Wilkinson, LS, and Emson, PC, *Somatostatin receptor 2 knockout/lacZ knockin mice show impaired motor coordination and reveal sites of somatostatin action within the striatum.* Eur J Neurosci, 2003. **17**: 1881-1895.
10. Moneta, D, Richichi, C, Aliprandi, M, Dournaud, P, Dutar, P, Billard, JM, Carlo, AS, Viollet, C, Hannon, JP, Fehlmann, D, Nunn, C, Hoyer, D, Epelbaum, J, and Vezzani, A, *Somatostatin receptor subtypes 2 and 4 affect seizure susceptibility and hippocampal excitatory neurotransmission in mice.* Eur J Neurosci, 2002. **16**: 843-849.
11. Viollet, C, Vaillend, C, Videau, C, Bluet-Pajot, MT, Ungerer, A, L'Heritier, A, Kopp, C, Potier, B, Billard, J, Schaeffer, J, Smith, RG, Rohrer, SP, Wilkinson, H, Zheng, H, and Epelbaum, J, *Involvement of sst2 somatostatin receptor in locomotor, exploratory activity and emotional reactivity in mice.* Eur J Neurosci, 2000. **12**: 3761-3770.
12. Hannon, JP, Petrucci, C, Fehlmann, D, Viollet, C, Epelbaum, J, and Hoyer, D, *Somatostatin sst2 receptor knock-out mice: localisation of sst1-5 receptor mRNA and binding in mouse brain by semi-quantitative RT-PCR, in situ hybridisation histochemistry and receptor autoradiography.* Neuropharmacology, 2002. **42**: 396-413.
13. Radosevic-Stasic, B, Trobonjaca, Z, Lucin, P, Cuk, M, Polic, B, and Rukavina, D, *Immunosuppressive and antiproliferative effects of somatostatin analog SMS 201-995.* Int J Neurosci, 1995. **81**: 283-297.
14. Luca, S, White, JF, Sohal, AK, Filippov, DV, van Boom, JH, Grisshammer, R, and Baldus, M, *The conformation of neurotensin bound to its G protein-coupled receptor.* Proc Natl Acad Sci U S A, 2003. **100**: 10706-10711.

# INDEX

GPSR Compliance

The European Union's (EU) General Product Safety Regulation (GPSR) is a set of rules that requires consumer products to be safe and our obligations to ensure this.

If you have any concerns about our products, you can contact us on ProductSafety@springernature.com

In case Publisher is established outside the EU, the EU authorized representative is:

Springer Nature Customer Service Center GmbH
Europaplatz 3
69115 Heidelberg, Germany

**Batch number: 09636419**

Printed by Printforce, the Netherlands